COMPOSITE MATERIALS AND STRUCTURAL ANALYSIS

COMPOSITE MATERIALS AND STRUCTURAL ANALYSIS

N.G.R. Iyengar

MV Learning

London • New Delhi

MV Learning

3, Henrietta Street
London WC2E 8LU
UK

4737/23, Ansari Road,
Daryaganj, New Delhi 110 002
India

A Viva Books imprint

ISBN: 978-81-309-2808-1

Printed and bound in India.

SUB-0001/SSUB-0004

Preface

Fiber-reinforced composites are employed where savings in weight is required. They are largely used in aerospace structures, automobiles, ship structures, civil engineering structures and chemical industrial applications. Fiber reinforced composites have also been successfully used to increase the life-span of existing metal bridges. As compared to metals, the use of composites has been growing due to the latter's better mechanical properties and the ease of fabrication of complex shapes since composites can be easily moulded to any desired shape.

A good number of books exist which basically deal with composite material behavior, material characterization, fatigue and impact characteristics, micro mechanics and lamination theories. However, only a few books cover not only the topics mentioned above but also the structural applications to a limited extent.

Nowadays, numerical techniques like finite element technique are being extensively used for solving real-life problems. The present book covers to some extent micro mechanics, and behavior of laminated composites. Chapters 1 to 7 cover the basic material like characterization and to some extent the response of composite materials. An attempt has also been made at making the text more comprehensive by including composite beams (Chapter 8), analysis of composite laminates with large number of examples on classical, first order, and higher order theories (Chapter 9), short fiber composites (Chapter 10), and optimization of composite laminates for vibration, buckling, and structural weight (Chapter 11). Manufacturing techniques are covered in Chapter 12 and evaluation of composite properties in Chapter 13.

The book will serve the requirements of undergraduate and postgraduate courses. It will also be useful to the practicing engineers. In all the chapters, a number of solved examples are provided. At the end of each chapter, exercise problems are given for better understanding of the topics covered in the chapter.

I am indebted to my graduate students, M.K. Patra, P.J. Sony, S.P. Joshi, S. Sharma, K. Singh, C.A. Shankara, S. Gajbir, and K. Sivakumar, whose works have made this book useful to the students.

Finally, I am extremely thankful to my wife Leela, for the patience she showed while I was busy in writing this book. I am thankful to Viva Books Pvt. Ltd. for helping to publish this book.

N.G.R. IYENGAR

Contents

1. Why Composites and What They Are

1.1 Introduction

In very broad terms, a composite is a material which is made up of two or more components that do not interact chemically. In other words, they retain their individual characteristics throughout their useful life. In India, even now, houses in the villages are built with mud reinforced with bamboo shoots and straw. As individual components they do not serve the purpose. Straw or bamboo shoots basically prevent the mud from cracking. Similarly, the ancient Inca and Mayan civilizations used plant fibers to strengthen bricks and pottery. Swords and armor were plated to add strength. Eskimos use moss to strengthen ice in building igloos.

One of the most efficient structures is the human body. Muscles in the human body are a very good example of efficient use of fibrous composites. The muscles are present in a system consisting of fibers at different angles and in different concentrations. The system gives rise to a very strong, adaptable structure that is capable of taking all kinds of loads. Other naturally occurring fibrous composites include the wings of a bird, the fins of fish, grass, etc. It is well known that solid wood, which is a naturally occurring composite, has a lignin matrix reinforced with cellulose fibers. The solid block of wood has a number of defects called knots (which are weak spots) and the grains are oriented along one direction. This results in a material that has good properties along only one direction.

Engineering materials such as steel alloys and aluminum alloys also consist of a number of constituent elements. However, these are not considered as composites. The various elements constituting the alloy cannot be seen individually under normal conditions. One can see them only when they are heat treated at elevated temperatures, provided they have not chemically reacted. Composite material is one which consists of two or more components that do not interact chemically. One is called the reinforcing phase and the other, the binding or matrix phase. The purpose of the binding material, as the name suggests, is to hold the reinforcing phase together and also protect it from environmental effects. The reinforcing material may be in the form

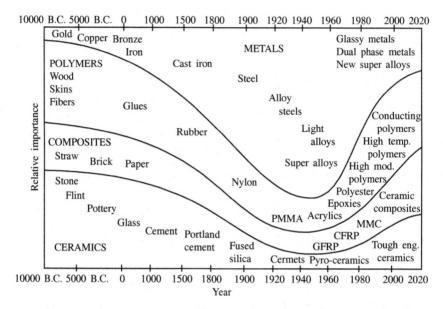

Fig. 1.1 Use of materials over the years.

of platelets, fibers, particles and flakes. We shall see later that both the reinforcing phase and the binding phase can be in the particulate form. Figure 1.1 shows the relative importance of different materials as a function of its growth over the years [1]. It can easily be seen that the use of composites and polymers has been gaining ground during the last 40 years or so.

It is well known that metallic materials in the bulk form have a large number of defects. The presence of these defects reduces the strength and stiffness of the components made from bulk material. If the size can be reduced, the number of defects can be minimized which in turn improves the material behavior. However, one should realize that reduction in cross-sectional dimensions reduces the defects. Two important parameters by which different metals and composite materials can be compared are (i) specific strength (σ_{ul}/ρ = ultimate tensile strength/density) and (ii) specific stiffness (E/ρ = elastic modulus/density). For a material, higher the values of specific strength and stiffness, the better it is for structural applications. For example, in order to improve ultimate tensile strength, the material has to be reduced from bulk to fiber form (diameter of the order of micrometers). Table 1.1 shows the properties of some fibers and composites generally employed for fabricating composite structural parts [2-6].

1.2 Definition and Some Applications of Composites

As mentioned before, a man-made composite material can be defined as a material which consists of two or more components that do not interact chemically — one is called the reinforcing phase and the other, the binding

Table 1.1 Properties of fibers

Fiber material	Volume fraction (V_f) (%)	Density (ρ) (g/cm^3)	Tensile strength σ_u (GPa)	Tensile modulus E (GPa)	Specific strength (σ_u/ρ)	Specific modulus (E/ρ)
E-glass	-	2.54	3.45	72.4	1.39	28.50
S-glass	-	2.49	4.30	86.9	1.72	34.90
Kevlar-49	-	1.45	3.62	131	2.52	90.34
Kevlar-149	-	1.47	3.45	179	2.20	121.7
Carbon (high modulus)	-	1.90	2.1	390	1.1	205.3
Carbon (high strength)	-	1.90	2.5	240	1.3	126.3
E-glass-epoxy	57	1.97	0.57	21.5	0.26	10.91
E-glass-polyester	55	1.8	0.75	30	0.417	16.67
Carbon-epoxy	62	1.6	1.5	140	0.937	87.5
Kevlar-149-epoxy	60	1.38	1.28	87	0.927	63.04
Aluminum-2024	-	2.7	0.41	73	0.152	27.03
Mild steel	-	7.8	0.45	210	0.058	26.92

or matrix phase. The purpose of the binding material, as the name suggests, is to hold the reinforcing phase together and also protect it from environmental effects. The reinforcing material may be in the form of sheets, fibers, particles and flakes. We shall see later that both the reinforcing phase and the binding phase can be in the particulate form. Thus, in general, a composite is heterogeneous. The heterogeneity depends on the scale of interest. If one looks at the micro level, these are heterogeneous. However, at the macro level, they can be treated as homogenous. Depending on the nature of reinforcements, composites can be isotropic, orthotropic (or transversely isotropic) and anisotropic.

Some of the properties of a material that can be improved upon by forming a composite are: (i) strength, (ii) stiffness, (iii) corrosion resistance, (iv) wear resistance, (v) fatigue resistance, (vi) temperature dependent behavior, (vii) thermal insulation, (viii) thermal conductivity, (ix) electrical conductivity, and (x) weight and many more.

All these properties may not be realized in a single material. They can be obtained by a proper choice of the reinforcing and matrix material. The reinforcing and matrix phases can be metallic or non-metallic. Some of the materials in common use are:

(a) Non-metallic in non-metallic
(i) Concrete-aggregate and sand in cement: Used largely in civil engineering construction.
(ii) Mica in glass: Used extensively in the electrical industry because of good insulating and machining qualities.

(b) Metal in non-metal
(i) Rocket propellant: Aluminum powder and perchlorate oxidizer in flexible organic binder, e.g. polyurethane.
(ii) Metal flake in suspension: Aluminum flakes suspended in paint. On application, the flakes orient themselves parallel to the surface giving a good finish. This is largely used for painting with silver finish.

(c) Non-metal in metal: Non-metallic particles such as ceramics can be suspended in metal matrix. The resulting composite is called cermet. They are of two kinds. Oxide based cermets can be oxide particles in a metal matrix. Tungsten carbide in cobalt matrix is used in machine parts requiring very high hardness such as dies, valves, etc.

In the examples mentioned above, both the reinforcing material and the binding material are in the form of particulates. In many situations, the laminated form is used. Some common examples of laminated composites are:

(i) Copper clad stainless steel is economically used for wall panels, rain drainage, windows and doorframes.
(ii) Toughened glass consisting of glass sheets on either side of a plastic sheet is used in aircraft windshields and automobile windows. Plastic cannot be used outside as it loses its transparency under ultraviolet light and becomes opaque. Glass retains its transparency but being brittle, shatters into sharp pieces when it breaks. Plastic provides the necessary ductility to the laminate, so that when the laminate breaks, it shatters the laminate into powder form.
(iii) Aluminum alloy gets corroded when exposed to the atmosphere. To improve its corrosion resistance properties, the alloy is cladded with pure aluminum.
(iv) Aluminum wire clad with 10% copper is introduced as a replacement for copper wire in electrical wiring. Aluminum wire, though it overheats easily and is difficult to connect, is light weight and economical. On the other hand, copper stays cool and can be connected easily, but it is expensive and relatively heavy.

Composites in Automobiles
Composite materials are being used increasingly in automobiles. Early applications were mostly plastics. However, the use of fiber reinforced plastics (FRP) have found increasing applications in automobile parts. Glass

fiber reinforced plastics have been largely used in fabricating automobile parts. The advantages of using FRPs are:

(i)   Lightening of vehicles: A reduction of mass of 1 kg induces a final reduction of 1.5 kg taking into account the consecutive lightening of mechanical components.

(ii)   Cost reduction: This is due to the reduction in number of parts required for a certain component.

(iii) Better corrosion resistance.

(iv)   Suspension components: Springs; a glass/resin spring absorbs 5 to 7 times more energy than a steel spring of the same mass. It is possible to integrate many functions in one particular system.

(v)   Composite transmission system: These are used in competition vehicles; they allow high speeds of rotation with low inertia.

(vi)   Parts of the engine where composites are used are rocker arms, pistons, and engine blocks.

(vii) Short randomly oriented glass fiber reinforced composites have been extensively used as surface molding compounds (SMCs) and bulk molding compounds (BMCs) in body panels and blocks.

Composites in Sporting Goods

(i)   One of the major advantages of using FRP is the reduction in overall weight. Tennis rackets and snow skies have carbon fiber composites as the skin and a soft core which makes the structure light weight without any decrease in stiffness. The use of FRP also increases the damping capacity.

(ii)   FRPs are also used in fishing rods, bicycle frames, sail-boats, hockey sticks, golf clubs, kayaks, etc.

Composites in Aerospace

Military aircraft designers were the first to take advantage of the high specific strength and high specific stiffness of composites as compared to metallic materials. Unlike in metallic structures where mechanical fasteners like bolts and rivets are used, the composite construction results in an aerodynamically smooth surface devoid of bolts and rivets. They are generally joined together with the help of adhesives. This results in a continuous connection between the two parts. The use of composites has also resulted in novel design concepts. Further, there is a reduction in the number of parts as large components can be molded into any desired shape. Composite structural elements like horizontal and vertical stabilizers, wing skin and various control surfaces have been employed in F-14, F-15 and F-16 military aircrafts resulting in a weight saving of almost 20% of the structural weight. The AV-8B fighter aircraft has a number of composite parts. The composite components correspond to 26% of the aircraft structural weight [7]. Boeing 757 and 767 are the first commercial aircraft to make extensive use of composites. In a Boeing 757, the weight of composites used is approximately 1350 kg. In India, the newly developed two-seater HANSA trainer aircraft makes use of composites in its primary structure. Laminated composites with carbon or Kevlar fibers in polymer matrices are used extensively in Light Combat Aircraft (LCA) airframe constituting 45% by weight of the aircraft. The wing is made up of carbon fiber composite and the radome of Kevlar. The Advance Light Helicopter (ALH) makes use of large quantities of composites like Kevlar, carbon, Kevlar-carbon and glass-carbon for the airframe and other structural components.

In the Airbus 310, the use of composites is of the order of 13.8% of the mass of the structure, while in an Airbus 320, it is of the order of 21.5% of the mass of the structure. Approximately 30% of the external surface area of Boeing 767 consists of composites.

Some other facts on the use of composites in aerospace structures are: a reduction of 1 kg in the mass of the structure of an aircraft leads to the reduction of fuel consumption of around 120 l/yr. A 60-tons aircraft weighs 150 tons with 250 passengers on board. If 1600 kg of high performance composite is used, this leads to the capacity to carry an extra 16 passengers along with their luggage.

The use of composite materials is generally more in military aircraft as compared to civilian aircrafts. Composites are also used in the manufacture of helicopter blades. Tailoring of the material can control the dynamic frequencies of the rotor blades. The shapes of the blades can easily be manufactured without any additional cost unlike in metal blades whose shapes are limited. As the blades are molded, one can obtain an optimized profile (variable chord and thickness, nonsymmetrical profile, nonlinear twist). The cost of fabrication is reduced by almost 50%.

Composites result in light weight missile structures, which in turn result in increased range for the same fuel consumption or a higher payload. However, the structures are generally made of advanced composites of graphite or carbon. One of the requirements for a space shuttle structure is dimensional stability. Graphite and carbon fibers are largely used in space structures as they possess high specific stiffness and low thermal coefficient of expansion as well as good damping characteristics. Thus, composites are extremely useful for spacecraft applications.

Composites in Ocean Engineering [8]

The use of composites in ocean engineering has tremendous advantages:

(i)   Resistance to environment, freedom from corrosion and other forms of degradation.

(ii)   Seamless molding, non-leaking structure of complex shapes and ability to reproduce these shapes easily.

(iii) Ability to orient strength to suit the load directions.
(iv) The structure will generally weigh about one half of the steel structure.
(v) Low maintenance and ease of repairs.
(vi) Excellent durability and long life for components.

Some of the applications of composites are:

(i) Pleasure boats: New types of hull forms have come up because of fiber reinforced composites.
(ii) Sail boats: Sandwich construction for the shell.
(iii) Fishing trawlers: The use of composites results in light weight structures which require low maintenance.

The advantages of using fiber reinforced composites over conventional wooden structures are low costs, low maintenance cost and aesthetics. Glass reinforced composites have been used in military and civil hovercrafts. They are also used as sheathing to protect wooden hulls from leakage and rot.

1.3 Characteristics of Composite Materials

A composite generally consists of two or more materials of which, one or more are in discontinuous phases embedded in a continuous phase. The discontinuous phase is generally referred to as reinforcing material, while the continuous phase is the matrix material or the binding material. In general, the reinforcements are much stronger and stiffer than the matrix. In the composite, both the phases retain their individual characteristics. However, the behavior of the composite is very different. Although metallic alloys and polymer blends consist of different phases, they are not called composites as they are formed at the microscopic level. In the case of man-made composites, the distinct phases can be identified at the macro level and the two phases do not chemically interact.

In general, a material is stronger and stiffer in fiber form than in bulk form. This is due to the large number of defects (flaws) that are present in a bulk material. The flaws are the sources of crack initiation. Griffith [9] measured the tensile strength of glass rods and glass fibers of different sizes. He observed that as the rods and fibers got thinner, they became stronger. This is because of the reduction in the number of defects as the size reduces. The use of fibers has the advantage of maximizing the tensile strength and stiffness of the material. However, there are many disadvantages. The fibers cannot resist compressive load and have very poor resistance in the transverse direction. Therefore, fibers by themselves cannot be used as structural material. They are enclosed in a matrix or a binder. The binder helps in load transfer as well as to protect the fibers from external damage and environmental exposure. Transverse properties can be improved by orienting the fibers at different angles according to the stress field.

The physical and mechanical properties of composites are dependent on the geometry, fiber orientation, concentration and distribution of the constituents. Increasing the fiber volume content will result in an increase in strength and stiffness. However, there are some limitations. As the total volume is fixed, an increase in fiber volume fraction would result in a corresponding decrease in the volume fraction of the matrix material. This may result in fiber being exposed to the external environment. Hence, there are many factors to be considered when designing with composite materials. Unlike in the case of conventional materials, the anticipated external, thermal loads and operating environment must be taken into account for design.

Fiber concentration is measured in terms of volume or weight fraction. This is defined as the volume or weight of the fiber divided by the total volume or weight of the composite. Fiber concentration affects the behavior of the composite. Higher the fiber volume fraction, greater is the strength. Another important parameter is the fiber distribution. This is a measure of homogeneity of the material. It indicates how the mechanical properties change from point to point in the material. The performance of the composite depends on this parameter.

1.4 Classification of Composites

Composite materials have been developed basically to improve the mechanical properties like strength, stiffness, fatigue behavior, toughness, etc. The performance of the composite depends on the type of reinforcement, size and distribution and binding or matrix material. Composites are broadly classified into: (i) fibrous, (ii) particulate and (iii) laminated. All these forms of composites are in use in structures.

Fibrous Composites

The fibrous composite consists of long or continuous fibers or short fibers along with a matrix material. If the fiber length-to-diameter ratio (L/d) of a composite is more than 1000, then it is classified as a long or continuous fiber composite. However, if the length-to-diameter ratio is of the order of 100, then the composite is classified as a short or discontinuous fiber composite. In the continuous and short fiber composite, the diameter of the fiber is of the order of 3-200 μm. Whiskers also belong to the class of short fibers but the diameter of the fibers are in the range 0.02-10 μm.

Particulate Composites

In a particulate composite, the reinforcement material and matrix material are in a particulate form. The particulates can be of any shape and size. Well-known examples of particulate composites are: (i) cement concrete and (ii) particle board. The particles in a particulate composite inhibit the plastic deformation of the matrix because they are relatively harder than the matrix.

The particles also carry the load, however, to a smaller extent when compared to the fibers. Thus, the particles increase the stiffness of the composite but do not offer the potential for more strengthening. Particle fillers are generally employed to modify the thermal and electrical conductivities, reduce friction, increase wear and abrasion resistance, increase surface hardness and reduce shrinkage. A particulate composite can be a combination of metallic/non-metallic materials. Composites with particles of tungsten or molybdenum, often are largely used for electrical contacts. Oxide based cermets are used as tool materials for high-speed cutting. Tungsten carbide in a cobalt matrix is used in machine parts requiring high surface hardness, such as cutting tools and valve parts. Chromium carbide in a cobalt matrix is highly resistant to corrosion and abrasion. Titanium carbide in a nickel or cobalt matrix is used for high temperature applications such as turbine parts [10].

There are two sub-classes of particulate composites: flake and filled/skeletal.

Flake

A flake composite generally consists of flakes with large ratios planform, area-to-thickness ratio, suspended in a material matrix. Examples of this kind are aluminum flakes suspended in oil medium used in paints and mica flakes used in electrical and heat insulating applications. Silver flakes are generally used where good conductivity is required. The flakes being two-dimensional, impart equal strength in the plane as compared to unidirectional fibers.

Filled/Skeletal

The composite is composed of a continuous skeletal matrix filled by a second material. For example, a honeycomb core filled with an insulating material. Figure 1.2 shows schematically the various types of fibrous composites.

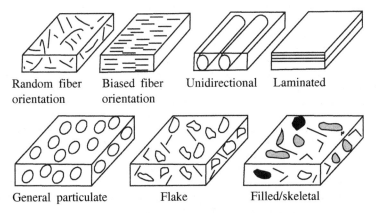

Fig. 1.2 Different types of fibrous composites.

Composites can further be classified on the basis of the type of reinforcement and alignment of the reinforcing material. The thickness of a single layer/ply/lamina of a reinforced composite is 0.1-0.125 mm. This thin lamina cannot be directly used as a structural member for carrying loads other than in-plane tensile loads. The lamina have to be stacked together to arrive at the required thickness to withstand the other forms of loads. These are termed as *laminated composites* or *multilayer composites*. They are also classified on the basis of the orientation of the fibers with respect to the geometric mid-surface of the laminate. Figure 1.3 shows the classification generally employed to describe the laminated composite.

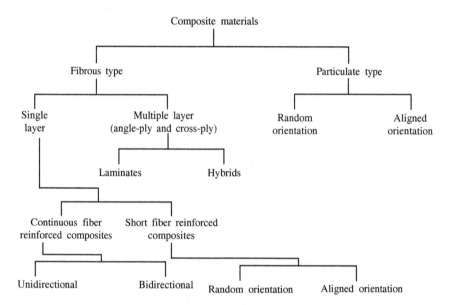

Fig. 1.3 Classification of laminated composites.

Table 1.1 shows the comparison of mechanical properties of the fibers and composites used in structural applications. As stated earlier, the fibers are very strong along the axis. This is due to the fact that the large flaws present in a bulk material have been removed or minimized. As seen from Table 1.1, the specific strength and stiffness of such materials is very much higher than the bulk material. However, one cannot take advantage of very high strength of fibers as they cannot be used directly for structural applications. One has to enclose these fibers in a matrix. This brings down the strength of the composite enormously, but it is still higher than the bulk material. It is this property, which makes composites potential structural material. As stated earlier, the thin layer cannot be used directly. They have to be stacked together to get the required thickness. A single layer or ply composite will invariably have a number of layers, with each layer or ply having the same orientation of fibers and material properties. In this case, the laminate will be a thick lamina or a single layer laminate. If however, the fiber orientation of each layer is different, then it is called a multilayered laminate. If the properties of each layer of the laminate are not the same, then it is called a *hybrid laminate*. For example, one can think of a combination of lay ups of glass fiber in one layer and the second layer could be of carbon. This is to make sure that the laminate can withstand the loads applied in different directions. It is possible that a laminate with a single material may not be able to withstand the load. One can also build a *hybrid laminate* by varying the volume fraction of a single material in each layer. It is not practical and is impossible to have a mixture of fibers in each layer. A hybrid provides certain advantages over the conventional laminates. One reason for using hybrid laminates is economics. The cost of manufacturing can be reduced by mixing less expensive fibers (glass) with more expensive fibers (graphite). For example, a mixture of 20% (by volume) of graphite fibers with glass fibers can produce a composite with 75% of the strength and stiffness, with about 30% of the cost of an all graphite composite. Combining different materials in a single lamina produces an *interlaminar* or *intraply hybrid*.

A single layer composite or a lamina represents a building block for structural applications.

For primary load carrying members in aerospace and other structures where the intensity of load is high, one prefers to use long or continuous fiber composites (*L/d* ratio > 1000). Such composite elements are termed as *continuous fiber reinforced composites*. However, for secondary structures where the load intensity is low or moderate, one employs discontinuous or short fibers. Such composites are termed as *discontinuous* or *short fiber reinforced composites*.

In *short fiber composites*, if the fibers are aligned along one direction only, then they are called *aligned short fiber composites*. Short fiber composites are generally used in high volume applications due to low manufacturing cost. They are also used in situations where the product involves large curved surfaces and it is not always possible to bend the continuous fibers. In such cases, short fibers provide a better solution. If the fibers are randomly oriented in a given lamina, then they are termed as *random short fiber composite lamina*. This, as will be seen later in Chapter 10 and Section 10.6, depends on the manufacturing technique used. The *random short fiber composite* behaves like an isotropic material. Short fibers when mixed with matrix material will result in a reinforced molding compound. Depending on the molding technique employed, the fiber may orient themselves in a preferred direction or continue to remain random.

Another class of composites in which continuous fibers are employed is woven *fiber composites*. In this case, the fibers are oriented along two orthogonal directions. If the amount of fibers in both the directions is the same, the lamina is called a *balanced lamina*. In many practical applications, the number of fibers in both the directions is generally not the same. For such a lamina, the mechanical properties are different in two orthogonal directions. Such a lamina is called an *orthotropic lamina*. It is used in situations where the external load acts in two orthogonal directions.

Another common composite construction is the sandwich construction which consists of composite face sheets bonded to a honey comb core. These have high flexural strength and are widely used in aerospace structures. Aircraft wings, fuselage floor panels are examples of sandwich construction. Figure 1.4 shows the various types of fiber reinforced composites [11].

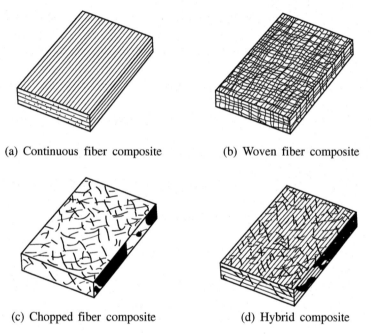

(a) Continuous fiber composite   (b) Woven fiber composite

(c) Chopped fiber composite   (d) Hybrid composite

Fig. 1.4 Various types of fiber reinforced composites.

Before going further, we shall list out the advantages and disadvantages of fiber reinforced composites vis-à-vis conventional metallic materials.

Advantages
- Light weight due to high specific strength and specific stiffness.
- Better fatigue and corrosion resistance.
- Ability to tailor the material to get the required directional strength and stiffness.
- Capability to mold a large structure in less time. This results in reducing the number of parts, which consequently reduces the number of fasteners.
- Capability to maintain dimensional alignment stability.

Disadvantages

*   Laminated structure with weak interfaces, has poor resistance to out-of-plane tensile load.
*   Susceptibility to impact damage. The probability of internal damage may go unnoticed.
*   Moisture absorption resulting in degradation of the material properties.
*   Extensive qualification requirements.
*   Need for new inspection techniques.
*   Material cost is still quite high.

Inspite of these disadvantages, composites are finding more applications in the design of primary and secondary structural components in aerospace, mechanical and civil engineering structures.

References

1. Ashby, M.F., 1987, "Technology in the 1990s: Advanced Materials and Predictive Design", *Philosophical Transactions of the Royal Society of London*, A322, pp. 393-407.
2. Agarwal, B.D., Broutman, L.J. and Chandrashekhara, K., 2006, *Analysis and Performance of Fiber Reinforced Composites*, Third Edition, John Wiley and Sons, Inc.
3. Daniel, I.M. and Ishai, O., 1994, *Engineering Mechanics of Composite Materials*, Oxford University Press.
4. Mallick, P.K., 1993, *Fiber Reinforced Composites*, 2nd Edition, Marcel Dekker Inc., USA.
5. Johnson, A.F., 1978, *Engineering Design Properties of GRP*, Pub. No. 215/1, British Plastic Federation.
6. Christiansen, R.M., 1991, *Mechanics of Composite Materials*, Krieger Pub. Co.
7. Reinhart, T.J. (Ed.), 1987, *Engineered Materials Handbook, Vol. 1: Composites*, ASM International, Materials Park, Ohio.
8. Smith, C.S., 1990, *Design of Marine Structures in Composite Materials*, Elsevier Applied Science.
9. Griffith, A.A., 1920, "The Phenomena of Rupture and Flow in Solids", *Philosophical Transactions of the Royal Society*, 221A, pp. 163-198.
10. Jones, R.M., 1975, *Mechanics of Composite Materials*, Scripta Book Company.
11. Gibson, R.F., 1994, *Principles of Composite Material Mechanics*, International Edition, McGraw-Hill Book Co., Singapore.

2. Fibers and Matrix Materials

In this chapter, we will discuss:

- The different types of natural and synthetic reinforcing fibers.
- Their forms and production techniques.
- Different types of polymer matrix and fillers used in making composites.

2.1 Introduction

In a composite, the reinforcing phase (fiber) is a stronger material and hence is less ductile, while the binding or matrix phase is a weaker material but more ductile. A combination of the two in proper proportion gives the desired property. Both fibers and the matrix material can be metals or non-metals. The other constituents that make up the composites are fillers, coatings and coupling agents. The filler material is generally mixed with the matrix material during manufacturing. Fillers do not increase the strength of the composite but improve some other properties. For example, clay or mica particles are used to reduce the cost, while carbon black particles are used for protection against ultraviolet light [1]. Coupling agents and coatings are applied on fibers to improve the adhesion of the matrix to the fiber and to reduce environmental effects.

2.2 Fibers

Bulk materials contain a large number of flaws/defects and hence, their strength and stiffness is low. One way of improving the mechanical properties is to minimize the defects in the bulk material. This can be achieved if the cross-sectional dimensions are reduced. By reducing the cross-sectional dimension of the fibers, the strength increases but the ductility reduces. Fibers are generally of small cross-section with fiber diameter in the range of 8-12 μm. We saw in Table 1.1 the properties of typical fibers along with the bulk material. As can be seen, the specific modulus (E/ρ) and specific strength (σ_{ul}/ρ) are very much higher for fibers as compared to the bulk material. However, one must realize that fibers alone cannot be used for making a component. These properties – high specific modulus and strength – reduce once they are embedded in a matrix or a binding material but they will still be higher than the bulk material. This is also observed from Table 1.1. Figure 2.1 shows the stress-strain behavior of the fibers. Their behavior is linearly elastic up to failure. Further, it can be seen that the failure strain is much lower compared to normal metals. The failure strain can be improved by embedding the fibers in a ductile matrix material.

Fibers are available in various forms. They could be continuous or discontinuous (chopped). Continuous fibers are good for withstanding tensile loads along the direction of the fiber. If, however, the loads act in different directions, one has to use chopped strand mats or woven fabrics. Continuous fibers can also be used; however, the fibers in the lamina or ply have to be oriented with respect to the reference axes to take the loads.

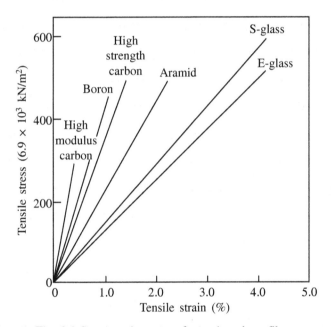

Fig. 2.1 Stress-strain curves for various bare fibers.

The following factors with regard to the fibers contribute to the stiffness and strength of the composite.

Length
As stated earlier, the fibers could either be long and continuous or short and discontinuous. Long fibers are easy to align while in the case of short fibers, their orientation cannot be controlled easily. It can be controlled by the manufacturing techniques employed.

Orientation
Fibers oriented in a particular direction give high strength along that direction and lower strength normal to this direction. Woven fabrics provide strength and stiffness in both the directions.

Cross-section
The most common cross-sectional shape is circular. The other kind of shapes like square, elliptical, platelets give higher fiber packing in a given volume. However, these outweigh the difficulty in processing.

Material
The material chosen for the fiber directly affects the strength and stiffness of the composite. Fibers generally possess high modulus and high tensile strength. The choice depends on the cost and the application.

2.2.1 Natural Fibers
There is a large variety of natural fibers, such as jute, kenaf, hemp, flax, pineapple, banana-leaf, banana-stem (in the form of fibers) and sisal. These are used with thermoplastics and thermosets as car door panels, seat backs, headliners, package trays, dashboards, doors, furniture manufacturing, building panels etc. Natural fibers benefit from the perception that they are 'green' or eco-friendly. They have the ability to provide stiffness enhancement and sound damping at a lower cost and density less than glass fibers. Natural fiber materials have better strength-to-weight ratio to replace glass-filled plastics. Natural fibers have specific gravities of 1.25-1.5 g/cc as compared to 2.5 g/cc for glass. Fibers such as kenaf, hemp and jute are very fine and easily dispersed. They have higher elongation and flexural strengths. This makes them prime candidates for fiber material for compression molding of complex curves and intricate parts with different radii. In contrast, sisal and banana-leaf fibers are coarser and stiffer, which makes them suitable for flat parts. Both polypropylene and polyurethane are employed as matrix materials for making the composite. Natural fibers can vary in quality, depending on where they come from and how they are processed. These natural fibers are not capable of carrying large loads. Man-made fibers are to a large extent defect free and are therefore stronger and stiffer in the fibrous form. Fibers with a large aspect ratio (length-to-diameter ratio) are very effective in transferring the load via the matrix material. This is the principal reason that fiber reinforced composites are being increasingly used in aerospace and automotive parts. Reinforced fibers used for carrying structural loads are: glass fibers, carbon fibers, graphite fibers, Kevlar fibers, boron fibers, silicon carbide fibers and a combination of these. In what follows, we shall describe the characteristics of these fibers.

2.2.2 Synthetic Fibers

Glass Fibers

Glass fiber is the most common fiber used in conjunction with polymeric matrix material. Among all other fibers like carbon, Kevlar etc., it is the cheapest. The advantages of glass fibers include high strength, low chemical resistance and very good insulating properties. The disadvantages include low elastic modulus and higher density. Adhesion of glass fiber to the polymer matrix is poor particularly in the presence of moisture. Therefore, the fiber surface has to be treated in order to improve the adhesion properties. Although glass fibers can be used at higher temperatures, the temperature up to which it can be used depends on the polymer.

There are two types of glass fibers available for structural applications. These are E-glass and S-glass fibers. S-glass fibers are superior to E-glass fibers in terms of mechanical properties. However, the E-glass fiber is cheaper than the S-glass fiber. The letter E in E-glass stands for electrical as it was originally used for electrical applications. It is now employed for structural applications as well. The letter S in S-glass stands for high content of silica. Table 2.1 shows the main elements in the composition of E-glass and S-glass fibers. Table 2.2 shows the properties of the glass fibers. From Table 2.2, it is observed that both modulus and strength of S-glass fibers are higher than those of the E-glass fibers. The tensile strength of S-glass is roughly 30% greater than E-glass, while the modulus is 20% higher than E-glass. Further, specific modulus and specific strength are also higher. This makes S-glass superior for design of structural elements that are to carry more loads. In addition, the thermal coefficient of expansion being lower, the thermal stresses that get generated at the junction of dissimilar materials will be lower. In India, E-glass fiber is largely used for structural applications.

Table 2.1 Chemical compositions of E-glass and S-glass fibers [2]

Material	E-glass (% weight)	S-glass (% weight)
Silicon oxide	54.3	64.20
Aluminum oxide	15.2	24.80
Ferrous oxide	-	0.21
Calcium oxide	17.2	0.01
Magnesium oxide	4.7	10.27
Sodium oxide	0.6	0.27
Boron oxide	8.0	0.01
Barium oxide	-	0.20
Others	-	0.03

Table 2.2 Properties of E-glass and S-glass fibers [2]

Property	E-glass	S-glass
Density (g/cm^3)	2.54	2.49
Range of diameter (μm)	3-20	8-13
Tensile strength (MPa)	3.45	4.30
Tensile modulus (GPa)	72.4	85.5
% Elongation (ultimate)	4.8	5.7
Coefficient of thermal expansion (10^{-6}/°C)	5.0	2.9
Specific strength	1.39	1.72
Specific modulus	28.5	34.33

Glass fibers are manufactured in a refractory furnace from a melt consisting of sand, limestone and alumina. It can also be manufactured by using glass marbles. The temperature of the furnace is maintained at 1400°C. The melt is passed through a heated platinum or ceramic bushing containing multiple holes with a diameter of the order of 10 μm. The molten glass comes out of these bushings, gathers together and is mechanically brought to the required size. In order to prevent the breaking of the fibers by rubbing against each other, a lubricant is applied. The fibers are then gathered into a bundle and wound on to a cylindrical spool at speeds of up to 50 m/s. These are dried in an oven to remove any water or sizing solution.

Passing a jet of air across the fibers coming out of the bushing produces short or discontinuous fibers. These fibers are collected on a rotating vacuum drum, sprayed with a matrix material. In order to prevent the fibers from breaking or crazing, chemical treatments are carried out during the drawing process. This process is called *sizing*. Sizings are of two types: (i) *temporary sizing* and (ii) *compatible sizing*. Temporary sizing is done to minimize the degradation of the fibers and its properties due to rubbing of the fibers against one another and to bind the fibers for easy handling. The material often used for temporary sizing is starch oil. Sometimes, oil lubricant is also employed. Temporary sizing has to be removed before impregnating with resin. These sizing materials inhibit good adhesion between the fiber and the matrix making load transfer inefficient. Temporary sizing can be removed by heating the fiber in an air-circulating oven at about 340°C for 15-20 h. Compatible sizing is done to improve the adhesion properties. The materials used for sizing are also called *coupling agents*. The most common coupling agents are silanes. The purpose of the silane is to provide good adhesion between the fiber and the matrix and also to maintain the bond strength of the interface even in the presence of moisture.

Glass fibers are available in various forms. These can be customized based on the application. The various forms are:

(i)   Roving: These are continuous fibers of diameters varying from 9 to 13 μm and typically have 20 strands. They are employed in making continuous fiber composites by making prepregs and filament winding.

(ii)   Woven roving: These are fibers which are woven in the form of a fabric. They are employed when a rapid build of thickness is required as in fiber glass boats etc.

(iii) Chopped strand mat: Chopped strand mat is a non-woven glass fiber which consists of short or discontinuous fibers randomly distributed in the horizontal plane. The fibers are bound together by a chemical binder.

(iv)   Continuous strand mat: This consists of continuous fibers interlocked in a spiral fashion like a rope.

(v)   Woven fabrics: These are once again continuous fibers woven into a fabric like a cloth. The properties of the fabric depends on the fabric constructions i.e. number of fibers in each direction, weave pattern (satin weave or plain weave).

(vi)   Surfacing mat or veil: This is a very thin mat of single continuous fibers randomly distributed. It is used as a surface layer to provide a smooth surface.

Carbon Fibers

Carbon fibers contain 93 to 95% of carbon content, while graphite fibers contain more than 99% carbon. The raw material used for making carbon and graphite fibers is the same. However, the temperature at which carbon fibers are drawn is lower than that for graphite fibers. Graphite fibers have higher modulus and strength as compared to carbon. This is due to the higher percentage of carbon content and also due to its microstructure.

Graphite fibers are manufactured by using one of the three precursor materials: polyacrylonitrile (PAN) fibers, rayon fibers and pitch. Carbon fibers produced by pitch are the cheapest, while carbon fibers produced by rayon are the most expensive. Carbon fibers have a wide range of tensile modulus ranging from 207 to 802 GPa. Both high strength and high modulus forms are available.

Carbon fibers when graphitized result in graphite fibers. Graphitization is carried out at 3000°C under inert atmospheres. The process improves the crystalline structure and preferred orientation of the graphite like crystalline within each individual fiber. After the heat treatment process, the fiber surface is treated in order to improve the adhesion with the matrix. This improves the interfacial shear strength.

Carbon fibers are characterized as high strength and high modulus fibers. Depending on the loading conditions, one can choose either of the two. However, the high modulus type is more expensive. The high strength, low modulus fibers have low density, lower cost and high ultimate elongation.

Like glass fibers, carbon and graphite fibers are available in various forms. Table 2.3 gives the properties of carbon fibers obtained from different precursors.

Table 2.3 Properties of carbon fibers [2, 3]

Properties	Precursor		
	PAN	Pitch	Rayon
Tensile strength (MPa)	3654-5033	1724-4060	2070-2760
Tensile modulus (GPa)	228-595	170-980	415-550
Density (g/cm^3)	1.77-1.96	2.0-2.2	1.7
Maximum elongation (%)	0.4-1.2	0.25-0.7	-
Coefficient of thermal expansion ($10^{-6}/°C$)	-0.75 to -0.4	-1.6 to -0.9	-
Fiber diameter (μm)	5-8	10-11	6.5

Kevlar Fibers

Du Pont produces the aramid polymer fiber under the trade name 'Kevlar®'. Kevlar is made of carbon, hydrogen, oxygen and nitrogen. The fibers are made by poly condensation of diamines and diacid halides at low temperatures. The polymers are spun from strong acid solution by a dry-jet wet spinning process. They are made by adding a diacid chloride to a cool amine solution while stirring. The clean polymer mixed with a strong acid is extruded from spinnerets at an elevated temperature of about 100°C. More details can be obtained from Agarwal et al. [2].

There are two main types of Kevlar fibers, Kevlar-29 and Kevlar-49. The density of both types of fibers is the same. However, Kevlar-49 has higher specific stiffness. Kevlar-29 is used for making bulletproof vests, rope and cables. Kevlar-49 has both high strength and stiffness. Both the fibers have low compressive strength leading to difficulty in machining. In situations where bending is predominant, Kevlar alone cannot be used. It has to be used in conjunction with glass or carbon as a hybrid. Kevlar resists tensile stresses, while carbon or glass resists compressive stresses.

Boron Fibers

Boron fibers possess very high tensile modulus compared to other fibers. The tensile modulus varies in the range 379-414 GPa, while the tensile strength varies in the range 2750-3445 MPa. The diameters of the boron fibers are of the order 100-200 μm. This is one order of magnitude higher than the other fibers. Boron fibers have much higher strength and stiffness as compared to graphite. They also have higher density of the order 2.57 g/cm^3. In view of the increased diameter, these fibers can withstand compressive stresses.

Boron fibers are produced by chemical vapor deposition of boron trichloride on a substrate of tungsten or carbon. The substrate is heated to a temperature of 1260°C and pulled continuously through a reactor which results in the required coating thickness. They are widely used for aerospace components manufacture. The high cost of the material prevents its use in other structures.

Ceramic Fibers

These fibers are employed for the manufacture of components that are subjected to a high temperature environment. They retain strength and stiffness over a large range of temperature and are also free from environmental degradation. Alumina and ceramic fibers are among important high temperature fibers. Alumina has inherent resistance to oxidation and, is therefore, employed in gas turbine blades.

Table 2.4 gives the properties of some of the fibers [2-4].

Table 2.4 Properties of the fibers

Properties	Material				
	Carbon	Kevlar-29	Kevlar-49	Boron	Silicon carbide
Diameter (μm)	7-9	12	12	100	10-140
Density (g/cm^3)	1.99	1.44	1.47	2.6	2.6-3.3
Elastic modulus (GPa)	250-390	65	125	410	180-430
Tensile strength (MPa)	2200-2700	2800	2800-3600	3450	2000-3500

2.3 Matrix or Binding Material

As stated earlier, fibers cannot transmit loads from one fiber to another. They have to be embedded in a matrix or a binding material to make them a composite capable of carrying the load. The matrix binds the fibers together, helps in transfer of loads between fibers and also protects them from environmental effects. The choice of the binding material depends on the melting and curing temperatures of the matrix, viscosity, variation of viscosity with temperature and adhesion with fibers. The bond between the fiber and the binding material should be good. This requires that the matrix must be capable of developing a good mechanical or chemical bond with the fiber. Further, the fiber and the matrix material must be compatible so that no undesirable reaction takes place at the interface.

2.3.1 Polymeric Material

Polymers are extensively used as matrix material. The main advantages are low cost, good chemical resistance, high elongation and low density. However, they cannot be used at high temperatures as the properties degrade. The useful temperature is in the range of 100 to 300°C.

Polymeric materials are divided into two categories: *thermosetting* and *thermoplastic*. *Thermosetting* polymers are largely used as matrix material for fiber reinforced composites. They do not soften but decompose on heating. Once the curing is complete, they cannot be reshaped. Common examples of thermosetting polymers are epoxies, polyesters, phenolics, silicones and polyimides. *Thermoplastic* polymers soften and melt on heating. Softening and solidification of these polymers are reversible. This has an advantage that products can be reshaped by the application of heat and pressure. Thus, in the case of thermoplastics, the material can be formed first and then the product. The temperature to which the thermoplastics can be used depends on whether it is semi-crystalline or amorphous. In the case of thermosetting polymers, the material and the product are formed at the same time. Thermoplastic polymers exhibit considerable strain at low stresses. Some of the common polymers are: polyethylene, polystyrene, polycarbonate, polyether-ether-ketone (PEEK), polysulfone etc.

We shall describe in detail some of the matrix materials generally employed for composites.

Epoxies

Epoxy resins are mostly used for aerospace applications with a variety of fibers. These can be cured at room temperature. The curing process can be accelerated by the application of pressure and temperature. One can improve the characteristics of epoxies by adding hardeners, plasticizers and fillers. Hardeners are added to improve curing of composites. Plasticizers are introduced to improve toughness, flexibility, processibility and ductility of polymers. Fillers are added to improve strength, surface texture and ultraviolet absorption of the polymer. The prime reason for using epoxy as a matrix material is due to its excellent mechanical properties and better adhesion to fibers. The epoxy resin cannot be used beyond 150°C.

Polyester

This is the most commonly used resin in glass reinforced composites. It is brittle and results in shrinkage of as much as 8% during curing. This implies that the flat surface before curing may turn out to be curved after curing. Calcium carbonate is used as the filler material. This not only brings down the cost, but also reduces shrinkage.

Phenolic Resins

These have excellent fire resistance properties. Hence, generally used for high temperature applications. Compared to epoxy and polyester, phenolic resins have inferior properties. However, they have better flame retardant and toxic gas emission characteristics. They exhibit high strength characteristics. Composite products result in high void content.

Vinyl Esters

Vinyl esters have good water resistance properties. They can be used at elevated temperatures, as there is very little deterioration in their strength and stiffness. The cost of the vinyl ester lies in between phenolic and epoxy resins. In view of the water resistance properties, they are used in marine industries.

Polyimides

Polyimides are employed for high temperature range applications. The operating temperature is in the range of 250-300°C. They possess excellent chemical resistance. However, they are generally very brittle.

Table 2.5 shows some of the properties of the thermosetting resins which are normally employed for making fiber reinforced composites.

Table 2.5 Properties of thermosetting polymers [5]

Properties	Thermosetting polymers				
	Epoxy	Vinyl ester	Polyester	Phenolic	Polyimides
Density (g/cm^3)	1.2-1.3	1.12-1.32	1.1-1.5	1.30	1.46
Tensile modulus (GPa)	2.75-6	3-4	1.1-4.5	2.7-4.1	3.5-4.5
Tensile strength (MPa)	55-130	65-90	40-90	50-75	120
Compressive strength (MPa)	100-200	127	90-250	200	206
Elongation (%)	1-8.5	1-5	2.5	2	2.0
Water absorption (%)	0.08-0.15	-	0.1-0.3	0.1-0.2	0.3
Coefficient of thermal expansion (10^{-6}/°C)	45-70	53	60-200	-	90

Thermoplastics

PEEK and polyphenylene sulfides (PPS) are finding several applications in view of their toughness, low moisture absorption properties and high temperature performance. PPS and PEEK have melting temperatures in the range of 315-370°C. As stated earlier, the elastic modulus variation depends on whether the material is amorphous or semi-crystalline. An amorphous thermoplastic material exhibits a sudden drop in the value of the elastic modulus near the *glass transition temperature* (T_g), while the semi-crystalline materials have a moderate drop of the modulus at the transition temperature. Depending on the application, the melting temperature governs the choice of the material. Table 2.6 shows the properties of some thermoplastic materials.

Table 2.6 Properties of thermoplastic polymers

Properties	Thermoplastic polymers			
	PEEK	PET	PC	PPS
Density (g/cm)	1.3	1.35	1.2	1.32
Tensile modulus (GPa)	3.7	2.8	2.3	3.3
Tensile strength (MPa)	92	80	60	70
Failure strain (%)	50	80	100	-

PC: Polycarbonate, PET: Polyethylene terephthalate

2.4 Fillers

Fillers are added to the polymeric matrix material to change certain characteristics of the material. They are generally added to the matrix material at the time of manufacture. These, however, do not change the mechanical properties. They can be employed to reduce shrinkage, viscosity and also increase the smoothness of the surface.

The most common filler employed with polyester and vinyl ester is calcium carbonate ($CaCO_3$) for cost reduction and shrinkage. This results in high surface smoothness. The other types of fillers are china clay, mica, silica and glass microspheres. China clay is used to increase the viscosity of resins as well as its fire resistance characteristics. Microspheres are used to reduce the density of the matrix material.

It is well known that ultraviolet light affects the appearance of the composite. To prevent this effect, ultraviolet absorbents like acetyl salicylic acid and benzotriazoles hydroxphenyl are used.

Natural silica is employed in thermosetting polymers for dimensional stability, good electrical resistance and improved thermal conductivity.

For more detailed information on fillers, the readers can refer to [6-8].

Summary

- A composite consists of two or more distinct phases: a reinforcing phase and a binding phase.
- Fibers are generally of circular cross-section with fiber diameter in the range of 8-12 μm.
- The length, orientation, cross-section and material of the fiber contribute to the strength and stiffness of the composite.
- Natural fibers have the ability to provide stiffness enhancement and sound damping at lower cost and density than glass fibers.
- Glass fibers are the most common synthetic fiber used in conjunction with polymeric matrix material. It is the cheapest and has high strength, low chemical resistance and very good insulating properties.

References

1. Katz, H.S. and Milewski, J.V. (Eds.), 1978, *Handbook of Fillers and Reinforcements for Plastics*, Van Nostrand Reinhold Co., New York.
2. Agarwal, B.D., Broutman, L.J. and Chandrashekhara, K., 2006, *Analysis and Performance of Fiber Composites*, 3rd Edition, John Wiley and Sons, Inc.
3. Gibson, R.F., 1994, *Principles of Composite Materials Mechanics*, International Edition, McGraw-Hill Book Co.
4. Stabb, G.H., 1999, *Laminar Composites*, Butterworth-Heinemann, MA.
5. Mukhopadhyay, M., 2004, *Mechanics of Composite Materials and Structures*, Universities Press (India) Private Ltd., Hyderabad.
6. Mazumdar, S.K., 2002, *Composites Manufacturing*, CRC Press, Boca Raton, Florida, USA.

7. Astrom, B.T., 2002, *Manufacturing of Polymer Composites*, Nelson Thornes, Cheltenham, UK.
8. Campbell, F.C., 2004, *Manufacturing Processes for Advanced Composites*, Elsevier, New York.

3. Lamina Constitutive Equations

In this chapter, we will:

- Derive the constitutive equations (stress-strain) for a 3D elastic body.
- Reduce the equations for a 2D case as the fiber reinforced lamina is basically a two-dimensional structure.
- Derive the transformation relations for stresses and strains from the material coordinate axes to reference orthogonal axes.

3.1 Introduction

As stated earlier, a single lamina is too thin to carry different types of loads. It can only support the tensile load along the directions of the fibers. Hence, a composite laminate consisting of a number of plies or lamina stacked together through an adhesive is generated to give the required thickness to withstand the loads applied. The mechanical properties of the fiber and the fiber orientation in the lamina are chosen to meet the load requirements. The mechanical properties of the laminate will depend on the combined effect of all the lamina properties. At the microscopic level, the composite is heterogeneous and anisotropic. However, at the macroscopic level, one can treat it as homogenous, that is, the properties do not vary from point to point. Further, for evaluating the properties of the lamina, it is convenient to consider a unidirectional lamina with parallel continuous fibers. The off axis properties can be obtained from the transformation relations. In this chapter, the constitutive relations for a unidirectional lamina with continuous fibers will be derived assuming that the bonding between the fiber and the matrix is perfect.

3.2 Stiffness Coefficients for 3D Composites

Consider an element with a three-dimensional state of stress. The stress at a point in the body can be described by nine stress components σ_{ij} $(i, j = 1, 2, 3)$ as shown in Fig. 3.1. Correspondingly, there are nine components of strain ε_{ij} $(i, j = 1, 2, 3)$. It is very important to distinguish between the 'tensor' strain ε_{ij} and the engineering strain γ_{ij}.

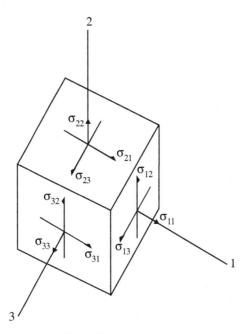

Fig. 3.1 Three-dimensional state of stress.

In the case of normal strain, the tensorial strain is the same as the engineering strain. However, for shear strain, the engineering strain γ_{ij} is twice the tensorial strain. The most general stress-strain relationship at a point in an elastic material can be written as

$$\{\sigma_{ij}\} = [Q_{ijkl}] \{\varepsilon_{kl}\} \tag{3.1}$$

$[Q_{ijkl}]$ is a 9×9 matrix of stiffness or elastic coefficients having 81 constants [1]. The first two subscripts correspond to the stresses and the latter two subscripts correspond to strains. Equation (3.1) represents a 3D anisotropic material. In the absence of body forces, both stresses and strains matrices are symmetric. Thus, there are only six independent stress components and six independent strain components. This implies that the stiffness coefficients must be symmetric with respect to the first two subscripts and with respect to the last two subscripts (i.e. $Q_{ijkl} = Q_{jikl}$ and $Q_{ijkl} = Q_{ijlk}$). The number of independent elastic constants reduces to 36. In view of this, Eq. (3.1) can be written in a simplified form as

$$\{\sigma\} = [Q] \{\varepsilon\} \tag{3.2}$$

The stiffness coefficient matrix $[Q]$ has 36 components and $\{\sigma\}$ and $\{\varepsilon\}$ are column vectors each having six elements, where

$$\sigma_1 = \sigma_{11}, \ \sigma_2 = \sigma_{22}, \ \sigma_3 = \sigma_{33}, \ \sigma_4 = \sigma_{23} = \sigma_{32}, \ \sigma_5 = \sigma_{13} = \sigma_{31}, \ \sigma_6 = \sigma_{12} = \sigma_{21}$$

$\varepsilon_1 = \varepsilon_{11}, \varepsilon_2 = \varepsilon_{22}, \varepsilon_3 = \varepsilon_{33}, \varepsilon_4 = \gamma_{23} = \gamma_{32} = 2\varepsilon_{23} = 2\varepsilon_{32}$
$\varepsilon_5 = \gamma_{31} = \gamma_{13} = 2\varepsilon_{23} = 2\varepsilon_{32}, \varepsilon_6 = \gamma_{12} = \gamma_{21} = 2\varepsilon_{12} = 2\varepsilon_{21}$

In a similar manner, the inverse form of the relation (3.2) can be written as

$$\{\varepsilon\} = [S]\{\sigma\} \tag{3.3}$$

in which $[S]$ is the compliance coefficient matrix.

Relation (3.2) can be further simplified as a consequence of the existence of strain energy density function W [2, 3]. This reduces the number of independent coefficients further. The work done per unit volume is given by

$$W = \frac{1}{2}\sigma_i \varepsilon_i (i = 1, 2, \ldots, 6) \tag{3.4}$$

Substituting for σ_i from Eq. (3.2) into Eq. (3.4) gives

$$W = \frac{1}{2}Q_{ij} \varepsilon_j \varepsilon_i \tag{3.5}$$

Taking second derivative of W, we obtain

$$\frac{\partial^2 W}{\partial \varepsilon_i \partial \varepsilon_j} = Q_{ij} \tag{3.6}$$

If the order of differential is changed, we obtain

$$\frac{\partial^2 W}{\partial \varepsilon_j \partial \varepsilon_i} = Q_{ij} \tag{3.7}$$

Since the order of differentiation is irrelevant, the result has to be the same

$$Q_{ij} = Q_{ji} \tag{3.8}$$

As a consequence, the number of independent constants reduces from 36 to 21. Thus, the response of an isotropic homogenous material to the applied forces can be estimated with the help of 21 constants.

Further, in most materials, some symmetry exists with regard to certain planes. The material properties do not change when the direction of the axis perpendicular to this plane is reversed. Consider the plane 1-2 to be the plane of symmetry. The symmetry requires that the material properties do not change when the properties of a point below and above the plane of symmetry are measured. This reduces the number of independent stiffness coefficients from 21 to 13. Such a material is termed as a *monoclinic material*.

A 3D orthotropic material has three planes of symmetry. In Fig. 3.2, the coordinate axes, 1, 2 and 3 are chosen along the fiber directions. They are generally referred to as the material axes. For such a material, the stiffness coefficient matrix has 9 independent material constants. The material is called a *specially orthotropic material*.

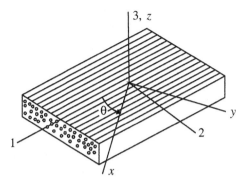

Fig. 3.2 Orthotropic lamina with orthogonal coordinate system.

The coefficient matrix is written as

$$[Q] = \begin{bmatrix} Q_{11} & Q_{12} & Q_{13} & 0 & 0 & 0 \\ Q_{12} & Q_{22} & Q_{23} & 0 & 0 & 0 \\ Q_{13} & Q_{23} & Q_{33} & 0 & 0 & 0 \\ 0 & 0 & 0 & Q_{44} & 0 & 0 \\ 0 & 0 & 0 & 0 & Q_{55} & 0 \\ 0 & 0 & 0 & 0 & 0 & Q_{66} \end{bmatrix} \tag{3.9}$$

If the properties are the same in any of the directions perpendicular to the fibers (i.e the properties are same along directions 2 and 3, the material is called *transversely isotropic*. For such a material, $Q_{22} = Q_{33}$, $Q_{12} = Q_{13}$, $Q_{55} = Q_{66}$ and Q_{44} would not be independent of other stiffnesses. Thus, the stiffness matrix for an especially orthotropic or transversely isotropic material is of the form

$$[Q] = \begin{bmatrix} Q_{11} & Q_{12} & Q_{12} & 0 & 0 & 0 \\ Q_{12} & Q_{22} & Q_{23} & 0 & 0 & 0 \\ Q_{12} & Q_{23} & Q_{22} & 0 & 0 & 0 \\ 0 & 0 & 0 & (Q_{22}-Q_{23})/2 & 0 & 0 \\ 0 & 0 & 0 & 0 & Q_{66} & 0 \\ 0 & 0 & 0 & 0 & 0 & Q_{66} \end{bmatrix} \tag{3.10}$$

The 2-3 plane and all other parallel planes are assumed to be planes of isotropy. Thus, we need five independent stiffness coefficients to characterize the material.

For isotropic material like metals, where all three coordinate axes represent the axes of symmetry, we have the stiffness coefficient matrix in the form

$$[Q] = \begin{bmatrix} Q_{11} & Q_{12} & Q_{12} & 0 & 0 & 0 \\ Q_{12} & Q_{11} & Q_{12} & 0 & 0 & 0 \\ Q_{12} & Q_{12} & Q_{11} & 0 & 0 & 0 \\ 0 & 0 & 0 & (Q_{11} - Q_{12})/2 & 0 & 0 \\ 0 & 0 & 0 & 0 & (Q_{11} - Q_{12})/2 & 0 \\ 0 & 0 & 0 & 0 & 0 & (Q_{11} - Q_{12})/2 \end{bmatrix} \qquad (3.11)$$

Thus, we observe from relation (3.11) that only two independent stiffness coefficients are required to characterize the material.

3.3 Stiffness Coefficients for 2D Composites

For a fiber reinforced lamina which is two-dimensional in its behavior (as the thickness of the lamina is very small compared to the other two dimensions), all the terms associated with axis 3, may be dropped to simplify the $[Q]$ matrix as

$$[Q] = \begin{bmatrix} Q_{11} & Q_{12} & 0 \\ Q_{12} & Q_{22} & 0 \\ 0 & 0 & Q_{66} \end{bmatrix} \qquad (3.12)$$

Thus, there are four independent stiffness coefficients to be evaluated to characterize an orthotropic lamina completely. The constitutive relation for such a lamina in terms of the stiffness coefficient can be written as

$$\begin{Bmatrix} \sigma_1 \\ \sigma_2 \\ \tau_{12} \end{Bmatrix} = \begin{bmatrix} Q_{11} & Q_{12} & 0 \\ Q_{12} & Q_{22} & 0 \\ 0 & 0 & Q_{66} \end{bmatrix} \begin{Bmatrix} \varepsilon_1 \\ \varepsilon_2 \\ \gamma_{12} \end{Bmatrix} \qquad (3.13)$$

The constitutive relation given by Eq. (3.13) relates stress to strain. In many situations, we may need to express stresses in terms of strains. Thus, the inverse form of Eq. (3.13) is written as

$$\begin{Bmatrix} \varepsilon_1 \\ \varepsilon_2 \\ \gamma_{12} \end{Bmatrix} = \begin{bmatrix} S_{11} & S_{12} & 0 \\ S_{12} & S_{22} & 0 \\ 0 & 0 & S_{66} \end{bmatrix} \begin{Bmatrix} \sigma_1 \\ \sigma_2 \\ \tau_{12} \end{Bmatrix} \qquad (3.14)$$

For an isotropic material, as stated earlier, there will be only two independent coefficients, either in terms of Q_{ij} or S_{ij}.

3.4 Relations Between Stiffness Coefficients and Engineering Constants

As mentioned before, a composite is a man-made material. Therefore, it needs to be characterized before being employed as a load carrying structural members. The relations mentioned in the earlier section are valid for several important classes of materials. When a material is characterized experimentally, we generally get the data on elastic modulus, Poisson's ratio as well as shear modulus, which will be a function of fiber volume fraction in the case of composites. Engineering constants are largely used in analysis and design as they relate simply the state of stress to the state of strain. It is possible to find relations between stiffness coefficient and engineering constants though it is not simple. Generally, it is convenient to work with stiffness coefficients for the purposes of analysis. For design, one needs to know the actual values of the material properties.

3.4.1 Especially Orthotropic Lamina

Consider an especially orthotropic lamina as shown in Fig. 3.3 with the material axes *L-T* coinciding with the direction of the fibers and normal to the fibers. The *L* axis is along the longitudinal direction and the *T* axis coincides with the transverse direction. We define the elastic modulus along the *L* and *T* directions as E_L and E_T, respectively, the in-plane shear modulus as G_{LT}, the major Poisson's ratio as v_{LT} and minor Poisson's ratio as v_{TL}. The major Poisson's ratio v_{LT} (ratio of lateral strain to longitudinal strain) is due to the longitudinal stress, while the minor Poisson's ratio v_{TL} (ratio of longitudinal strain to lateral strain) is caused by the transverse stress.

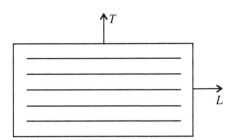

Fig. 3.3 Material axes for orthotropic lamina.

We assume that the deformations are small and all the fibers are intact during the application of the stress. As a consequence, the stress-strain relation is linear. We can, therefore, apply one stress at a time and find the resulting strains. We then add up all the strains due to the application of all the possible in-plane stresses.

Case 1
Only σ_L is the non-zero stress ($\sigma_T = \tau_{LT} = 0$).
   The strains developed due to this stress are ε_L, ε_T with $\gamma_{LT} = 0$

$$\varepsilon_L = \frac{\sigma_L}{E_L}, \quad \varepsilon_T = -v_{LT}\,\varepsilon_L = -v_{LT}\frac{\sigma_L}{E_L}, \gamma_{LT} = 0 \qquad (3.15)$$

Case 2

Only σ_T is the non-zero stress ($\sigma_L = 0$, $\tau_{LT} = 0$).

The strains developed due to this stress are ε_T, ε_L and $\gamma_{LT} = 0$.

$$\varepsilon_T = \frac{\sigma_T}{E_T}, \quad \varepsilon_L = -\nu_{TL}\,\varepsilon_T = -\nu_{TL}\frac{\sigma_T}{E_T}, \gamma_{LT} = 0 \tag{3.16}$$

Case 3

Only τ_{LT} is the non-zero stress ($\sigma_L = \sigma_T = 0$).

The strain produced by this stress is γ_{LT} with $\varepsilon_L = \varepsilon_T = 0$ $\qquad$ (3.17)

Superposition of all the three cases gives the state of the strains developed due to the application of all the possible in-plane stresses. Combining Eqs. (3.15), (3.16) and (3.17) gives the following three relations:

$$\varepsilon_L = \frac{\sigma_L}{E_L} - \nu_{TL}\frac{\sigma_T}{E_T}$$

$$\varepsilon_T = \frac{\sigma_T}{E_T} - \nu_{LT}\frac{\sigma_L}{E_L} \tag{3.18}$$

$$\gamma_{LT} = \frac{\tau_{LT}}{G}$$

Equations (3.18) are the strain-stress relations for an especially orthotropic lamina.

Equations (3.18) appear to have four independent material constants, E_L, E_T, G_{LT}, ν_{LT} and ν_{TL}. However, from Eq. (3.13), we have only four stiffness coefficients. It is shown subsequently that there exist only four independent engineering constants. E_L, E_T, G_{LT} and ν_{LT} or ν_{TL}.

Consider once again the same lamina as shown in Fig. 3.3 and apply the state of stress as considered in Cases 1 to 3.

For Case 1, with $\sigma_L \neq 0$, $\sigma_T = \tau_{LT} = 0$, we have from Eq. (3.13)

$$\sigma_L = Q_{11}\varepsilon_L + Q_{12}\varepsilon_T$$

$$0 = Q_{12}\varepsilon_L + Q_{22}\varepsilon_T \tag{3.19}$$

$$0 = Q_{66}\gamma_{LT}$$

We can solve for ε_L and ε_T from the first two equations of (3.18).

$$\varepsilon_L = \frac{Q_{22}}{(Q_{11}Q_{22} - Q_{12}^2)}\sigma_L \tag{3.20a}$$

$$\varepsilon_T = -\frac{Q_{22}}{(Q_{11}Q_{22} - Q_{12}^2)}\sigma_L \tag{3.20b}$$

$$\gamma_{LT} = 0 \tag{3.20c}$$

For Case 2, with $\sigma_T \neq 0$, and $\sigma_L = \tau_{LT} = 0$, we get

$$0 = Q_{11}\,\varepsilon_L + Q_{12}\,\varepsilon_T$$
$$\sigma_T = Q_{12}\,\varepsilon_L + Q_{22}\,\varepsilon_T$$
$$0 = Q_{66}\,\gamma_{LT}$$

Solving for ε_L and ε_T, we obtain

$$\varepsilon_T = \frac{Q_{11}}{(Q_{11}Q_{22} - Q_{12}^2)}\sigma_T \tag{3.21a}$$

$$\varepsilon_L = -\frac{Q_{12}}{(Q_{11}Q_{22} - Q_{12}^2)}\sigma_T \tag{3.21b}$$

$$\gamma_{LT} = 0 \tag{3.21c}$$

For Case 3, with $\tau_{LT} \neq 0$, and $\sigma_L = \sigma_T = 0$, we get

$$0 = Q_{11}\,\varepsilon_L + Q_{12}\,\varepsilon_T \tag{3.22a}$$

$$0 = Q_{12}\,\varepsilon_L + Q_{22}\,\varepsilon_T \tag{3.22b}$$

$$\tau_{LT} = Q_{66}\,\gamma_{LT} \tag{3.22c}$$

From Eqs. (3.20a) and (3.20b), we can obtain the longitudinal elastic modulus E_L and Poisson's ratio ν_{LT} as

$$E_L = \frac{\sigma_L}{\varepsilon_L} = \frac{Q_{11}Q_{22} - Q_{12}^2}{Q_{22}} \tag{3.23a}$$

Substituting for σ_L from Eq. (3.20a) in Eq. (3.20b) and simplifying, we get

$$\nu_{LT} = -\frac{\varepsilon_T}{\varepsilon_L} = \frac{Q_{12}}{Q_{22}} \tag{3.23b}$$

From Eqs. (3.21a) and (3.21b), we obtain the transverse elastic modulus E_T and Poisson's ratio ν_{TL} as

$$E_T = \frac{\sigma_T}{\varepsilon_T} = \frac{Q_{11}Q_{22} - Q_{12}^2}{Q_{11}} \tag{3.23c}$$

$$\nu_{TL} = -\frac{\varepsilon_L}{\varepsilon_T} = \frac{Q_{12}}{Q_{11}} \tag{3.23d}$$

From Eq. (3.22c), we obtain G_{LT} as

$$G_{LT} = \frac{\tau_{LT}}{\gamma_{LT}} = Q_{66} \tag{3.23e}$$

Using Eqs. (3.23a) to (3.23e), it is possible to obtain the relations between the individual stiffness coefficients Q_{ij} in terms of E_L, E_T, G_{LT}, ν_{LT} and ν_{TL}. These are

$$Q_{11} = \frac{E_L}{1 - \nu_{TL}\nu_{LT}}, \quad Q_{22} = \frac{E_T}{1 - \nu_{TL}\nu_{LT}}$$

$$Q_{12} = \frac{\nu_{TL}E_L}{1 - \nu_{LT}\nu_{TL}} = \frac{\nu_{LT}E_T}{1 - \nu_{LT}\nu_{TL}} \tag{3.24}$$

$$Q_{66} = G_{LT}$$

Since Q_{12} is the same irrespective of which relation is used, it is observed that there are only four independent elastic constants to define the material. The fifth constant is constrained by the relation

$$\nu_{LT} E_T = \nu_{LT} E_L \tag{3.25}$$

Equation (3.14), which relates the strains to stresses, can also be written in terms of engineering constants. These are also obtained in a similar manner by applying one stress at a time.

$$S_{11} = \frac{1}{E_L}, \quad S_{22} = \frac{1}{E_T}, \quad S_{12} = -\frac{\nu_{LT}}{E_L} = -\frac{\nu_{TL}}{E_T}, \quad S_{66} = \frac{1}{G_{LT}} \tag{3.26}$$

The elastic constants derived in this section are valid only when the reinforcement is continuous and along the X-axis. If, however, the fibers are along the Y-axis, E_L and E_T get interchanged.

A special class of lamina in which the amount of fiber reinforcement is the same along 0° and 90° is termed as a balanced orthotropic lamina. For this case, the number of independent elastic constants reduces to 3 because of the symmetry of the properties with respect to the L and T axes. Therefore, for a balanced lamina, we have $E_L = E_T$, $S_{11} = S_{22}$ and $Q_{11} = Q_{22}$. This is true provided the volume fraction of the fiber along both the directions is same.

3.5 Constitutive Relations for a Generally Orthotropic Lamina

For composite components, which are expected to carry out-of-plane loads as well, several laminae have to be stacked together to get the required stiffness. This is obtained by stacking several unidirectional laminae with specified sequence of fiber orientation. Therefore, it is necessary to obtain the constitutive equations for an off-axis lamina. That is, the principal directions of the lamina do not coincide with the reference axes or the structural axes. Such a lamina is generally referred to as a *generally orthotropic lamina*. The sign convention for the fiber orientation angle θ is shown in Fig. 3.4. The fiber orientation θ is positive when the axis is rotated in the anti-clockwise direction from the reference axis to the material axis (principal axis). The stress/strain relationships are found by making use of the strain

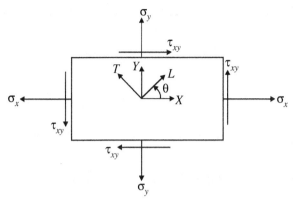

Fig. 3.4 2D orthotropic element of the lamina with possible in-plane stresses.

and stress transformations from the material axes (*L-T*) to the reference axes (*X-Y*). In many books, the material axes are also called 1-2 axes. The direction 1 is parallel to the fibers and the direction 2 is transverse to the fibers. Relations for the transformation of stresses and strains from the *L-T* coordinate system to the *X-Y* coordinate or vice versa can be derived by writing the equations of static equilibrium for the differential element shown in Fig. 3.4.

3.5.1 Transformation of Stresses

Considering the equilibrium of all the forces along a direction *L*, we get

$$\sum F_L = (\sigma_L - \sigma_x \cos^2\theta - \sigma_y \sin^2\theta - 2\tau_{xy} \sin\theta \cos\theta)\, dA = 0 \quad (3.27)$$

Since $dA \neq 0$, the quantity inside the bracket should be zero to satisfy the equation. Hence

$$\sigma_L = \sigma_x \cos^2\theta + \sigma_y \sin^2\theta + 2\tau_{xy} \sin\theta \cos\theta \qquad (3.28a)$$

Similarly, by summing up all the forces in direction *T*, we obtain

$$\tau_{LT} = -\sigma_x \sin\theta \cos\theta + \sigma_y \sin\theta \cos\theta + \tau_{xy}(\cos^2\theta - \sin^2\theta) \qquad (3.28b)$$

Similarly, by making a cut at an angle θ normal to direction *T*

$$\sigma_T = \sigma_x \sin^2\theta + \sigma_y \cos^2\theta - 2\tau_{xy} \sin\theta \cos\theta \qquad (3.28c)$$

Equations (3.28a) to (3.28c) can be rewritten in the matrix form as

$$\begin{Bmatrix} \sigma_L \\ \sigma_T \\ \tau_{LT} \end{Bmatrix} = \begin{bmatrix} \cos^2\theta & \sin^2\theta & 2\sin\theta\cos\theta \\ \sin^2\theta & \cos^2\theta & -2\sin\theta\cos\theta \\ -\sin\theta\cos\theta & \sin\theta\cos\theta & \cos^2\theta - \sin^2\theta \end{bmatrix} \begin{Bmatrix} \sigma_x \\ \sigma_y \\ \tau_{xy} \end{Bmatrix} \qquad (3.29)$$

The 3×3 matrix in Eq. (3.29) which relates the stresses in the *L-T* coordinate system to the *X-Y* coordinate system is called the transformation matrix [*T*].

Inverting this square matrix, we obtain the stresses in the *X-Y* coordinate system in terms of the *L-T* coordinates. $[T]^{-1}$ is written as

$$[T]^{-1} = \begin{bmatrix} \cos^2\theta & \sin^2\theta & -2\sin\theta\cos\theta \\ \sin^2\theta & \cos^2\theta & 2\sin\theta\cos\theta \\ \sin\theta\cos\theta & -\sin\theta\cos\theta & \cos^2\theta-\sin^2\theta \end{bmatrix} \qquad (3.30)$$

Thus, the stresses in the reference coordinate *X-Y* are related to those in the *L-T* coordinate as

$$\begin{Bmatrix} \sigma_x \\ \sigma_y \\ \tau_{xy} \end{Bmatrix} = \begin{bmatrix} \cos^2\theta & \sin^2\theta & -2\sin\theta\cos\theta \\ \sin^2\theta & \cos^2\theta & 2\sin\theta\cos\theta \\ \sin\theta\cos\theta & -\sin\theta\cos\theta & \cos^2\theta-\sin^2\theta \end{bmatrix} \begin{Bmatrix} \sigma_L \\ \sigma_T \\ \tau_{LT} \end{Bmatrix} \qquad (3.31)$$

3.5.2 Transformation of Strains

As stated earlier, the material axes are *L-T* and the reference axes are *X-Y*. Assuming small deformations, the strain-displacement relationships in the *X-Y* axes can be written as

$$\varepsilon_x = \frac{\partial u}{\partial x}, \varepsilon_y = \frac{\partial v}{\partial y}, \gamma_{xy} = \frac{\partial u}{\partial y} + \frac{\partial v}{\partial x}$$

u, *v* are the in-plane displacements along the *X* and *Y* directions. Transforming the strains in terms of the *L-T* coordinates, we get

$$\varepsilon_L = \varepsilon_x \cos^2\theta + \varepsilon_y \sin^2\theta + \gamma_{xy} \sin\theta\cos\theta$$
$$\varepsilon_T = \varepsilon_x \sin^2\theta + \varepsilon_y \cos^2\theta - \gamma_{xy} \sin\theta\cos\theta \qquad (3.32)$$
$$\gamma_{LT} = -2\varepsilon_x \sin\theta\cos\theta + 2\varepsilon_y \sin\theta\cos\theta + \gamma_{xy} (\cos^2\theta - \sin^2\theta)$$

In matrix form, Eq. (3.32) is written as

$$\begin{Bmatrix} \varepsilon_L \\ \varepsilon_T \\ \gamma_{LT}/2 \end{Bmatrix} = \begin{bmatrix} \cos^2\theta & \sin^2\theta & 2\sin\theta\cos\theta \\ \sin^2\theta & \cos^2\theta & -2\sin\theta\cos\theta \\ -\sin\theta\cos\theta & \sin\theta\cos\theta & \cos^2\theta-\sin^2\theta \end{bmatrix} \begin{Bmatrix} \varepsilon_x \\ \varepsilon_y \\ \gamma_{xy}/2 \end{Bmatrix} \qquad (3.33)$$

It is observed that the 3 × 3 matrix in Eq. (3.33) is the same matrix as that in Eq. (3.29).

Inverting Eq. (3.33) gives

$$\begin{Bmatrix} \varepsilon_x \\ \varepsilon_y \\ \gamma_{xy}/2 \end{Bmatrix} = \begin{bmatrix} \cos^2\theta & \sin^2\theta & -2\sin\theta\cos\theta \\ \sin^2\theta & \cos^2\theta & 2\sin\theta\cos\theta \\ \sin\theta\cos\theta & -\sin\theta\cos\theta & \cos^2\theta-\sin^2\theta \end{bmatrix} \begin{Bmatrix} \varepsilon_L \\ \varepsilon_T \\ \gamma_{LT}/2 \end{Bmatrix} \qquad (3.34)$$

It can be verified that $[T]\,[T]^{-1}$ is a unit matrix. It should be understood that Eqs. (3.33) and (3.34) show that the *tensor strains* transform the same way as the stresses. However, in most engineering applications, we require the relation between stresses and *engineering strains* with respect to the reference coordinates $(X\text{-}Y)$.

3.5.3 Transformed Stiffness Coefficients

We are interested in finding the relation between stresses and engineering strains with respect to the reference coordinate. This is represented as

$$\begin{Bmatrix} \sigma_x \\ \sigma_y \\ \tau_{xy} \end{Bmatrix} = \begin{bmatrix} \overline{Q}_{11} & \overline{Q}_{12} & \overline{Q}_{16} \\ \overline{Q}_{12} & \overline{Q}_{22} & \overline{Q}_{26} \\ \overline{Q}_{16} & \overline{Q}_{26} & \overline{Q}_{66} \end{bmatrix} \begin{Bmatrix} \varepsilon_x \\ \varepsilon_y \\ \gamma_{xy} \end{Bmatrix} \tag{3.35}$$

in which $\overline{Q}_{ij}$ are the components of the transformed lamina stiffness coefficients.

We define a transformation matrix $[R]$ as

$$[R] = \begin{bmatrix} 1 & 0 & 0 \\ 0 & 1 & 0 \\ 0 & 0 & 2 \end{bmatrix} \tag{3.36}$$

This matrix is used to transform the tensorial strain to engineering strain.

$$\begin{Bmatrix} \varepsilon_x \\ \varepsilon_y \\ \gamma_{xy} \end{Bmatrix} = \begin{bmatrix} 1 & 0 & 0 \\ 0 & 1 & 0 \\ 0 & 0 & 2 \end{bmatrix} \begin{Bmatrix} \varepsilon_x \\ \varepsilon_y \\ \gamma_{xy}/2 \end{Bmatrix} \tag{3.37}$$

Similarly, we define the engineering strains in terms of the L-T coordinates.

Equation (3.13) is refined in terms of the L-T coordinates as

$$\begin{Bmatrix} \sigma_L \\ \sigma_T \\ \tau_{LT} \end{Bmatrix} = \begin{bmatrix} Q_{11} & Q_{12} & 0 \\ Q_{12} & Q_{22} & 0 \\ 0 & 0 & Q_{66} \end{bmatrix} \begin{Bmatrix} \varepsilon_L \\ \varepsilon_T \\ \gamma_{LT} \end{Bmatrix} \tag{3.38}$$

Making use of $[R]$ and $[Q]$ matrices, Eq. (3.38) is rewritten as

$$\begin{Bmatrix} \sigma_x \\ \sigma_y \\ \tau_{xy} \end{Bmatrix} = [R]^{-1}[Q] \begin{Bmatrix} \varepsilon_L \\ \varepsilon_T \\ \gamma_{LT} \end{Bmatrix} \tag{3.39}$$

Equation (3.34) can be rewritten as

$$\begin{Bmatrix} \varepsilon_L \\ \varepsilon_T \\ \gamma_{LT} \end{Bmatrix} = [R][T][R]^{-1} \begin{Bmatrix} \varepsilon_x \\ \varepsilon_y \\ \gamma_{xy} \end{Bmatrix} \tag{3.40}$$

Substituting Eq. (3.40) in Eq. (3.39), we get

$$\begin{Bmatrix} \sigma_x \\ \sigma_y \\ \tau_{xy} \end{Bmatrix} = [T]^{-1}[Q][R][T][R]^{-1} \begin{Bmatrix} \varepsilon_x \\ \varepsilon_y \\ \gamma_{xy} \end{Bmatrix} \tag{3.41}$$

Substituting the various matrices in Eq. (3.41), we obtain

$$\begin{Bmatrix} \sigma_x \\ \sigma_y \\ \tau_{xy} \end{Bmatrix} = \begin{bmatrix} \overline{Q}_{11} & \overline{Q}_{12} & \overline{Q}_{16} \\ \overline{Q}_{12} & \overline{Q}_{22} & \overline{Q}_{26} \\ \overline{Q}_{16} & \overline{Q}_{26} & \overline{Q}_{66} \end{bmatrix} \begin{Bmatrix} \varepsilon_x \\ \varepsilon_y \\ \gamma_{xy} \end{Bmatrix} \tag{3.42}$$

Equation (3.42) gives the constitutive equation of the lamina along the X-Y axes in terms of the transformed stiffness coefficients. The elements of the $\left[\overline{Q} \right]$ matrix in the expanded form is

$$\overline{Q}_{11} = Q_{11} \cos^4 \theta + Q_{22} \sin^4 \theta + 2(Q_{12} + 2Q_{66}) \sin^2 \theta \cos^2 \theta$$

$$\overline{Q}_{22} = Q_{11} \sin^4 \theta + Q_{22} \cos^4 \theta + 2(Q_{12} + 2Q_{66}) \sin^2 \theta \cos^2 \theta$$

$$\overline{Q}_{12} = (Q_{11} + Q_{22} - 4Q_{66}) \cos^2 \theta \sin^2 \theta + Q_{12} (\cos^4 \theta + \sin^4 \theta)$$

$$\overline{Q}_{66} = (Q_{11} + Q_{22} - 2Q_{12} - 2Q_{66}) \cos^2 \theta \sin^2 \theta + Q_{66} (\cos^4 \theta + \sin^4 \theta) \tag{3.43}$$

$$\overline{Q}_{16} = (Q_{12} - Q_{22} - 2Q_{66}) \cos^3 \theta \sin \theta - (Q_{22} - Q_{12} - 2Q_{66}) \sin^3 \theta \cos \theta$$

$$\overline{Q}_{26} = (Q_{12} - Q_{22} - 2Q_{66}) \cos \theta \sin^3 \theta - (Q_{22} - Q_{12} - 2Q_{66}) \sin \theta \cos^3 \theta$$

Thus, it can be seen that the transformed stiffness matrix consists of six coefficients. They are functions of Q_{11}, Q_{12}, Q_{22}, Q_{66} and fiber orientation θ. Further, $\overline{Q}_{11}, \overline{Q}_{22}, \overline{Q}_{12}, \overline{Q}_{66}$ are functions of even powers of θ. This implies that the magnitude of these quantities does not change if the fiber orientation θ is changed to $- \theta$. On the other hand, $\overline{Q}_{16}$ and $\overline{Q}_{26}$ are functions of odd powers of θ. This means that these quantities change when θ changes to $- \theta$. The magnitude remains the same but the sign changes. Further, in the X-Y coordinate system, the lamina with fiber orientation other than $0°$ and $90°$ exhibit coupling between normal stress and shear strain, and between shear stress and normal strain. This implies that if an angle-ply is subjected to a uniform load along the X direction, it will not only have uniform strain along the X direction but also give rise to shear strain and vice versa.

Equation (3.42) can be written in inverse form as

$$\begin{Bmatrix} \varepsilon_x \\ \varepsilon_y \\ \gamma_{xy} \end{Bmatrix} = \begin{bmatrix} \overline{S}_{11} & \overline{S}_{12} & \overline{S}_{16} \\ \overline{S}_{12} & \overline{S}_{22} & \overline{S}_{26} \\ \overline{S}_{16} & \overline{S}_{26} & \overline{S}_{66} \end{bmatrix} \begin{Bmatrix} \sigma_x \\ \sigma_y \\ \tau_{xy} \end{Bmatrix} \tag{3.44}$$

where S_{ij} are the elements of the transformed reduced compliance coefficient matrix. These are expressed as

$$\overline{S}_{11} = S_{11}\cos^4\theta + S_{22}\sin^4\theta + (2S_{12} + S_{66})\sin^2\theta\cos^2\theta$$

$$\overline{S}_{22} = S_{11}\sin^4\theta + S_{22}\cos^4\theta + (2S_{12} + S_{66})\sin^2\theta\cos^2\theta$$

$$\overline{S}_{12} = (S_{12} + S_{22} - S_{66})\cos^2\theta\sin^2\theta + S_{12}(\cos^4\theta + \sin^4\theta)$$

$$\overline{S}_{66} = 2(2S_{11} + 2S_{22} - 4S_{12} - S_{66})\cos^2\theta\sin^2\theta + S_{66}(\cos^4\theta + \sin^4\theta) \tag{3.45}$$

$$\overline{S}_{16} = (2S_{11} - 2S_{12} - S_{66})\cos^3\theta\sin\theta - (2S_{22} - 2S_{12} - S_{66})\sin^3\theta\cos\theta$$

$$\overline{S}_{26} = (2S_{11} - 2S_{12} - S_{66})\cos\theta\sin^3\theta - (2S_{22} - 2S_{12} - S_{66})\sin\theta\cos^3\theta$$

It is observed that the transformation is similar but the coefficients are not the same.

3.5.4 Transformed Engineering Constants

The lamina or ply engineering constants, E_L, E_T, G_{LT} and ν_{LT} in the *L-T* coordinate (material axes) system can also be transformed into the *X-Y* (reference axes) coordinate system. In many situations, we need this transformation. Equation (3.18) gives the constitutive equation for the lamina in terms of engineering constants referred to the *L-T* axes.

Consider once again a thin lamina with fibers oriented at an angle θ to the *X-Y* axes as shown in Fig. 3.4. Let us further assume that only the axial stress σ_x is acting on the lamina. The normal and shear stresses acting along the *L-T* direction can be obtained by making use of Eq. (3.29). These are

$$\sigma_L = \sigma_x\cos^2\theta$$
$$\sigma_T = \sigma_x\sin^2\theta \tag{3.46}$$
$$\tau_{LT} = -\sigma_x\sin\theta\cos\theta$$

Substituting for σ_L, σ_T and τ_{LT} from Eq. (3.46) in Eq. (3.18), we get

$$\varepsilon_L = \sigma_x\left(\frac{\cos^2\theta}{E_L} - \nu_{TL}\frac{\sin^2\theta}{E_T}\right)$$

$$\varepsilon_T = \sigma_x\left(\frac{\sin^2\theta}{E_T} - \nu_{LT}\frac{\cos^2\theta}{E_L}\right) \tag{3.47}$$

$$\gamma_{LT} = -\sigma_x\frac{\sin\theta\cos\theta}{G_{LT}}$$

Substituting for ε_L, ε_T and $\gamma_{LT}/2$ from Eq. (3.47) into Eq. (3.34), we obtain

$$\varepsilon_x = \sigma_x \left[\frac{\cos^4 \theta}{E_L} + \frac{\sin^4 \theta}{E_T} + \frac{1}{4}\left(\frac{1}{G_{LT}} - \frac{2\nu_{LT}}{E_L} \right) \sin^2 2\theta \right]$$

$$\varepsilon_y = \sigma_x \left[\frac{-\nu_{LT}}{E_L} + \frac{1}{4}\left(\frac{1}{E_L} + \frac{1}{E_T} - \frac{1}{G_{LT}} + \frac{2\nu_{LT}}{E_L} \right) \sin^2 2\theta \right] \qquad (3.48)$$

$$\gamma_{xy} = \sigma_x \left[\frac{-\nu_{LT}}{E_L} - \frac{1}{E_T} + \frac{1}{2G_{LT}} + \left(\frac{1}{E_L} + \frac{2\nu_{LT}}{E_L} + \frac{1}{E_T} - \frac{1}{G_{LT}} \right) \cos^2 \theta \right] \sin 2\theta$$

Defining

$$E_x = \frac{\sigma_x}{\varepsilon_x} \quad \text{(elastic modulus along the X direction)}$$

the first relation from Eq. (3.48) gives

$$\frac{1}{E_x} = \frac{\cos^4 \theta}{E_L} + \frac{\sin^4 \theta}{E_T} + \frac{1}{4}\left(\frac{1}{G_{LT}} - \frac{2\nu_{LT}}{E_L} \right) \sin^2 2\theta \qquad (3.49)$$

Poisson's ratio ν_{xy} is defined as the strain produced in the Y direction as a consequence of applied stress in the X direction only.

$$\nu_{xy} = -\frac{\varepsilon_x}{\varepsilon_y}$$

Substituting for ε_x and ε_y from Eq. (3.48), we get

$$\nu_{xy} = \left[\frac{\nu_{LT}}{E_L} - \frac{1}{4}\left(\frac{1}{E_L} + \frac{1}{E_T} + \frac{2\nu_{LT}}{E_L} \frac{1}{G_{LT}} \right) \sin^2 2\theta \right] E_x \qquad (3.50)$$

From the third relation of Eq. (3.48), it is observed that the normal stress σ_x gives rise to shear strain γ_{xy}. The presence of shear strain indicates the shear coupling. The amount of shear coupling is obtained by defining shear coupling coefficient or mutual influence coefficient. We define as in Ref. [4], a cross coefficient η_x that relates the shearing strain to the normal stress σ_x in the following manner:

$$\eta_x = -\frac{E_L \gamma_{xy}}{\sigma_x} \qquad (3.51)$$

Comparing Eq. (3.51) with the third relation of Eq. (3.48) gives

$$\eta_x = \left[\nu_{LT} + \frac{E_L}{E_T} - \frac{E_L}{2G_{LT}} - \cos^2 \theta \left(1 + 2\nu_{LT} + \frac{E_L}{E_T} - \frac{E_L}{G_{LT}} \right) \right] \sin 2\theta \qquad (3.52)$$

If we now consider the loading such that $\sigma_x = \tau_{xy} = 0$ with $\sigma_y \neq 0$, we get from Eq. (3.29)

$$\sigma_L = \sigma_y \sin^2 \theta$$

$$\sigma_T = \sigma_y \cos^2 \theta \qquad (3.53)$$

$$\tau_{LT} = \sigma_y \sin \theta \cos \theta$$

Making use of Eq. (3.53), the strains in the L and T directions are obtained from Eq. (3.18) as

$$\varepsilon_L = \sigma_y \left(\frac{\sin^2 \theta}{E_L} - v_{TL} \frac{\cos^2 \theta}{E_T} \right)$$

$$\varepsilon_T = \sigma_y \left(\frac{\cos^2 \theta}{E_T} - v_{LT} \frac{\sin^2 \theta}{E_L} \right) \qquad (3.54)$$

$$\gamma_{LT} = \sigma_y \frac{\sin \theta \cos \theta}{G_{LT}}$$

The strains in the X and Y directions are obtained from Eq. (3.34) as

$$\varepsilon_x = -\sigma_y \left[\frac{v_{TL}}{E_T} - \frac{1}{4} \left(\frac{1}{E_L} + \frac{1}{E_T} + \frac{2v_{LT}}{E_L} - \frac{1}{G_{LT}} \right) \sin^2 2\theta \right]$$

$$\varepsilon_y = \sigma_y \left[\frac{\sin^4 \theta}{E_L} + \frac{\cos^4 \theta}{E_T} + \frac{1}{4} \left(\frac{1}{G_{LT}} - \frac{2v_{LT}}{E_L} \right) \sin^2 2\theta \right] \qquad (3.55)$$

$$\gamma_{xy} = -\sigma_y \left[\frac{v_{LT}}{E_L} + \frac{1}{E_T} - \frac{1}{G_{LT}} - \sin^2 \theta \left(1 + 2v_{LT} + \frac{1}{E_T} - \frac{1}{G_{LT}} \right) \right] \sin 2\theta$$

One can now obtain E_y from the second relation of Eq. (3.55) as

$$\frac{1}{E_y} = \frac{\sin^4 \theta}{E_L} + \frac{\cos^4 \theta}{E_T} + \frac{1}{4} \left(\frac{1}{G_{LT}} - \frac{2v_{LT}}{E_L} \right) \sin^2 2\theta \qquad (3.56)$$

Equation (3.56) can also be obtained by replacing θ with $\theta + 90°$ in Eq. (3.49). Poisson's ratio v_{yx} is obtained from the first relation of Eq. (3.55).

$$v_{yx} = \left[\frac{v_{TL}}{E_T} - \frac{1}{4} \left(\frac{1}{E_L} + \frac{1}{E_T} - \frac{1}{G_{LT}} + \frac{2v_{LT}}{E_L} \right) \sin^2 2\theta \right] E_y \qquad (3.57)$$

Define a cross coefficient η_y that relates the shearing strain to normal stress σ_y as

$$\eta_y = -\frac{E_L \gamma_{xy}}{\sigma_y} \qquad (3.58)$$

Making use of the third relation of Eq. (3.55), we get

$$\eta_y = \left[\nu_{LT} + \frac{E_L}{E_T} - \frac{E_L}{2G_{LT}} - \left(1 + 2\nu_{LT} + \frac{E_L}{E_T} - \frac{E_L}{G_{LT}} \right) \right] \sin 2\theta \qquad (3.59)$$

considering that only $\tau_{xy} \neq 0$, $\sigma_x = 0$ and $\sigma_y = 0$. From Eq. (3.29), we have

$$\sigma_L = 2\tau_{xy} \sin\theta \cos\theta$$

$$\sigma_T = -2\tau_{xy} \sin\theta \cos\theta \qquad (3.60)$$

$$\tau_{LT} = \tau_{xy}(\cos^2\theta - \sin^2\theta)$$

Substituting for σ_L, σ_T and τ_{LT} from Eq. (3.60) into Eq. (3.18), the strains ε_L, ε_T and γ_{LT} are given as

$$\varepsilon_L = 2\tau_{xy} \left(\frac{1}{E_L} + \frac{\nu_{TL}}{E_T} \right) \sin\theta \cos\theta$$

$$\varepsilon_T = -2\tau_{xy} \left(\frac{1}{E_T} + \frac{\nu_{LT}}{E_L} \right) \sin\theta \cos\theta \qquad (3.61)$$

$$\gamma_{LT} = \frac{1}{G_{LT}} \tau_{xy}(\cos^2\theta - \sin^2\theta)$$

The strains ε_x, ε_y and γ_{xy} are obtained from Eq. (3.34). The relation for shear strain γ_{xy} is given by

$$\gamma_{xy} = 4\tau_{xy} \sin^2\theta \cos^2\theta \left(\frac{1}{E_L} + \frac{\nu_{TL}}{E_T} \right) + 4\tau_{xy} \sin^2\theta \cos^2\theta \left(\frac{1}{E_T} + \frac{\nu_{LT}}{E_L} \right)$$

$$+ \tau_{xy} \frac{1}{G_{LT}} (\cos^2\theta - \sin^2\theta)$$

$$= \tau_{xy} \left[\left(\frac{1}{E_L} + \frac{1}{E_T} + \frac{2\nu_{LT}}{E_L} \right) \sin^2 2\theta + \frac{1}{G_{LT}} \cos^2 2\theta \right]$$

$$= \tau_{xy} \left[\left(\frac{1}{E_L} + \frac{1}{E_T} + \frac{2\nu_{LT}}{E_L} \right)(1 - \cos^2 2\theta) + \frac{1}{G_{LT}} \cos^2 2\theta \right]$$

$$= \tau_{xy} \left[\left(\frac{1}{E_L} + \frac{1}{E_T} + \frac{2\nu_{LT}}{E_L} \right) - \left(\frac{1}{E_L} + \frac{1}{E_T} + \frac{2\nu_{LT}}{E_L} - \frac{1}{G_{LT}} \right) \cos^2 2\theta \right]$$

Therefore

$$\frac{1}{G_{xy}} = \left[\left(\frac{1}{E_L} + \frac{1}{E_T} + \frac{2\nu_{LT}}{E_L} \right) - \left(\frac{1}{E_L} + \frac{1}{E_T} + \frac{2\nu_{LT}}{E_L} - \frac{1}{G_{LT}} \right) \cos^2 2\theta \right] \qquad (3.62)$$

As stated earlier, τ_{xy} gives rise to normal strains ε_x and ε_y along the X and Y directions. These can be obtained from the first two relations of Eq. (3.34). As the reciprocal relation exists, these can be written as

$$\varepsilon_x = -\eta_x \frac{\tau_{xy}}{E_L}$$

$$\varepsilon_y = -\eta_y \frac{\tau_{xy}}{E_L}$$

(3.63)

η_x and η_y are given by Eq. (3.52) and Eq. (3.59), respectively.

The superposition of all these loadings results in the following equation:

$$\begin{Bmatrix} \varepsilon_x \\ \varepsilon_y \\ \gamma_{xy} \end{Bmatrix} = \begin{bmatrix} \dfrac{1}{E_x} & -\dfrac{v_{yx}}{E_y} & -\dfrac{\eta_x}{E_L} \\[2mm] -\dfrac{v_{xy}}{E_x} & \dfrac{1}{E_y} & -\dfrac{\eta_y}{E_L} \\[2mm] -\dfrac{\eta_x}{E_L} & -\dfrac{\eta_y}{E_L} & \dfrac{1}{G_{xy}} \end{bmatrix} \begin{Bmatrix} \sigma_x \\ \sigma_y \\ \tau_{xy} \end{Bmatrix}$$

(3.64)

Equation (3.64) is the constitutive relation for a lamina in the *X-Y* coordinate system; the fibers are oriented at an angle θ to the *X-Y* axes. It is observed from Eq. (3.64) that E_x, E_y, G_{xy}, v_{xy}, η_x and η_y is a complicated function of E_L, E_T, G_{LT}, v_{LT} and fiber orientation θ. Further, from the symmetry of the compliance matrix

$$\frac{v_{yx}}{E_x} = \frac{v_{xy}}{E_y}$$

(3.65)

Example 3.1

For a fiber reinforced composite lamina, the material properties along the *L-T* directions are E_L = 150 GPa, E_T = 20 GPa, G_{LT} = 5 GPa and v_{LT} = 0.2. Obtain the elements of the stiffness matrix [*Q*].

Solution

$$v_{TL} = v_{LT} \frac{E_T}{E_L} = 0.2 \times \frac{20}{150} = 0.0267$$

Hence $1 - v_{TL} \, v_{LT} = 1 - 0.2 \times 0.0267 = 0.9947$

Making use of Eq. (3.24)

$$Q_{11} = \frac{E_L}{1 - v_{LT}v_{TL}} = \frac{150}{0.9947} = 150.81\,\text{GPa}$$

$$Q_{22} = \frac{E_T}{1 - v_{LT}v_{TL}} = \frac{20}{0.9947} = 20.11\,\text{GPa}$$

$$Q_{12} = v_{TL}Q_{11} = 0.0267 \times 150.81 = 4.027 \text{ GPa}$$

$$Q_{66} = G_{LT} = 5\,\text{GPa}$$

Example 3.2

For a 45° lamina (material properties are given in Example 3.1), obtain the elements of the transformed reduced stiffness matrix $[\bar{Q}]$.

Solution

$$\cos 45° = 0.7071,\ \sin 45° = 0.7071$$

From Example 3.1, $Q_{11} = 150.81$ GPa, $Q_{22} = 20.11$ GPa, $Q_{12} = 4.027$ GPa, $Q_{66} = 5$ GPa. Substituting these values in Eq. (3.43) with $\cos^4 \theta = 0.25$, $\cos^3 \theta = 0.3535$, $\cos^2 \theta = 0.5$ and $\cos \theta = 0.7071$, we get

$\bar{Q}_{11} = 150.81 \times 0.25 + 20.11 \times 0.25 + 2 \times (4.027 + 2 \times 5) \times 0.5 \times 0.5$
$\quad = 49.8335$ GPa

$\bar{Q}_{22} = 150.81 \times 0.25 + 20.11 \times 0.25 + 2 \times (4.027 + 2 \times 5) \times 0.5 \times 0.5$
$\quad = 49.8335$ GPa

$\bar{Q}_{12} = (150.81 + 20.11 - 4 \times 5) \times 0.5 \times 0.5 + 4.027 \times (0.25 + 0.25)$
$\quad = 39.8625$ GPa

$\bar{Q}_{66} = (150.81 + 20.11 - 2 \times 4.027 - 2 \times 5.0) \times 0.5 \times 0.5 + 5.0 \times (0.25$
$\quad + 0.25)$
$\quad = 40.6265$ GPa

$\bar{Q}_{16} = (150.81 - 4.027 - 2 \times 5.0) \times 0.3535 \times 0.7071 - (20.11 - 4.027 - 2$
$\quad \times 5.0) \times 0.7071 \times 0.3535$
$\quad = 35.7554$ GPa

$\bar{Q}_{26} = (150.81 - 4.027 - 2 \times 5.0) \times 0.7071 \times 0.3535 - (20.11 - 4.027 - 2$
$\quad \times 5.0) \times 0.3535 \times 0.7071$
$\quad = 35.7554$ GPa

Example 3.3

Obtain the stresses in the L-T coordinate system for a lamina with ply angle $\theta = 30°$. The applied stresses are $\sigma_x = 100$ GPa, $\sigma_y = -50$ GPa and $\tau_{xy} = 25$ GPa.

Solution

$$\cos 30° = 0.866,\ \sin 30° = 0.5$$

From Eq. (3.29), we get

$$
\begin{Bmatrix} \sigma_L \\ \sigma_T \\ \tau_{LT} \end{Bmatrix} =
\begin{bmatrix}
\cos^2 \theta & \sin^2 \theta & 2\sin \theta \cos \theta \\
\sin^2 \theta & \cos^2 \theta & -2\sin \theta \cos \theta \\
-\sin \theta \cos \theta & \sin \theta \cos \theta & \cos^2 \theta - \sin^2 \theta
\end{bmatrix}
\begin{Bmatrix} \sigma_x \\ \sigma_y \\ \tau_{xy} \end{Bmatrix}
$$

Substituting the values of $\sin \theta$, $\cos \theta$ and the applied stresses, we have

$$
\begin{Bmatrix} \sigma_L \\ \sigma_T \\ \tau_{LT} \end{Bmatrix} =
\begin{bmatrix}
0.75 & 0.25 & 0.866 \\
0.25 & 0.75 & -0.866 \\
-0.433 & 0.433 & 0.500
\end{bmatrix}
\begin{Bmatrix} 100 \\ -50 \\ 25 \end{Bmatrix} \text{GPa}
$$

$$\left\{ \begin{array}{c} \sigma_L \\ \sigma_T \\ \tau_{LT} \end{array} \right\} = \left[\begin{array}{c} 84.15 \\ -34.5 \\ -52.45 \end{array} \right] GPa$$

Example 3.4

The material properties of a lamina in the *L-T* coordinate system are $E_L =$ 20 GPa, $E_T = 2.0$ GPa, $G_{LT} = 0.7$ GPa and $\nu_{LT} = 0.35$. Obtain the material properties E_x, E_y, G_{xy} and ν_{xy} for $\theta = 45°$.

Solution

From Eq. (3.25)

$$\nu_{TL} = \nu_{LT} \frac{E_L}{E_T} = 0.35 \frac{2}{20} = 0.035$$

Given

$$\cos 45° = 0.7071, \ \sin 45° = 0.7071$$

From Eqs. (3.49), (3.50), (3.56) and (3.57), we can obtain the mechanical properties.

$$\frac{1}{E_x} = \frac{\cos^4 45}{20} + \frac{\sin^4 45}{2} + \frac{1}{4} \left(\frac{1}{0.7} - \frac{2 \times 0.35}{14} \right) \sin^2 90$$
$$= 0.0125 + 0.125 + 0.445 = 0.5825$$

$$E_x = 1.7167 \ GPa$$

$$\frac{1}{E_y} = \frac{\sin^4 45}{20} + \frac{\cos^4 45}{2} + \frac{1}{4} \left(\frac{1}{0.7} - \frac{2 \times 0.35}{14} \right) \sin^2 90$$

$$E_y = 1.7167 \ GPa$$

$$\frac{1}{G_{xy}} = \frac{1}{20} + \frac{2 \times 0.35}{20} + \frac{1}{2} - \left(\frac{1}{20} + \frac{1}{2} - \frac{1}{0.7} + \frac{2 \times 0.35}{20} \right) \cos^2 90$$
$$= 0.05 + 0.035 + 0.5 = 0.585$$

$$G_{xy} = 1.7094 \ GPa$$

$$\nu_{xy} = \left[\frac{0.35}{20} - \frac{1}{4} \left(\frac{1}{20} + \frac{2 \times 0.35}{20} + \frac{1}{2} - \frac{1}{0.7} \right) \sin^2 90 \right] \times 1.7167$$
$$= [0.0175 - 0.25 (0.05 + 0.035 + 0.5 - 1.428)] \times 1.7167$$

$$\nu_{xy} = -0.2377$$

$$v_{yx} = \left[\frac{0.035}{2} - \frac{1}{4}\left(\frac{1}{20} + \frac{0.35 \times 2}{20} + \frac{1}{2} - \frac{1}{0.7}\right)\sin^2 90\right] \times 1.7167$$

$$= [0.0175 - 0.25 \ (0.05 + 0.035 + 0.5 - 1.428)] \times 1.7167$$

$$v_{yx} = -0.2377$$

Example 3.5

Calculate the values of E_x, E_y, v_{xy}, η_x and η_y for a 60° lamina with the following properties in the L-T coordinate system: $E_L = 15$ GPa, $E_T = 15$ GPa, $G_{LT} = 2.5$ GPa and $v_{LT} = 0.20$.

Solution

Since $\dfrac{E_L}{v_{LT}} = \dfrac{E_T}{v_{TL}}$, for this lamina, $v_{TL} = 0.2$. Such a lamina is called a balanced lamina. Equations (3.49), (3.50), (3.52), (3.56), (3.57) and (3.59), can be used to obtain the material properties in the X-Y coordinate system.

$$\frac{1}{E_x} = \frac{\cos^4 60°}{15} + \frac{\sin^4 60°}{15} + \frac{1}{4}\left(\frac{1}{2.5} - \frac{2 \times 0.2}{15}\right)\sin^2 120°$$

$$= 0.00417 + 0.0374 + 0.07$$

$$E_x = 8.9623 \text{ GPa}$$

$$\frac{1}{E_y} = \frac{\sin^4 60°}{15} + \frac{\cos^4 60°}{15} + \frac{1}{4}\left(\frac{1}{2.5} - \frac{2 \times 0.2}{15}\right)\sin^2 120°$$

$$= 0.0374 + 0.00417 + 0.07$$

$$E_y = 8.9623 \text{ GPa}$$

$$\frac{1}{G_{xy}} = \frac{1}{15} + \frac{1}{15} + \frac{2 \times 0.2}{15} - \left(\frac{1}{15} + \frac{1}{15} - \frac{1}{2.5} + \frac{2 \times 0.2}{15}\right)\cos^2 120°$$

$$= 0.0666 + 0.0666 + 0.0266 - (0.0666 + 0.0666 - 0.4 + 0.0266) \ 0.25$$

$$G_{xy} = 4.539 \text{ GPa}$$

$$v_{xy} = \left[\frac{0.2}{15} - \frac{1}{4}\left(\frac{1}{15} + \frac{1}{15} - \frac{1}{2.5} + \frac{2 \times 0.2}{15}\right)\sin^2 120°\right]E_x$$

$$= [0.0133 - 0.25 \ (0.0666 + 0.0666 - 0.4 + 0.0266) \ 0.75] \ E_x$$

$$= 0.0583 \ E_x$$

$$v_{xy} = 0.5225$$

$$\eta_x = \left[0.2 + 1.0 - \frac{15}{2 \times 2.5} - \left(1 + 0.4 + 1 - \frac{15}{2.5}\right)\cos^2 60°\right]\sin 120°$$

$$= [0.2 + 1.0 - 3.0 + 0.9] \ 0.866$$

$$\eta_x = -0.779$$

$$\eta_y = \left[0.2 + 1.0 - \frac{15}{2 \times 2.5} - \left(1 + 2 \times 0.2 + 1.0 - \frac{15}{2.5}\right)\sin^2 60°\right]\sin 120°$$

$$= [0.2 + 1.0 - 3.0 + 2.7]\ 0.866$$

$$\eta_y = 0.779$$

Example 3.6

For a 30° lamina, the stresses in the reference coordinate system are given as $\sigma_x = 10$ MPa, $\sigma_y = 5$ MPa and $\tau_{xy} = -5$ MPa. Obtain the stresses in the L-T coordinate system.

Solution

The stresses are given by Eq. (3.29)

$$\begin{Bmatrix} \sigma_L \\ \sigma_T \\ \tau_{LT} \end{Bmatrix} = \begin{bmatrix} \cos^2\theta & \sin^2\theta & 2\sin\theta\cos\theta \\ \sin^2\theta & \cos^2\theta & -2\sin\theta\cos\theta \\ -\sin\theta\cos\theta & \sin\theta\cos\theta & \cos^2\theta - \sin^2\theta \end{bmatrix} \begin{Bmatrix} \sigma_x \\ \sigma_y \\ \tau_{xy} \end{Bmatrix}$$

$$\sigma_L = \sigma_x \cos^2\theta + \sigma_y \sin^2\theta + 2\tau_{xy} \sin\theta\cos\theta$$
$$= 10 \cos^2 30° + 5 \sin^2 30° - 2 \times 5 \sin 30° \cos 30°$$
$$= 7.5 + 1.25 - 4.33$$
$$= 3.92 \text{ MPa}$$

$$\sigma_T = \sigma_x \sin^2 30° + \sigma_y \cos^2 30° - 2\tau_{xy} \cos 30° \sin 30°$$
$$= 10 \sin^2 30° + 5 \cos^2 30° + 10 \cos 30° \sin 30°$$
$$= 2.5 + 2.5 + 4.33$$
$$= 9.33 \text{ MPa}$$

$$\tau_{LT} = -\sigma_x \sin 30°\cos 30° + \sigma_y \sin 30°\cos 30° + \tau_{xy} (\cos^2 30° - \sin^2 30°)$$
$$= -10 \sin 30° \cos 30° + 5\sin 30° \cos 30° - 5 (\cos^2 30° - \sin^2 30°)$$
$$= -4.33 + 2.165 - 1.25$$
$$= -3.145 \text{ MPa}$$

Example 3.7

A high strength composite lamina has the following properties: $E_L = 145$ GPa, $E_T = 12$ GPa, $G_{LT} = 6$ GPa and $\nu_{LT} = 0.25$. Obtain the transformed stiffness matrix if $\theta = 60°$.

Solution

From Eq. (3.24), we have

$$Q_{11} = \frac{E_L}{1 - \nu_{TL}\nu_{LT}}, \quad Q_{22} = \frac{E_T}{1 - \nu_{TL}\nu_{LT}}$$

$$Q_{12} = \frac{v_{TL} E_L}{1 - v_{LT} v_{TL}} = \frac{v_{LT} E_T}{1 - v_{LT} v_{TL}}, \quad Q_{66} = G_{LT}$$

From the given data, $v_{TL} = v_{LT} \dfrac{E_T}{E_L} = 0.25 \dfrac{12}{145} = 0.0207$

$$Q_{11} = \frac{145}{1 - 0.25 \times 0.0207} = 145.75 \text{ GPa}, \quad Q_{22} = \frac{12}{1 - 0.25 \times 0.0207} = 12.063 \text{ GPa}$$

$$Q_{12} = \frac{0.0207 \times 145}{1 - 0.25 \times 0.0207} = 3.017 \text{ GPa}, \quad Q_{66} = G_{LT} = 6 \text{ GPa}$$

The reduced stiffness coefficients can be obtained from Eq. (3.43). For the given lamina

$\cos 60° = 0.5, \ \cos^2 60° = 0.25, \ \cos^3 60° = 0.125, \ \cos^4 60° = 0.0625$

$\sin 60° = 0.866, \ \sin^2 60° = 0.75, \ \sin^3 60° = 0.649, \ \sin^4 60° = 0.562$

$\bar{Q}_{11} = 145.75 \times 0.0625 + 12.063 \times 0.562 + 2 \times (3.017 + 2 \times 6) \times 0.75 \times 0.25$
$\quad = 9.1094 + 6.7794 + 5.631$
$\quad = 21.5198 \text{ GPa}$

$\bar{Q}_{22} = 145.75 \times 0.562 + 12.063 \times 0.0625 + 2 \times (3.017 + 2 \times 6) \times 0.75 \times 0.25$
$\quad = 81.9115 + 0.7539 + 5.6313$
$\quad = 88.2968 \text{ GPa}$

$\bar{Q}_{12} = (145.75 + 12.063 - 4 \times 6) \times 0.75 \times 0.25 + 3.017 \times (0.0625 + 0.562)$
$\quad = 25.0899 + 1.8841$
$\quad = 26.974 \text{ GPa}$

$\bar{Q}_{66} = (145.75 + 12.063 - 2 \times 3.017 - 2 \times 6) \times 0.75 \times 0.25 + 6 \times (0.562 + 0.0625)$
$\quad = 26.2085 + 3.747$
$\quad = 29.9555 \text{ GPa}$

$\bar{Q}_{16} = (145.75 - 3.017 - 2 \times 6) \times 0.125 \times 0.866 - (12.063 - 3.017 - 2 \times 6) \times 0.5 \times 0.649$
$\quad = 14.1518 + 0.9585$
$\quad = 15.1103 \text{ GPa}$

$\bar{Q}_{26} = (145.75 - 3.017 - 2 \times 6) \times 0.5 \times 0.649 - (12.063 - 3.017 - 2 \times 6) \times 0.125 \times 0.866$
$\quad = 42.4228 + 0.3197$
$\quad = 42.5477 \text{ GPa}$

Example 3.8
Obtain the strains ε_x, ε_y and γ_{xy} and the engineering constants E_x, E_y, G_{xy}, η_x and η_y for the lamina subjected to the in-plane stresses as shown in Fig. 3.4.

Given $E_L = 14$ GPa, $E_T = 3.5$ GPa, $G_{LT} = 4.2$ GPa, $v_{LT} = 0.4$, $v_{TL} = 0.1$. $\sigma_x = 3.0$ MPa, $\sigma_y = 7.0$ MPa, $\tau_{xy} = -1.5$ MPa, $\theta = 30°$.

Solution

The transformation matrix $[T]$ is obtained from Eq. (3.29). For $\theta = 30°$

$$[T] = \begin{bmatrix} 0.75 & 0.25 & 0.866 \\ 0.25 & 0.75 & -0.866 \\ -0.433 & 0.433 & 0.50 \end{bmatrix}$$

Therefore

$$\begin{Bmatrix} \sigma_L \\ \sigma_T \\ \tau_{LT} \end{Bmatrix} = \begin{bmatrix} 0.75 & 0.25 & 0.866 \\ 0.25 & 0.75 & -0.866 \\ -0.433 & 0.433 & 0.50 \end{bmatrix} \begin{Bmatrix} 3.0 \\ 7.0 \\ -1.5 \end{Bmatrix}$$

$$= \begin{Bmatrix} 2.701 \\ 7.299 \\ 0.982 \end{Bmatrix} \text{MPa}$$

Equations (3.49), (3.50), (3.52), (3.56), (3.57), (3.59) and (3.62) can be used to obtain the engineering constants

$E_x = 10.9$ GPa, $E_y = 5.024$ GPa, $G_{xy} = 2.701$ GPa, $v_{xy} = -0.049$, $v_{yx} = -0.023$, $\eta_x = 0.7633$ and $\eta_y = 1.833$

Substituting these values in Eq. (3.64), we get

$\varepsilon_x = 0.3889 \times 10^{-3}$, $\varepsilon_y = 1.603 \times 10^{-3}$, $\gamma_{xy} = -1.635 \times 10^{-3}$

Example 3.9

For the 30° lamina of graphite/epoxy shown in Fig. 3.4, obtain the reduced stiffness matrix $[\bar{Q}]$. The material properties for the lamina are: $E_L = 181$ GPa, $E_T = 10.3$ GPa, $G_{LT} = 7.17$ GPa, $v_{LT} = 0.28$. If the applied stress is $\sigma_x = 2$ MPa, $\sigma_y = -3$ MPa, $\tau_{xy} = 4$ MPa obtain the strains in the L-T coordinates, strains in the x-y coordinates, principal stresses and principal strains.

Solution

The transformed reduced stiffness matrix is obtained from Eq. (3.40) for $\theta = 30°$. For the given engineering constants, we first obtain the $[Q]$ matrix.

$$v_{TL} = \frac{E_T v_{LT}}{E_L} = \frac{10.3 \times 0.28}{181} = 0.016$$

$$Q_{11} = \frac{181}{1 - 0.28 \times 0.016} = 181.81 \text{ GPa}, \quad Q_{22} = \frac{10.3}{1 - 0.28 \times 0.016} = 10.35 \text{ GPa}$$

$$Q_{12} = \frac{0.28 \times 10.3}{1 - 0.28 \times 0.016} = 2.897 \text{ GPa}, \quad Q_{66} = 7.17 \text{ GPa}$$

$\cos 30° = 0.866$, $\cos^2 30° = 0.75$, $\cos^3 30° = 0.649$, $\cos^4 30° = 0.562$
$\sin 30° = 0.5$, $\sin^2 30° = 0.25$, $\sin^3 30° = 0.125$, $\sin^4 30° = 0.0625$

$\bar{Q}_{11} = 181.81 \times 0.562 + 10.35 \times 0.0625 + 2(2.897 + 2 \times 7.17) \times 0.25 \times 0.75$
$\quad = 109.311$ GPa
$\bar{Q}_{22} = 181.81 \times 0.0625 + 10.35 \times 0.562 + 2(2.897 + 2 \times 7.17) \times 0.25 \times 0.75$
$\quad = 23.644$ GPa
$\bar{Q}_{12} = (181.81 + 10.35 - 4 \times 7.17) \times 0.75 \times 0.25 + 2.897 \times (0.562 + 0.0625)$
$\quad = 32.461$ GPa
$\bar{Q}_{66} = (181.81 + 10.35 - 2 \times 2.897 - 2 \times 7.17) \times 0.75 \times 0.25 + 7.17 \times (0.562$
$\quad + 0.0625)$
$\quad = 36.733$ GPa
$\bar{Q}_{16} = (181.81 - 2.897 - 2 \times 7.17) \times 0.649 \times 0.5 - (10.35 - 2.897 - 2 \times$
$\quad 7.17) \times 0.866 \times 0.125$
$\quad = 54.149$ GPa
$\bar{Q}_{26} = (181.81 - 2.897 - 2 \times 7.17) \times 0.866 \times 0.125 - (10.35 - 2.897 - 2$
$\quad \times 7.17) \times 0.649 \times 0.5$
$\quad = 20.05$ GPa

From Eq. (3.29), we have

$$\begin{Bmatrix} \sigma_L \\ \sigma_T \\ \tau_{LT} \end{Bmatrix} = \begin{bmatrix} \cos^2 \theta & \sin^2 \theta & 2\sin\theta\cos\theta \\ \sin^2 \theta & \cos^2 \theta & 2\sin\theta\cos\theta \\ -\sin\theta\cos\theta & \sin\theta\cos\theta & \cos^2\theta - \sin^2\theta \end{bmatrix} \begin{Bmatrix} \sigma_x \\ \sigma_y \\ \tau_{xy} \end{Bmatrix}$$

Substituting for σ_x, σ_y, τ_{xy} and θ, and carrying the matrix multiplication, we get

$$\begin{Bmatrix} \sigma_L \\ \sigma_T \\ \tau_{LT} \end{Bmatrix} = \begin{bmatrix} 4.214 \\ -5.214 \\ -0.165 \end{bmatrix} \text{MPa}$$

Strains in the L and T directions are obtained from Eq. (3.14)

$$S_{11} = \frac{1}{E_L} = 0.552 \times 10^{-11}, \ S_{22} = \frac{1}{E_T} = 9.66 \times 10^{-11}$$

$$S_{12} = \frac{-\nu_{LT}}{E_L} = -0.155 \times 10^{-11}, \ S_{66} = \frac{1}{G_{LT}} = 0.0139 \times 10^{-11}$$

$\varepsilon_L = S_{11}\sigma_L + S_{12}\sigma_T = 0.522 \times 10^{-11} \times 2 \times 10^6 - 0.155 \times 10^{-11} \times (-3) \times 10^6$
$\quad = 0.5694 \times 10^{-5}$
$\varepsilon_T = S_{12}\sigma_L + S_{22}\sigma_T = -0.155 \times 10^{-11} \times 2 \times 10^6 - 9.66 \times 10^{-11} \times 3 \times 10^6$
$\quad = -29.9 \times 10^{-5}$
$\gamma_{LT} = S_{66}\tau_{LT} = 0.0139 \times 10^{-11} \times 4 \times 10^6 = 0.0556 \times 10^{-5}$

$$
\left\{
\begin{array}{c}
\varepsilon_x \\
\varepsilon_y \\
\gamma_{xy}/2
\end{array}
\right\} = [T]^{-1}
\left\{
\begin{array}{c}
\varepsilon_L \\
\varepsilon_T \\
\gamma_{LT}/2
\end{array}
\right\}
$$

$$
= \left[
\begin{array}{ccc}
\cos^2\theta & \sin^2\theta & -2\sin\theta\cos\theta \\
\sin^2\theta & \cos^2\theta & 2\sin\theta\cos\theta \\
\sin\theta\cos\theta & -\sin\theta\cos\theta & \cos^2\theta-\sin^2\theta
\end{array}
\right]
\left\{
\begin{array}{c}
\varepsilon_L \\
\varepsilon_T \\
\gamma_{LT}/2
\end{array}
\right\}
$$

$$
= \left[
\begin{array}{ccc}
0.75 & 0.25 & -0.866 \\
0.25 & 0.75 & 0.866 \\
-0.433 & 0.433 & 0.5
\end{array}
\right]
\left\{
\begin{array}{c}
0.5674\times10^{-5} \\
-29.9\times10^{-5} \\
0.278\times10^{-5}
\end{array}
\right\}
$$

$$
= \left\{
\begin{array}{c}
-6.809\times10^{-5} \\
-22.043\times10^{-5} \\
-13.063\times10^{-5}
\end{array}
\right\}
$$

$$
\left\{
\begin{array}{c}
\varepsilon_x \\
\varepsilon_y \\
\gamma_{xy}
\end{array}
\right\} = \left\{
\begin{array}{c}
-6.809\times10^{-5} \\
-22.043\times10^{-5} \\
-26.126\times10^{-5}
\end{array}
\right\}
$$

Principal stresses are given by

$$
\sigma_{1,2} = \frac{\sigma_x - \sigma_y}{2} \pm \sqrt{\left(\frac{\sigma_x - \sigma_y}{2}\right)^2 + 4\tau_{xy}}
$$

$$
= \frac{2-3}{2} \pm \sqrt{\left(\frac{2+3}{2}\right)^2 + 16}
$$

$$
= -0.5 \pm 4.717
$$

$$
= (4.217, -5.217)\ \text{MPa}
$$

Principal strains are given by

$$
\varepsilon_{1,2} = \frac{(\varepsilon_x + \varepsilon_y)}{2} \pm \sqrt{\left(\frac{\varepsilon_x - \varepsilon_y}{2}\right)^2 + \gamma_{xy}^2}
$$

$$
= \left(\frac{-6.809 - 22.043}{2} \pm \sqrt{\left(\frac{-6.809 + 22.043}{2}\right)^2 + (13.063)^2}\right)\times10^{-5}
$$

$$
= (-14.426 \pm 15.121)\times10^{-5}
$$

$$
= 0.6955\times10^{-5},\ -29.547\times10^{-5}
$$

3.6 Invariant Form of Stiffness and Compliance Coefficients

Equations (3.43) and (3.45) give the transformed stiffness and compliance matrix coefficients. As can be seen, these are products of the higher powers of cosine and sine functions of fiber orientation. It is not easy to visualize how the individual terms change with a change in fiber orientation θ, and their consequent contribution to the overall stiffness. In addition, the laminae are stacked together to form a laminate. The overall stiffness of the laminate has to be obtained by integrating the lamina stiffnesses over the laminate thickness. The integration of such terms becomes rather difficult. To overcome this difficulty, Tsai and Pagano [5] proposed a convenient invariant form of stiffness transformation equations. They showed that Eqs. (3.43) and (3.45) can be rewritten as

$$
\begin{aligned}
\overline{Q}_{11} &= U_1 + U_2 \cos 2\theta + U_3 \cos 4\theta \\
\overline{Q}_{22} &= U_1 - U_2 \cos 2\theta + U_3 \cos 4\theta \\
\overline{Q}_{12} &= U_4 - U_3 \cos 4\theta \\
\overline{Q}_{66} &= \frac{(U_1 - U_4)}{2} - U_3 \cos 4\theta \\
\overline{Q}_{16} &= \frac{U_2}{2} \sin 2\theta + U_3 \sin 4\theta \\
\overline{Q}_{26} &= \frac{U_2}{2} \sin 2\theta - U_3 \sin 4\theta
\end{aligned}
\tag{3.66}
$$

$$
\begin{aligned}
\overline{S}_{11} &= V_1 + V_2 \cos 2\theta + V_3 \cos 4\theta \\
\overline{S}_{22} &= V_1 - V_2 \cos 2\theta + V_3 \cos 4\theta \\
\overline{S}_{12} &= V_4 - V_3 \cos 4\theta \\
\overline{S}_{66} &= 2(V_1 - V_4) - 4V_3 \cos 4\theta \\
\overline{S}_{16} &= V_2 \sin 2\theta + 2V_3 \sin 4\theta \\
\overline{S}_{26} &= V_2 \sin 2\theta - 2V_3 \sin 4\theta
\end{aligned}
\tag{3.67}
$$

where

$$
\begin{aligned}
U_1 &= \frac{1}{8}[3Q_{11} + 3Q_{22} + 2Q_{12} + 4Q_{66}] \\
U_2 &= \frac{1}{2}[Q_{11} - Q_{22}] \\
U_3 &= \frac{1}{8}[Q_{11} + Q_{22} - 2Q_{12} - 4Q_{66}] \\
U_4 &= \frac{1}{8}[Q_{11} + Q_{22} + 6Q_{12} - 4Q_{66}]
\end{aligned}
\tag{3.68}
$$

$$V_1 = \frac{1}{8}[3S_{11} + 3S_{22} + 2S_{12} + S_{66}]$$

$$V_2 = \frac{1}{2}[S_{11} - S_{22}]$$

$$V_3 = \frac{1}{8}[S_{11} + S_{22} - 2S_{12} - S_{66}]$$

$$V_4 = \frac{1}{8}[S_{11} + S_{22} + 6S_{12} - S_{66}]$$

(3.69)

We illustrate below the method to obtain the coefficients with regard to $\bar{Q}_{11}$. Making use of the trigonometric relations, we can write

$$\cos^2\theta = \frac{1}{2}(1 + \cos 2\theta), \sin^2\theta = \frac{1}{2}(1 - \cos 2\theta)$$

$$\cos^2 2\theta = \frac{1}{2}(1 + \cos 4\theta), \sin^2 2\theta = \frac{1}{2}(1 - \cos 4\theta)$$

$$\cos^4\theta = \frac{1}{8}(3 + 4\cos 2\theta + \cos 4\theta), \sin^4\theta = \frac{1}{8}(3 - 4\cos 2\theta + \cos 4\theta)$$

$$\cos^3\theta\sin\theta = \frac{1}{8}(2\sin 2\theta + \sin 4\theta), \cos^2\theta\sin^2\theta = \frac{1}{8}(1 - \cos 4\theta)$$

$$\cos\theta\sin^3\theta = \frac{1}{8}(3 - 4\cos 2\theta + \cos 4\theta)$$

$$\bar{Q}_{11} = Q_{11}\cos^4\theta + Q_{22}\sin^4\theta + 2(Q_{12} + 2Q_{66})\sin^2\theta\cos^2\theta$$

$$= \frac{Q_{11}}{8}[3 + 4\cos 2\theta + \cos 4\theta] + \frac{Q_{22}}{8}[3 - 4\cos 2\theta + \cos 4\theta]$$

$$+ \frac{2}{8}(Q_{12} + 2Q_{66})(1 - \cos 4\theta)$$

$$= \frac{1}{8}[3Q_{11} + 3Q_{22} + 2Q_{12} + 4Q_{66}] + \frac{1}{2}(Q_{11} - Q_{22})\cos 2\theta$$

$$+ \frac{1}{8}[Q_{11} + Q_{22} - 2Q_{12} - 4Q_{66}]\cos 4\theta$$

$$= U_1 + U_2\cos 2\theta + U_3\cos 4\theta$$

One can derive similar equations by using trigonometric relations. This expression is useful when examining the consequences of changing lamina orientations to obtain the desired stiffness requirements. It is also very important while designing laminates to meet certain load or stress requirements.

Summary

- The constitutive relations for a 2D orthotropic lamina have been derived from a 3D elastic body for a fiber reinforced composite lamina.
- The stress-strain relations have been derived for a lamina having fibers along the material axes oriented at an angle with respect to the reference axes.
- A number of worked out examples are presented in the chapter.

Problems

3.1   The following are the material properties of a unidirectional lamina along the L-T directions

$$E_L = 140 \text{ GPa}, E_T = 10 \text{ GPa}, G_{LT} = 6 \text{ GPa}, \nu_{LT} = 0.25$$

Determine the material properties along a direction $+ 45°$ to the material axis.

3.2   The elastic properties of a unidirectional composite lamina along the material axes are

$$E_L = 120 \text{ GPa}, E_T = 12 \text{ GPa}, G_{LT} = 5 \text{ GPa}, \nu_{LT} = 0.28$$

Determine the elastic properties along the reference axis inclined at an angle α.

3.3   The in-plane stresses in an L-T system is given by $\{15\ 10\ 6\}$. Find the stress components with respect to x, y axes, where the x-axis is located at $\alpha = 45°$, and $30°$.

3.4   Express the compliance matrix $[S]$ in terms of the invariants U_1, U_2 etc.

3.5   A unidirectional lamina is subjected to a uniform stress σ_0 along the axial direction. The fibers are oriented along $30°$ anticlockwise to the loading direction. Calculate the displacements U, V along and perpendicular to the loading direction. The material properties are

$$E_L = 140 \text{ GPa}, E_T = 10 \text{ GPa}, G_{LT} = 6 \text{ GPa}, \nu_{LT} = 0.28, \sigma_0 = 350 \text{ MPa}$$

3.6   Show that $Q_{11} + 2Q_{12} + Q_{22}$ is invariant with respect to the rotation with respect to the Z-axis (out-of-plane axis).

References

1. Green, A.E. and Zerna, W., 1968, *Theoretical Elasticity*, 2nd Edition, Oxford University Press, New York.
2. Crandall, S.H., Dahl, N.C. and Lardner, T.J., 1978, *An Introduction to the Mechanics of Solids*, 2nd Edition, McGraw-Hill, Inc., New York.
3. Jones, R.M., 1975, *Mechanics of Composite Materials*, Scripta Book Co., Washington D.C.
4. Agarwal, B.D., Broutman, L.J. and Chandrashekhara, K., *Analysis and Performance of Fiber Composites*, 3rd Edition, John Wiley and Sons, Inc., USA.
5. Tsai, S.W. and Pagano, N.J., 1968, "Invariant Properties of Composite Materials", In *Composite Materials Workshop*, Tsai, S.W., Halpin, J.C. and Pagano, N.J. (Eds.), Technomic Publishing Co., Lancaster, Pennsylvania.

4. Analysis of Continuous Fiber Lamina for Strength and Stiffness

In this chapter, we will:

- Discuss the behavior of an orthotropic lamina when subjected to in-plane loads.
- Derive the relations for engineering constants for a 2D lamina and its in-plane strengths.
- Derive the relations for thermal and moisture coefficients of the lamina.
- Investigate the behavior of the lamina under axial compressive load.

4.1 Introduction

In Chapter 3, we discussed constitutive equations of the lamina in terms of stiffness coefficients and also the relation between stiffness and compliance coefficients with the engineering constants. For a lamina, which is orthotropic in its behavior, the number of independent engineering constants is four. They are E_L, E_T, G_{LT} and v_{LT}. These engineering constants depend on a number of variables such as the engineering constants of fiber and matrix, volume fraction of fiber and matrix, fiber distribution, packing geometry, method of manufacture etc. They can be obtained experimentally by performing several tests, such as measuring tension, compression, shear etc. It is well known that experimental evaluation of these constants is costly and time consuming. Further, there are a large numbers of parameters, which affect the experimental values. Hence, there is a need to develop analytical techniques to determine these engineering constants. In addition to the engineering constants, we have strength values like, longitudinal ultimate tensile and compressive strengths (σ_{LU}, σ'_{LU}), transverse ultimate tensile and compressive strengths (σ_{TU}, σ'_{TU}) and in-plane ultimate shear strength (τ_{LTU}).

In this chapter, some relationships for estimating stiffness and strength of the lamina in terms of the individual constituents of the composite, fiber volume fraction and packing geometry have been discussed. A number of methods have been suggested to estimate the mechanical and strength properties of continuous fiber composites [1]. They are:

(i) Strength of materials approach (rule of mixtures)
(ii) Theory of elasticity approach
(iii) Semi-empirical approach

In the strength of materials approach, certain simplistic assumptions are made regarding the behavior of fiber and matrix. These are easy to formulate, but the assumptions made may violate the elasticity assumptions.

In the theory of elasticity approach, the technique has been divided into three categories: (i) bounding techniques, (ii) exact approach and (iii) self-consistent model. In the bounding technique, the bounds on the modulus are obtained by using minimum complimentary energy and minimum potential energy principles. Paul [2] and Hashin and Rosen [3] have employed this technique. The exact method assumes that the fibers are arranged in a regular periodic array. The resulting elasticity equation has to be solved by an approximate technique. Adams and Doner [4] employed the finite difference technique to obtain an estimate of transverse properties. Chen and Lin [5] also used the finite element technique. In the self-consistent model, it is assumed that a cylinder of matrix material encloses the fiber. This in turn, is enclosed in a homogenous material. Hill [6] and other investigators employed this model for predicting the composite modulus. Readers can refer to review papers by Chamis and Sendeckyj [7], Christensen [8], Hashin [9] and Halpin [10] for more details. The empirical approach is generally a curve fitting technique to the experimental data and the theoretical data obtained from one of the analysis techniques. In what follows, the strength of materials model will be discussed in detail.

4.2 Volume and Weight Fractions

Properties like stiffness, strength, coefficient of thermal expansion and moisture absorption coefficient of the composite depend on the relative proportions of the matrix and fiber. The relative proportions can be expressed in terms of either weight or volume fractions. Weight fraction is used generally with experimental or manufacturing work as it is easy to obtain. However, for theoretical calculations, volume fraction is used. One can easily go from one to the other. The subscripts c, f, m denote the composite, fiber and matrix materials, respectively. The lamina properties are based on the assumption that fiber and matrix are isotropic and the composite is homogenous. Homogeneity is determined by taking X-ray photographs of several representative volume elements (RVE) and comparing them. These are generally statistically homogenous, although voids do exist in spite of the best possible quality control during manufacture (Fig. 4.1).

The total volume (v_c) and weight (w_c) of the RVE are: $w_c = w_f + w_m$ and $v_c = v_f + v_m + v_v$; v_v must be as small as possible as they affect the mechanical properties. Further, they should be randomly distributed in the

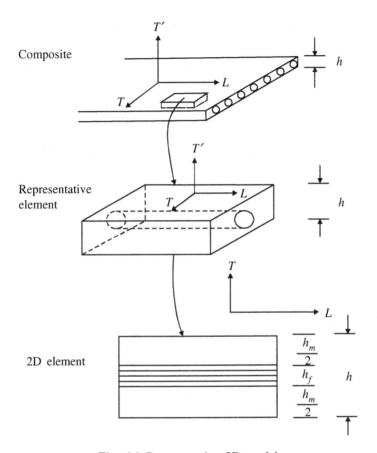

Fig. 4.1 Representative 2D model.

sample. Voids are the weakest spot in the specimen. The presence of voids results in lower fatigue life and greater possibility of water absorption. A good composite should have less than 1% voids in a given volume.

The volume and weight fractions are defined as

$$V_f = \frac{v_f}{v_c}, W_f = \frac{w_f}{w_c}, V_m = \frac{v_m}{v_c}, W_m = \frac{w_m}{w_c}, V_v = \frac{v_v}{v_c} \qquad (4.1)$$

If the void volume is negligible

$$v_c = v_f + v_m \qquad (4.2a)$$

and

$$w_c = w_f + w_m \qquad (4.2b)$$

Therefore, dividing the volume by total volume, and weight by total weight, we get

$$V_f + V_m = 1 \text{ and } W_f + W_m = 1 \qquad (4.3)$$

The density of the composite (ρ_c) can be expressed as

$$\rho_c v_c = \rho_f v_f + \rho_m v_m \qquad (4.4)$$

Dividing by v_c, and making use of Eq. (4.1), we get

$$\rho_c = \rho_f V_f + \rho_m V_m \qquad (4.5)$$

Making use of Eq. (4.2a) and replacing volume by weight terms, we get

$$\frac{w_c}{\rho_c} = \frac{w_f}{\rho_f} + \frac{w_m}{\rho_m} \qquad (4.6)$$

Dividing by w_c, and making use of Eq. (4.1), we obtain

$$\rho_c = \frac{1}{\dfrac{W_f}{\rho_f} + \dfrac{W_m}{\rho_m}} \qquad (4.7)$$

Equations (4.5) and (4.6) give the density of the composite in terms of volume and weight fractions, respectively.

If the number of constituents is more than two, Eqs. (4.5) and (4.7) can be generalized as

$$\rho_c = \sum_{i=1}^{N} (\rho_i V_i) \quad \text{and} \quad \rho_c = \frac{1}{\displaystyle\sum_{i=1}^{N}\left(\dfrac{W_i}{\rho_i}\right)} \qquad (4.8)$$

The density of the composite obtained theoretically using volume fraction may not be the same as that obtained experimentally. This could be due to the presence of voids. The voids contribute to the volume and not to the weight. Defining ρ_{cth} and ρ_{cex} as theoretical and experimental composite densities, the void volume fraction V_v is given by

$$V_v = \frac{\rho_{cth} - \rho_{cex}}{\rho_{cth}} \qquad (4.9)$$

It is to be pointed out here that autoclave cured and vacuum bagged composites generally result in void fractions of the order of 0.1-1.0%. Without vacuum bagging, the volatiles trapped in the specimen can cause void contents of the order of 5%.

As stated earlier, a composite of laminates is needed to arrive at a strength that is much higher than the matrix strength. This is only possible if one can get as high a volume fraction of the fiber in a given volume of the composite. It is, therefore, necessary to consider the possible packing geometries in a representative volume. Figure 4.2 shows two possible packing

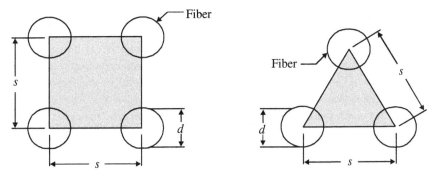

Fig. 4.2 Two possible stacking geometries [11].

geometries. In actual situations, the packing geometry may not exactly correspond to these idealized cases. Assuming the bare fiber diameter to be d and the fiber spacing to be s which does not change along the length of the specimen, the fiber volume fraction is the same as area fraction. For square packing, the area of the square, which corresponds to the matrix area, is s^2, and the area of the fiber is $\frac{\pi}{4}d^2$. Hence, the fiber volume fraction for this packing geometry is

$$V_f = \text{Area of fiber/Matrix area} = \frac{\pi d^2}{4s^2} \qquad (4.10)$$

If we assume very close packing, then $d = s$. Hence

$$V_{f(\text{max})} = 0.785$$

Similarly, one can obtain volume fractions for different packing geometries. Volume fraction is generally much lower because of processing limitations. Generally, fibers are coated with a chemical for better bonding between the fibers and the matrix. It is possible to get a volume fraction of the order of 0.8, if one uses prepreg tapes rather than continuous fibers. This helps in keeping the fibers unidirectional and does not allow the fiber to distort and break when composites are fabricated.

4.3 Modulus and Strength of the Continuous Fiber Lamina

As stated in Section 4.1, a lamina is characterized by the mechanical properties E_L, E_T, G_{LT} and ν_{LT}. The strength properties are σ_{LU}, σ_{TU}, σ'_{LU}, σ'_{TU} and τ_{LTU}. Various approaches for estimating these values have been suggested in literature. The complexities increase as the models improve. The nature of composite materials is very different from that of metallic materials.

4.3.1 Strength of Materials Approach

This is one of the simplest approaches to obtain an estimate of the modulus of the composite. The lamina consists of unidirectional fibers embedded in a matrix. It is assumed to be homogenous and exhibits orthotropic behavior in the plane of the lamina. Figure 4.1 shows a representative volume element. It is assumed that the matrix surrounds the bundle of fiber. The following assumptions are made regarding the fiber and the matrix material:

(i)   Both fiber and the matrix exhibit linear elastic behavior.
(ii)   Fiber and the matrix are assumed to be isotropic.
(iii) There is no slip between the fiber and the matrix i.e. perfect bonding exists till the failure.
(iv)   Fiber cross-section is uniform throughout the length of the specimen.
(v)   The fibers in the composite are perfectly aligned and remain parallel throughout.
(vi)   Voids in the specimen are negligible.
(vii) Lamina is macroscopically homogenous, linearly elastic and exhibits orthotropic behavior.
(viii) Lamina is free from residual stress.

Longitudinal Elastic Modulus and Strength

The representative volume element of length L is subjected to a uniform stress σ_c as shown in Fig. 4.3. The deformation of the element due to the application of the load is also shown. The displacements are measured from one end of the element. Since there is no slip between the fiber and the matrix, the axial strain of the fiber, matrix and the composite is same. This is expressed as

$$\varepsilon_f = \varepsilon_m = \varepsilon_c = \varepsilon_L \qquad (4.11)$$

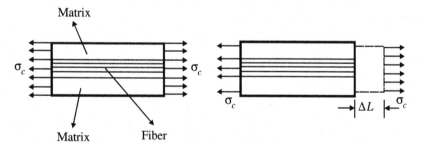

Fig. 4.3 Composite element under axial tensile load.

The total load $P_c = \sigma_c A_c$ is shared by the fibers and the matrix. Therefore, from equilibrium of the forces along the longitudinal direction

$$P_c = P_f + P_m \qquad (4.12)$$

Defining $P_f = A_f \, \sigma_f$ and $P_m = A_m \, \sigma_m$, Eq. (4.12) can be written as

$$\sigma_c \, A_c = \sigma_f \, A_f + \sigma_m \, A_m \tag{4.13}$$

Since A_c is not zero, on dividing by A_c, Eq. (4.13) can be rewritten as

$$\sigma_c = \sigma_f \, \frac{A_f}{A_c} + \sigma_m \, \frac{A_m}{A_c} \tag{4.14}$$

Since the length of the element is same for both fiber and the matrix, the area fractions can be replaced by volume fractions. Hence

$$\sigma_c = \sigma_f \, V_f + \sigma_m \, V_m \tag{4.15}$$

Equation (4.15) shows that the contribution of the fiber and matrix to the composite stress is directly proportional to the respective volume fractions.

Equation (4.11) defines the compatibility condition with respect to the strains. Differentiating Eq. (4.15) with respect to strain, we obtain

$$\frac{d\sigma_c}{d\varepsilon} = \frac{d\sigma_f}{d\varepsilon} V_f + \frac{d\sigma_m}{d\varepsilon} V_m \tag{4.16}$$

In view of the assumptions made regarding the linear elastic behavior of the fiber and the matrix, the composite also behaves in a linear elastic manner. The derivatives, which give the slope of the stress-strain curves, can be replaced by their corresponding moduli. Thus, Eq. (4.16) can be written as

$$E_c = E_L = E_f \, V_f + E_m V_m \tag{4.17}$$

Equation (4.17) shows that the modulus of the composite along the fiber direction is directly proportional to the respective volume fractions. Further, this equation shows that the longitudinal modulus (E_L) varies linearly with the volume fraction.

Equation (4.15) gives the stress in the composite for any volume fraction. However, the longitudinal strength of the composite (σ_{LU}) depends on the fiber volume fraction V_f. Since the fibers are strong, the load shared by them is generally substantial. In which case, if we assume that all the fibers break at the same strain (ε_{fu}), the composite will not be able to sustain the applied load and fails. The ultimate longitudinal tensile strength of the composite can then be obtained from Eq. (4.15).

$$\sigma_{cu} = \sigma_{LU} = \sigma_{fu} \, V_f + [\sigma_m | \varepsilon_{fu}](1 - V_f) \tag{4.18}$$

where σ_{fu} is the ultimate strength of the fiber and $[\sigma_m | \varepsilon_{fu}]$ is the matrix stress at the fiber fracture strain ε_{fu}.

The load shared by the fiber in a unidirectional lamina is given by

$$\frac{P_f}{P_c} = \frac{\sigma_f A_f}{\sigma_c A_c} = \frac{\sigma_f V_f}{\sigma_f V_f + \sigma_m V_m} = \frac{\sigma_f V_f}{\sigma_f V_f + \sigma_m (1 - V_f)} \tag{4.19}$$

Making use of Eq. (4.11) and replacing stresses in terms of modulus E_f and E_m, Eq. (4.19) is rewritten as

$$\frac{P_f}{P_c} = \frac{E_f V_f}{E_f V_f + E_m (1 - V_f)} \tag{4.20}$$

Figure 4.4 shows the plot of (P_f/P_c) vs. V_f with E_f/E_m as parameter. It is observed that higher the values of (E_f/E_m), the required fiber volume ratio reduces for a substantial load sharing by the fiber. Further, for a given value of (E_f/E_m), the fiber volume fraction, V_f has to be maximized if the fibers have to carry a major portion of the composite load. As stated earlier, fiber volume fraction cannot be greater than 0.8. This will result in poor bonding of the fibers to the matrix. As a consequence, the load transfer will get affected.

Whitney and Riley [12] have suggested a modification to Eq. (4.17) to take into account the misalignment of the fibers. This is given by

$$E_c = E_L = K E_f - (K E_f - E_m) V_m \tag{4.21}$$

where K is a correction factor that lies between $0 \leq K \leq 1$. For no misalignment, $K = 1$.

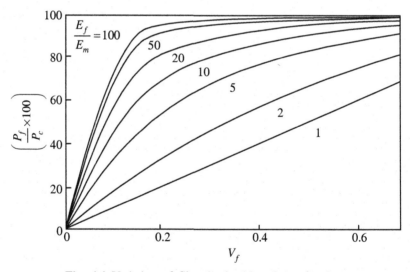

Fig. 4.4 Variation of fiber load with volume fraction.

Transverse Elastic Modulus and Strength

We once again consider the same representative volume element shown in Fig. 4.1. The specimen is subjected to a loading transverse to the fiber direction as shown in Fig. 4.5.

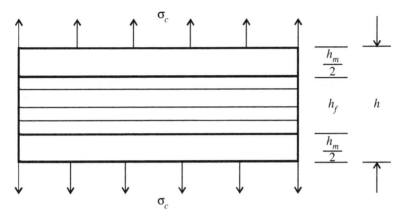

Fig. 4.5 Representative element with transverse loading.

The transverse tensile load is uniformly applied normal to the fibers. Since each layer carries the same load, from the continuity condition

$$\sigma_c = \sigma_f = \sigma_m \qquad (4.22)$$

The displacement of the composite specimen (δ_c) along the loading direction is the sum of displacements of the fiber (δ_f) and the matrix (δ_m). Thus

$$\delta_c = \delta_f + \delta_m \qquad (4.23)$$

The thickness of the matrix and the fiber is t_m and t_f, respectively. Therefore, the deformations can be written in terms of the strains and stiffness as

$$h\,\varepsilon_c = h_f\,\varepsilon_f + h_m\,\varepsilon_m \qquad (4.24)$$

Dividing Eq. (4.24) by h, we get

$$\varepsilon_c = \varepsilon_f\,\frac{h_f}{h} + \varepsilon_m\,\frac{h_m}{h} \qquad (4.25)$$

Since the length of the specimen is constant, the thickness ratio is replaced by volume fraction. Therefore

$$\varepsilon_c = \varepsilon_f\,V_f + \varepsilon_m\,V_m \qquad (4.26)$$

Making use of the constitutive relations for fiber and matrix, we get

$$\frac{\sigma_c}{E_c} = V_f\,\frac{\sigma_f}{E_f} + V_m\,\frac{\sigma_m}{E_m} \qquad (4.27)$$

In view of Eq. (4.22), Eq. (4.27) reduces to

$$\frac{1}{E_T} = \frac{1}{E_c} = \frac{V_f}{E_f} + \frac{V_m}{E_m} \qquad (4.28)$$

or

$$E_T = \frac{E_f E_m}{V_m E_f + V_f E_m} \qquad (4.29)$$

where E_c is the transverse modulus, also defined as E_T. Figure 4.6 shows the variation of transverse modulus E_L and E_T with fiber volume fraction V_f and E_c/E_m. It is observed that unlike the longitudinal modulus, the transverse modulus increases slightly with increase in volume fraction up to about 50%, beyond which it increases rapidly with increase in fiber volume fraction. This implies that the transverse modulus of the composite can be increased only by the presence of large fiber volume fraction.

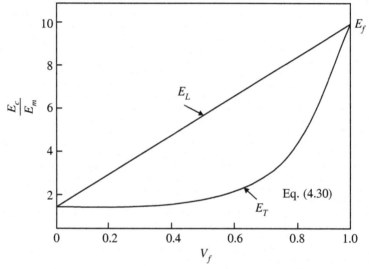

Fig. 4.6 Variation of E_T and E_L with fiber volume fraction V_f

Equation (4.21) indicates that the strength of the composite cannot be more than matrix strength and that it is independent of fiber volume fraction. Furthermore, there is a mismatch of strains at the interface as the material properties of the fiber and the matrix are different. Thus, the strength and the transverse modulus predicted by the strength of materials approach are approximate and do not conform to the experimental observations. Jones [1] has shown that the assumption of equal stresses in fibers and matrix is not justified. Therefore, one has to look for better models to take into account the continuity of displacements at the interface and the effect of fiber volume fraction on strength and modulus.

Hopkins and Chamis [13] have developed a better model for transverse and shear properties considering square fiber packing and dividing the RVE into subregions. The transverse modulus E_T is given by

$$E_T = E_m \left[\left(1 - \sqrt{V_f}\right) + \frac{\sqrt{V_f}}{1 - \sqrt{V_f}\,(1 - E_m / E_f)} \right] \qquad (4.30)$$

In-plane Shear Modulus and Strength

Figure 4.7 shows the representative volume element subject to in-plane shear stress τ. The shear stress acting on the matrix and the fiber are same. Hence

$$\tau_c = \tau_f = \tau_m \qquad (4.31)$$

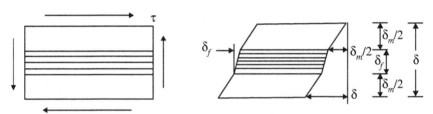

Fig. 4.7 Representative volume element for determining G_{LT}.

The total deformation due to the applied shear δ_c is due to the sum of shear deformation of the fiber (δ_f) and due to the matrix (δ_m)

$$\delta_c = \delta_f + \delta_m \qquad (4.32)$$

From Fig. 4.5

$$\delta_c = \gamma_c\, h, \ \delta_f = \gamma_f\, h_f, \ \delta_m = \gamma_m\, h_m \qquad (4.33)$$

Substituting Eq. (4.33) in Eq. (4.32), and dividing by h, we obtain

$$\gamma_c = \gamma_f \frac{h_f}{h} + \gamma_m \frac{h_m}{h} \qquad (4.34)$$

Since the specimen length is same, the thickness ratio is proportional to volume fraction.

Equation (4.34) can then be written as

$$\gamma_c = \gamma_f\, V_f + \gamma_m\, V_m \qquad (4.35)$$

Assuming a linear shear stress-shear strain behavior for the fiber and matrix, we get

$$\frac{1}{G_{LT}} = \frac{V_f}{G_f} + \frac{V_m}{G_m} \qquad (4.36)$$

or

$$G_{LT} = \frac{G_m G_f}{V_m G_f + V_f G_m} \qquad (4.37)$$

in view of Eq. (4.31). Equation (4.37) is similar to Eq. (4.28). Hence, the comments made with regard to the behavior of E_T are also valid for G_{LT}. We need to find better models for estimating the in-plane shear modulus. A modified relation similar to Eq. (4.30) has been obtained for the in-plane shear modulus (G_{LT}) by Hopkins and Chamis [13] and is given as

$$G_{LT} = G_m \left[\left(1 - \sqrt{V_f} \right) + \frac{\sqrt{V_f}}{1 - \sqrt{V_f} \, (1 - G_m / G_f)} \right] \qquad (4.38)$$

Major Poisson's Ratio

Consider once again the RVE shown in Fig. 4.3. It is subjected to a uniform tensile load along the longitudinal direction. As a result there will be contraction along the transverse direction. The major Poisson's ratio ν_{LT} is defined as

$$\nu_{LT} = -\frac{\varepsilon_T}{\varepsilon_L} \qquad (4.39)$$

The total displacement of the composite in the lateral direction is

$$\delta_c = \delta_f + \delta_m \qquad (4.40)$$

Equation (4.40) can be rewritten as

$$\varepsilon_{cT} \, h = \varepsilon_{fT} \, h_f + \varepsilon_{mT} \, h_m \qquad (4.41)$$

where ε_{fT}, ε_{mT} and ε_{cT} are the strains of fiber, matrix and composite along the lateral direction. These, in turn, can be written in terms of longitudinal strain ε_L.

$$\varepsilon_{fT} = -\nu_f \, \varepsilon_L, \; \varepsilon_{mT} = -\nu_m \, \varepsilon_L \text{ and } \varepsilon_{cT} = -\nu_{LT} \, \varepsilon_L \qquad (4.42)$$

Substituting Eq. (4.42) into Eq. (4.41), we obtain

$$-\nu_{LT} \, \varepsilon_L \, h = -\nu_f \, \varepsilon_L \, h_f - \nu_m \, \varepsilon_L \, h_m \qquad (4.43)$$

Equation (4.43) can further be simplified as

$$\nu_{LT} = \nu_f \, V_f + \nu_m \, V_m \qquad (4.44)$$

Thus, Poisson's ratio also follows the rule of mixtures. One can then find the minor Poisson's ratio by using the constraint relation given by Eq. (3.25).

Theory of Elasticity Approach

The elasticity approach satisfies the equations of equilibrium, compatibility and constitutive relations in three dimensions. An attempt is made here in brief to describe the composite cylinder assemblage (CCA) model presented by Hill [6]. In this model, it is assumed that the fibers are circular in cross-section and spread uniformly. The RVE consists of a solid cylinder of fiber encased in a cylinder of matrix of diameters a and b, respectively. The fiber volume fraction V_f is given as

$$V_f = a^2/b^2 \qquad (4.45)$$

The elastic modulus E_L, ν_{LT} and G_{LT} are obtained as

$$E_L = E_f V_f + E_m V_m + \frac{4(\nu_m - \nu_f)^2 V_f V_m}{\dfrac{V_f}{K_m} + \dfrac{V_m}{K_f} + \dfrac{1}{G_f}} \qquad (4.46a)$$

$$G_{LT} = G_f \frac{G_f V_f + G_m(1 + V_m)}{G_m V_f + V_f(1 + V_m)} \qquad (4.46b)$$

$$\nu_{LT} = \nu_f V_f + \nu_m V_m + \frac{(\nu_m - \nu_f)\left(\dfrac{1}{K_f} - \dfrac{1}{K_m}\right) V_f V_m}{\dfrac{V_f}{K_m} + \dfrac{V_m}{K_f} + \dfrac{1}{G_f}} \qquad (4.46c)$$

$$K_f = \frac{E_f}{3(1 - 2\nu_f)} \qquad (4.46d)$$

$$K_m = \frac{E_m}{3(1 - 2\nu_m)} \qquad (4.46e)$$

For more details on the method, the reader may refer to Hashin and Rosen [3].

Semi-empirical Approach

The transverse modulus E_T and in-plane shear modulus G_{LT} obtained by employing the strength of materials approach do not agree well with the experimental results as shown in Figs. 4.8 and 4.9, respectively. Halpin and Tsai [14] proposed a semi-empirical relation by curve fitting to results that are based on the elasticity approach. These equations are simple and are quite accurate. They involve two parameters ξ and η. ξ represents the degree of reinforcement and depends on (i) fiber geometry, (ii) packing geometry and (iii) loading conditions. The parameters depend on fiber, matrix moduli and ξ.

The transverse elastic modulus E_T is given by

$$\frac{E_T}{E_m} = \frac{1 + \zeta \eta V_f}{1 - \eta V_f} \qquad (4.47)$$

$$\eta = \frac{(E_f / E_m) - 1}{(E_f / E_m) + \xi} \qquad (4.48)$$

The in-plane shear modulus G_{LT} is given by

$$\frac{G_{LT}}{G_m} = \frac{1 + \zeta \eta V_f}{1 - \eta V_f} \tag{4.49}$$

$$\eta = \frac{(G_f / G_m) - 1}{(G_f / G_m) + \zeta} \tag{4.50}$$

For estimating the *transverse modulus*, ζ = 2.0, for circular cross-section fibers in a packing geometry of square array. ζ = 1.0 for rectangular fiber cross-section of length a and width b in a hexagonal array, $\zeta = 2(a/b)$, where a is in the direction of loading. Figure 4.8 shows the comparison of experimental values, strength of materials approach and Halpin-Tsai relations.

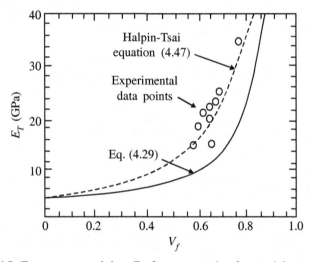

Fig. 4.8 Transverse modulus E_T from strength of materials approach and Halpin-Tsai equation [12].

For estimating in-plane shear modulus, ζ = 1.0 for circular cross-section fibers in a square array. For rectangular cross-section fibers of length a and width b in hexagonal array, $\zeta = \sqrt{3}\log_e(a/b)$, where length a is in the direction of loading. Figure 4.9 shows the comparison of experimental values, strength of material approach and Halpin-Tsai relations.

4.4 Hygrothermal Behavior of Lamina

The stress-strain relationships derived for the lamina in the earlier sections have been based on the assumption that the environmental conditions do not change. However, the environmental conditions do change during fabrication as well as during its lifetime. The effect of the environment is important especially for polymer based fiber reinforced composites as they absorb

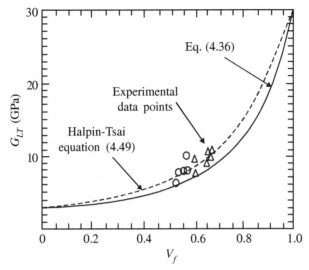

Fig. 4.9 Shear modulus G_{LT} from strength of materials approach and Halpin-Tsai equation [12].

moisture and will result in the swelling of the composites. Some reinforcing fibers also absorb moisture. Therefore, it is necessary to include the effect of such changes in the constitutive relations of the lamina.

Composite materials are generally processed at high temperatures and then cooled down to operating temperatures. For polymer-based composites, this difference in temperature is of the order of 200-300°C. The thermal coefficient of expansion of the fibers and the matrices are very different. Due to this mismatch, thermal stresses are generated when the temperature is brought down. Increase in moisture content causes swelling of the matrix, while reduced moisture content causes contraction of the matrix. This swelling and contraction is resisted by the fibers. The effects of temperature are usually referred to as 'thermal' effects, whereas those of moisture are referred to as 'hygroscopic' effects. If both the effects are present then it is referred to as 'hygrothermal'. In an actual situation, the laminae are stacked together to form a laminate. The fiber orientation of each lamina may be different. The neighboring lamina acts as a constraint and does not allow free expansion of the lamina. This results in hygrothermal stresses generated in the laminate even though there are no external loads applied to the laminate. Such stresses are called self-equilibrating stresses. Further, the strains developed in the transverse and longitudinal directions are not the same, since the elastic modulus along the two directions, in general, are different.

It is, therefore, necessary to incorporate their effect in the analysis, as they will affect the overall load carrying capacity of the laminate.

4.4.1 Coefficient of Thermal Expansion and Associated Thermal Strains
The coefficient of thermal expansion is defined as change in length per unit length per unit change in temperature. For isotropic materials such as aluminum, steel etc., the coefficient of thermal expansion α is the same in all directions. It varies from material to material. The associated linear strain is given by

$$\varepsilon^{Th} = \alpha \, \Delta T \tag{4.51}$$

where ΔT is the change in temperature. In general, the initial temperature T_0 is regarded as the temperature at which no thermal stresses exist, and T is the final temperature at which the component is required to operate. $\Delta T = T - T_0$. Equation (4.51) is based on the experimental observation that a temperature change induces only expansion or contraction and no distortion of the section [11].

For a unidirectional lamina, there are two thermal coefficients of expansion, one along the direction of the fiber and the other transverse to the direction of the fiber. They are defined as α_L and α_T, respectively. These are related to the thermal coefficients of expansion of the constituent materials. Gibson [11] derived the relation for α_L by using the strength of materials approach, similar to the one employed for longitudinal elastic modulus. In what follows, the method used by Gibson [11] is presented for estimating α_L.

The constitutive relation for a one-dimensional unidirectional composite, including the thermal strain, is given by

$$\varepsilon_L = \frac{\sigma_L}{E_L} + \alpha_L \Delta T \tag{4.52}$$

Rewriting Eq. (4.52), we get

$$\sigma_L = E_L \, (\varepsilon_L - \alpha_L \, \Delta T) \tag{4.53}$$

Substituting Eq. (4.53) in Eq. (4.15) for composite, fiber and matrix, we obtain

$$E_L \, (\varepsilon_L - \alpha_L \, \Delta_T) = E_f \, (\varepsilon_f - \alpha_f \, \Delta T) \, V_f + E_m \, (\varepsilon_m - \alpha_m \, \Delta T) \, V_m \tag{4.54}$$

where α_f and α_m are the thermal coefficients of expansion for fiber and matrix material, respectively. By making use of Eqs. (4.11) and (4.17) along with Eq. (4.54), we obtain

$$E_L \, \alpha_L = \alpha_f \, E_f \, V_f + \alpha_m \, E_m \, V_m \tag{4.55}$$

Therefore

$$\alpha_L = \frac{\alpha_f V_f E_f + \alpha_m V_m E_m}{E_f V_f + E_m V_m} \tag{4.56}$$

Equation (4.56) has also been derived by Schapery [15] by using energy principles.

The transverse thermal coefficient of expansion has been obtained by Schapery [15] as

$$\alpha_T = \alpha_f V_f (1 + v_f) + \alpha_m V_m (1 + v_m) - \alpha_L v_{LT} \qquad (4.57)$$

where v_f and v_m are Poisson's ratios of the fiber and matrix material, and v_{LT}, the major Poisson's ratio.

Rosen [16] has shown that for a typical graphite/epoxy composite having higher fiber volume fraction ($V_f \geq 0.6$), the predicted value of α_L is practically zero. Ishikawa et al. [17] have shown that for most of the practical cases α_T is greater than α_L.

The thermal strains generated due to a change in temperature in the material directions is given by

$$\begin{Bmatrix} \varepsilon_L^{Th} \\ \varepsilon_T^{Th} \\ \gamma_{LT}^{Th} \end{Bmatrix} = \begin{Bmatrix} \alpha_L \\ \alpha_T \\ 0 \end{Bmatrix} \Delta T \qquad (4.58)$$

Since thermal strains are also physical strains, they can be transformed into *X-Y* axes in a similar manner as the physical strains by employing Eq. (3.34). Hence

$$\begin{Bmatrix} \alpha_x \\ \alpha_y \\ \alpha_{xy}/2 \end{Bmatrix} = \begin{bmatrix} \cos^2\theta & \sin^2\theta & -2\sin\theta\cos\theta \\ \sin^2\theta & \cos^2\theta & 2\sin\theta\cos\theta \\ \sin\theta\cos\theta & -\sin\theta\cos\theta & (\cos^2\theta - \sin^2\theta) \end{bmatrix} \begin{Bmatrix} \alpha_L \\ \alpha_T \\ 0 \end{Bmatrix} \qquad (4.59)$$

where α_x, α_y and α_{xy} are called the apparent coefficients of thermal expansion. Multiplying both sides of Eq. (4.59), we can obtain the strains in the *X-Y* coordinate system in terms of the strains in the *L-T* coordinate system. The stresses associated with these strains can be obtained by making use of Eq. (3.42).

4.4.2 Coefficient of Moisture Absorption and Associated Strains

As stated earlier, the polymer absorbs moisture. This results in hygroscopic expansion or contraction similar to thermal strains. The moisture induced in isotropic materials causes only normal strains and no distortion of the cross-section. This can be represented as

$$\varepsilon^H = \beta \, \Delta C \qquad (4.60)$$

where β is the coefficient of moisture absorption. $\Delta C = (C - C_0)$ is the change in the moisture concentration. C_0 is the reference moisture condition.

For a composite lamina, which is orthotropic, there are two coefficients of moisture absorption along the material axes *L* and *T*, defined as β_L and β_T, respectively. The strains associated are defined as

$$\begin{Bmatrix} \varepsilon_L^H \\ \varepsilon_T^H \\ \gamma_{LT}^H \end{Bmatrix} = \begin{Bmatrix} \beta_L \\ \beta_T \\ 0 \end{Bmatrix} \Delta C \qquad (4.61)$$

Following the steps as indicated for the thermal strains, the coefficients β_L and β_T are written as

$$\beta_L = \frac{\beta_f E_f V_f + \beta_m E_m V_m}{E_f V_f + E_m V_m} \qquad (4.62)$$

$$\beta_T = \beta_f V_f (1 + \nu_f) + \beta_m V_m (1 + \nu_m) - \beta_L \nu_{LT} \qquad (4.63)$$

in which β_f and β_m are the coefficients of moisture absorption for fiber and matrix, respectively.

Making use of Eq. (3.34), one can write the apparent coefficient of moisture absorption along the X-Y direction as

$$\begin{Bmatrix} \beta_x \\ \beta_y \\ \beta_{xy}/2 \end{Bmatrix} = \begin{bmatrix} \cos^2\theta & \sin^2\theta & -2\sin\theta\cos\theta \\ \sin^2\theta & \cos^2\theta & 2\sin\theta\cos\theta \\ \sin\theta\cos\theta & -\sin\theta\cos\theta & (\cos^2\theta - \sin^2\theta) \end{bmatrix} \begin{Bmatrix} \beta_L \\ \beta_T \\ 0 \end{Bmatrix} \qquad (4.64)$$

Table 4.1 indicates the coefficient of thermal expansion and coefficient of moisture absorption for different composites.

Table 4.1 Coefficient of thermal expansion and coefficient of moisture absorption for some typical composites

Material	Coefficient of thermal expansion (10^{-6} m/m)		Coefficient of moisture expansion (10^{-6} m/m)	
	α_L	α_T	β_L	β_T
AS graphite/epoxy	0.88	31.0	0.09	0.30
E-glass/epoxy	6.30	20.0	0.014	0.29
Carbon/epoxy	− 0.90	24.0	-	-
Kevlar/epoxy	− 4.0	79.0	-	-

4.5 Strength and Failure of a Lamina

4.5.1 Longitudinal Tensile Strength

Equation (4.15) gives the longitudinal stress for a unidirectional composite specimen subjected to axial tensile load. Equation (4.18) gives the ultimate tensile strength of the same specimen. While writing Eq. (4.18), it is assumed that the matrix failure strain ε_{mu} is greater than the fiber failure strain. This is true for many polymer matrix composites. However, in the case of ceramic matrix composites, the fiber failure strain ε_{fu} is greater than ε_{mu}.

In polymer matrix, unidirectional composites subjected to a uniaxial tensile load, failure is initiated when the fiber strain reaches the ultimate value. The following assumptions are made for deriving the strength of the unidirectional composite: (i) fibers are of equal strength, (ii) stress-strain relation for the fiber is linear up to failure and (iii) longitudinal strain in the fiber, matrix and the composite are equal.

If the volume fraction of the fiber in the composite is sufficiently high, the matrix may not be able to sustain the load when all the fibers break. Theoretically, if the matrix could support the applied load after fiber failure, the strain could be increased to the matrix failure. For all practical purposes, fiber failure means composite failure. Under these conditions, the ultimate longitudinal tensile strength of the composite is equal to the composite stress at the failure strain of the fiber. Equation (4.18) gives the ultimate tensile strength of the composite.

$$\sigma_{cu} = \sigma_{fu} V_f + [\sigma_m | \varepsilon_{fu}](1 - V_f) \tag{4.18}$$

where σ_{cu} is the ultimate longitudinal strength of the composite, σ_{fu} is the ultimate strength of the fibers, and $\sigma_m | \varepsilon_{fu}$ is the matrix stress at the failure strain of the fiber. Equation (4.18) predicts whether the composite strength is higher or lower than the matrix strength depending on the fiber volume fraction. The main purpose of the composite is to have composite strength higher than matrix strength. In order to achieve this, it is necessary to have the fiber fraction greater than a certain critical value. A critical fiber volume fraction V_{fcric} is defined as the one that should be exceeded in order to have a composite strength higher than the matrix strength

$$\sigma_{cu} = \sigma_{fu} V_f + [\sigma_m | \varepsilon_{fu}](1 - V_f) \geq \sigma_{mu} \tag{4.65}$$

For a limiting case beyond which strengthening takes place

$$\sigma_{fu} V_{fcric} + [\sigma_m | \varepsilon_{fu}](1 - V_{fcric}) = \sigma_{mu} \tag{4.66}$$

Therefore

$$V_{fcric} = \frac{\sigma_{mu} - (\sigma_m)\varepsilon_{fu}}{\sigma_{fu} - (\sigma_m)\varepsilon_{fu}} \tag{4.67}$$

If the volume fraction of the fibers is less than minimum fiber volume V_{fmin}, the matrix takes most of the applied load and it is stretched. Since the fiber ultimate strain is smaller than the matrix, it gets stretched beyond its ultimate value and all the fibers break. The composite fails when the maximum stress of the composite is equal to σ_{mu}. Thus, the ultimate strength of the composite with the fiber volume fraction less than V_{fmin} is given by

$$\sigma_{cu} = \sigma_{mu} (1 - V_f) \tag{4.68}$$

The minimum value of fiber volume fraction above which the fiber starts taking the load is obtained by equating Eq. (4.68) and Eq. (4.18). Thus

$$\sigma_{fu} \, V_{fmin} + [\sigma_m \mid \varepsilon_{fu}](1 - V_{fmin}) = \sigma_{mu}(1 - V_{fmin}) \tag{4.69}$$

From which

$$V_{fmin} = \frac{\sigma_{mu} - (\sigma_m)\varepsilon_{fu}}{\sigma_{fu} + \sigma_{mu} - (\sigma_m)\varepsilon_{fu}} \tag{4.70}$$

Figure 4.10 shows the plot of ultimate composite strength as a function of fiber volume fraction V_f. It is observed that for a fiber volume fraction less than V_{fmin}, the composite strength is always less than the matrix strength. For a fiber volume fraction lying in between V_{fmin} and V_{fcric}, the composite strength is still less than the matrix strength, while for a fiber volume fraction greater than V_{fcric}, the composite strength is always greater than the matrix strength. The volume fractions obtained from the above relations depend strongly on the assumptions made. One of the weakest assumptions is that all the fibers are of uniform strength. It is generally observed that the fibers fail much before their ultimate value. The fiber strength decreases with increase in fiber length as the defects in the fibers generally increase because of the manufacturing process. Readers are referred to Rosen [16] for a review of various statistical models.

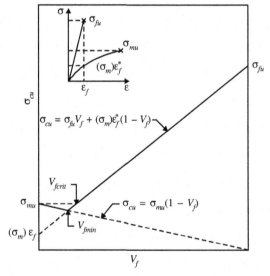

Fig. 4.10 Variation of ultimate longitudinal stress with fiber volume fraction [18].

The assumption of linear elastic behavior up to failure is not valid if the fibers are ductile. However, the errors introduced by this assumption are negligible.

Different types of failures occur in composites depending on the properties of constituents. Failure in unidirectional composites are generally initiated with fiber breakage at the weakest cross-sections. Breaking of fibers is generally a random process in view of the manufacturing process. If a number of fibers break at the same location, the cross-section will not be able to withstand the load resulting in complete failure.

The micro cracks that develop at different sections may join up resulting in debonding of the fiber from the matrix along their length or by shear failure of the matrix. The sequence in which different types of failure behavior occurs is strongly dependent on the constituent properties and also the manufacturing process.

4.5.2 Longitudinal Compressive Strength

The failure modes of a composite subjected to axial compressive loading are different from the composite subjected to tensile loading. It is well known that an isotropic specimen subjected to axial compressive loading buckles. However, in case composite columns form before buckling occurs, micro fiber buckling is likely to occur. In this case, the structural form remains unchanged but fibers would have buckled internally. The nature of buckling depends on the characteristics of the material and the fiber volume fraction. The initial investigation into fiber micro buckling was carried out by Rosen [19] based on the work of Timoshenko for columns supported on elastic foundation. The analysis presented is based on the work of Rosen [18] and is also given in Jones [1]. Fiber buckling is defined as the fiber instability which decreases the ability of the fiber to carry load. Two types of fiber micro buckling is possible: (i) symmetric flexural mode and (ii) anti-symmetric flexural mode. In the first case, the matrix undergoes extensional deformation and, is therefore, referred to as extensional mode of buckling, while in the second case, the matrix is subjected to shearing deformation and hence is referred to as shear mode of buckling. The possible mode of buckling depends on the fiber characteristics, matrix characteristics and fiber volume fraction.

Figure 4.11(a) shows the model considered for both types of buckling. It is assumed that the fibers are uniformly placed. It is further assumed that the fiber takes the entire load and the load is distributed uniformly across all the fibers. The fibers are of thickness w separated by a matrix of width $2d$. The thickness of the specimen is taken to be unity along the z-direction. The load acting is taken as P. The energy approach is employed to obtain the buckling load. It is further assumed that the stresses in the matrix and the fiber are within the elastic limit. The fiber primarily undergoes bending and since the cross-section is very small compared to the length of the fiber, the contribution to the strain energy will be mainly due to flexure. However, in the case of matrix, it undergoes either the extensional mode or the shear mode.

Extensional Mode

Figures 4.11(b) and 4.12 show the deformation of the fiber and matrix in the extensional mode. In the energy approach, the sum of the change in the

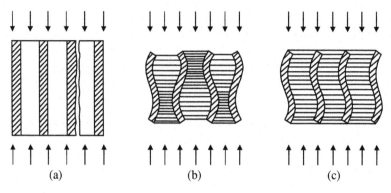

Fig. 4.11 Modes of failure for a lamina subjected to axial compressive load [1].

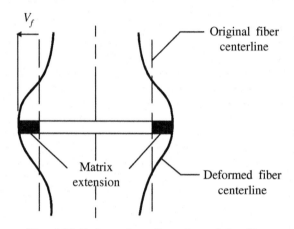

Fig. 4.12 Deformed configuration of the fiber.

bending energy of the fiber and the extensional energy of the matrix is equal to the work done by the external load P acting on the fiber. This can be stated mathematically as

$$\Delta U_f + \Delta U_m = \Delta W_e \qquad (4.71)$$

It is well known that the buckling load estimated by this approach is an upper bound to the exact value since a solution is assumed to be satisfying only the geometric boundary conditions specified for the problem. Assuming the column to be simply supported at the two ends, the admissible function satisfying the boundary conditions is given by

$$v = \sum_{n=1}^{\infty} A_n \sin \frac{n\pi x}{L} \qquad (4.72)$$

where v is the displacement of the fiber in the transverse direction and L is the length of the fiber.

The strain energy due to bending of the fiber is expressed as

$$\Delta U_b = \frac{E_f I}{2} \int_0^L \frac{d^2 v}{dx^2} \, dx \qquad (4.73)$$

where E_f is the elastic modulus of the fiber and I, the second moment of area of the fiber.

Substituting for v from Eq. (4.72) and integrating, we obtain

$$\Delta U_b = \frac{E_f I \pi^4}{4L^3} \sum_{n=1}^{\infty} n^4 A_n^2 \qquad (4.74)$$

The second moment of area of the fiber is $I = \dfrac{h^3}{12}$. Substituting the value of I in Eq. (4.74), we get

$$\Delta U_b = \frac{E_f h^3 \pi^4}{48 L^3} \sum_{n=1}^{\infty} n^4 A_n^2 \qquad (4.75)$$

The change in the extensional energy of the matrix is obtained as follows: It is assumed that there is no debonding of the fiber and the matrix. Hence, the displacement of the fiber and the matrix at any given point along the length of the fiber is the same. Therefore, the change in the strain energy of the matrix is

$$\Delta U_m = \frac{1}{2} \int_0^L \varepsilon_y \sigma_y \, dx \qquad (4.76)$$

The displacement of the matrix at any point is $2v$ over a length $2c$. Hence, the strain ε_y is

$$\varepsilon_y = \frac{2v}{2c} \quad \text{(the width } 2c \text{ is chosen for convenience)} \qquad (4.77)$$

Therefore, the stress in the matrix is

$$\sigma_y = E_m \frac{v}{c} \qquad (4.78)$$

Substituting for ε_y, σ_y and v in Eq. (4.76) and integrating, we get

$$\Delta U_m = \frac{E_m L}{2c} \sum_{n=1}^{\infty} A_n^2 \qquad (4.79)$$

The work done by the external applied load P is

$$\Delta W_e = \frac{P}{2} \int_0^L \left(\frac{dv}{dx} \right)^2 dx \qquad (4.80)$$

Substituting the value of (dv/dx) and subsequent integration leads to

$$\Delta W_e = \frac{P\pi^2}{4L} \sum_{n=1}^{\infty} n^2 A_n^2 \tag{4.81}$$

Substituting the expressions in Eq. (4.71), the fiber micro buckling load P is

$$P = \frac{\pi^2 E_f h^3}{12L^2} \left(\frac{\displaystyle\sum_{n=1}^{\infty} n^4 A_n^2 + \frac{24L^4 E_m}{\pi^4 ch^3 E_f} \sum_{n=1}^{\infty} A_n^2}{\displaystyle\sum_{n=1}^{\infty} n^2 A_n^2} \right) \tag{4.82}$$

The load P acting on the fiber is

$$P = \sigma_f (h \times 1) \tag{4.83}$$

where n defines the number of half sine waves at which the fiber buckles. Let us assume that the buckling will take place at a particular value of n which is p. We do not know the value of m at which the fiber buckles. The value of p is obtained by taking the derivative of σ_{cr} with respect to m. This gives the minimum value of m when the fiber buckles. The value depends on the spacing of the fibers, fiber and matrix modulus and the fiber dimension.

$$\sigma_{cr} = \frac{\pi^2 E_f h^2}{12L^2} \left[p^2 + \frac{24L^4 E_m}{\pi^4 ch^3 E_f} \left(\frac{1}{p^2} \right) \right] \tag{4.84}$$

As stated earlier, the minimum value is obtained by taking the derivative of Eq. (4.84) with respect to m and putting it equal to zero i.e. $\partial \sigma_{cr}/\partial p = 0$, subject to the condition $\partial^2 \sigma_{cr}/\partial p^2 > 0$. However, it has to be borne in mind that m is an integer.

$$\frac{\partial \sigma_{cr}}{\partial p} = \left[2p - \frac{24L^4 E_m}{\pi^4 ch^3 E_f} \frac{2}{p^3} \right] = 0 \tag{4.85}$$

This gives

$$p^2 = \left(\frac{24L^4 E_m}{\pi^4 ch^3 E_f} \right)^{1/2} \tag{4.86}$$

Substituting this value of p in Eq. (4.84), we get

$$\sigma_{cr} = \sqrt{\frac{2E_f E_m h}{3c}} \tag{4.87}$$

Equation (4.87) can be rewritten in terms of fiber volume fraction V_f, which is defined as

$$V_f = \frac{h}{(h+2c)} \tag{4.88}$$

Equation (4.88) can be written in terms of volume fraction of the fiber as

$$\sigma_{cr} = 2\sqrt{\frac{V_f E_f E_m}{3(1-V_f)}} \tag{4.89}$$

For the extensional mode, the number of half sine waves is generally very large (this has been observed experimentally).

If we assume that the stress developed in the matrix is very small, then the maximum stress of the composite is given as

$$\sigma_{cul} = V_f \sigma_{cr} = 2V_f \sqrt{\frac{V_f E_f E_m}{3(1-V_f)}} \tag{4.90}$$

Since there is no debonding between the fiber and the matrix, the strain in the fiber and matrix is same. Hence, σ_m stress in the matrix at σ_{cr} is given by

$$\sigma_m = E_m \left(\frac{\sigma_{cr}}{E_f} \right) \tag{4.91}$$

Following Eq. (4.15), the ultimate stress of the composite is written as

$$\sigma_{cul} = V_f \, \sigma_{cr} + (1 - V_f) \, \sigma_m \tag{4.92}$$

Making use of Eqs. (4.89) and (4.91) in Eq. (4.92), we get

$$\sigma_{cul} = 2 \left[V_f + (1-V_f)\frac{E_m}{E_f} \right] \sqrt{\frac{V_f E_m E_f}{3(1-V_f)}} \tag{4.93}$$

Equation (4.93) assumes that the void volume is negligible.

Shear Mode
Figures 4.11(c) and 4.13 show the shear mode of fiber buckling. Here the fiber buckles in the anti-symmetric mode in flexure. The matrix deformation is predominantly shear as shown in Fig. 4.13. We shall once again employ the energy method to get an estimate of the micro fiber buckling load. We will once again assume that there is no debonding between the fiber and the matrix. The flexural energy of the fiber remains the same and is given by Eq. (4.75). However, the contribution to the energy from the matrix will be the shear energy due to shear deformation. We neglect the extensional energy of the matrix as the matrix deformation is predominantly shear.

The shear strain in the matrix is

$$\gamma_{xy} = \frac{\partial u}{\partial y} + \frac{\partial v}{\partial x} \tag{4.94}$$

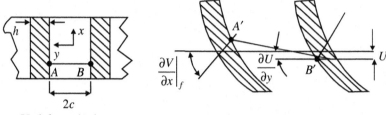

| Undeformed plate | Deformed shape of fibers |

Fig. 4.13 Deformed configuration of the fiber in shear mode [1].

The displacement v is along the y-direction and u is along the x-direction. Since the transverse is independent of y

$$\left(\frac{\partial v}{\partial x}\right)_{matrix} = \left(\frac{\partial v}{\partial x}\right)_{fiber} \tag{4.95}$$

Since the shear strain is independent of y

$$\frac{\partial u}{\partial y} = \frac{1}{2c}[u(c) - u(-c)] \tag{4.96}$$

From Fig. 4.13

$$u(c) = \frac{h}{2}\left(\frac{\partial v}{\partial x}\right)_{fiber} \tag{4.97}$$

Substituting Eq. (4.97) in Eq. (4.96), we get

$$\frac{\partial u}{\partial y} = \frac{h}{2c}\left(\frac{\partial v}{\partial x}\right)_{fiber} \tag{4.98}$$

Substituting Eq. (4.98) in Eq. (4.94) and making use of Eq. (4.95), we obtain

$$\gamma_{xy} = \left(1 + \frac{h}{2c}\right)\left(\frac{\partial v}{\partial x}\right)_{fiber} \tag{4.99}$$

The change in shear strain energy of the matrix is given as

$$\Delta U_m = \frac{1}{2}\int_0^L\int_A \tau_{xy}\gamma_{xy}\,dA\,dx \tag{4.100}$$

Assuming a linear relationship between shear and shear strain for the matrix as

$$\tau_{xy} = G_m\,\gamma_{xy} \tag{4.101}$$

Eq. (4.100) can be rewritten as

$$\Delta U_m = \frac{1}{2} \int_0^L \int_A G_m \gamma_{xy}^2 \, dA \, dx \tag{4.102}$$

Substituting for γ_{xy} from Eq. (4.99), and for v from Eq. (4.72) and subsequent integration with respect to the volume, we get

$$\Delta_m = G_m c \left(1 + \frac{h}{2c}\right)^2 \frac{\pi^2}{2L} \sum_{n=1}^{\infty} n^2 A_n^2 \tag{4.103}$$

The change in the bending energy of the fiber is still given by Eq. (4.75) and the work done by the applied compressive load P is given by Eq. (4.81). Making use of Eq. (4.71), the critical load can be expressed as

$$P = \frac{G_m c \left(1 + \dfrac{h}{2c}\right)^2 \dfrac{\pi^2}{2L} \displaystyle\sum_{n=1}^{\infty} n^2 A_n^2 + \dfrac{\pi^4 E_f h^3}{48 L^3} \displaystyle\sum_{n=1}^{\infty} h^4 A_n^2}{\dfrac{\pi^2}{4L} \displaystyle\sum_{n=1}^{\infty} n^2 A_n^2} \tag{4.104}$$

Assuming that a particular value of n equal to p gives the critical load, we get the critical load as

$$P_{cr} = 2 G_m c \left(1 + \frac{h}{2c}\right)^2 + \frac{E_f h^3 p^2 \pi^2}{12 L^3} \tag{4.105}$$

Therefore

$$\sigma_{fcr} = \frac{2 G_m c}{h}\left(1 + \frac{h}{2c}\right)^2 + \frac{E_f h^2 p^2 \pi^2}{12 L^2} \tag{4.106}$$

Making use of Eq. (4.88), we can rewrite Eq. (4.106) as

$$\sigma_{fcr} = \frac{G_f}{V_f (1 - V_f)} + \frac{\pi^2 E_f}{12}\left(\frac{ph}{L}\right)^2 \tag{4.107}$$

In general, the buckled wavelength for the shear mode is very much larger than for the extensional mode. The contribution of the second term to the critical load is negligible. Therefore, the critical stress for shear mode can be represented as

$$\sigma_{fcr} = \frac{G_m}{V_f (1 - V_f)} \tag{4.108}$$

In order to assess which of these two modes of fiber micro buckling is critical, let us consider a glass/epoxy lamina with the following properties: Elastic modulus of a typical epoxy = 3.793 GPa, E-glass modulus = 85 GPa. Assuming the matrix to be elastic, G_m = 1.405 GPa. The critical buckling stress for the extension and shear modes can be obtained from Eqs. (4.89) and (4.108).

Extension $\qquad \sigma_{cr} = 2\sqrt{\dfrac{V_f E_f E_m}{3(1-V_f)}} = 20.733\sqrt{\dfrac{V_f}{(1-V_f)}}$

Shear $\qquad \sigma_{cr} = \dfrac{G_m}{V_f(1-V_f)} = \dfrac{1.405}{V_f(1-V_f)}$

From Table 4.2, it is observed that for low values of fiber volume fraction, the extensional mode governs the buckling. However, for higher volume fraction, the shear mode is critical.

Table 4.2 Fiber buckling loads at different fiber volume fraction

	Fiber volume fraction V_f	
Mode	Extensional	Shear
0.05	4.7	29.6
0.1	6.9	15.6
0.2	10.4	8.78
0.4	16.9	8.58
0.6	25.4	8.58
0.8	41.5	8.78
0.9	62.2	15.6

On the basis of the experimental and theoretical studies of graphite fibers in various matrices, Greszczuk [20] concluded: (i) micro buckling in the shear mode occurs for low modulus resins and (ii) extensional mode of buckling for high modulus resins.

Summary
- We observed that the lamina has four independent engineering constants and five independent in-plane strengths.
- We also discussed the behavior of the lamina under in-plane tensile or compressive loadings.

Problems

4.1 Assume that there are circular fibers of diameter d embedded in a matrix material. The two fiber packing arrangements with separation distance D are shown. Obtain fiber volume fraction V_f for the two arrangements.

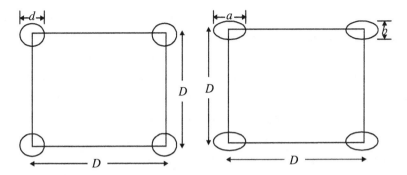

4.2 If the fibers are of rectangular cross-section of length a and width b and separation distance D for the two packing shown, obtain the fiber volume fraction, V_f.

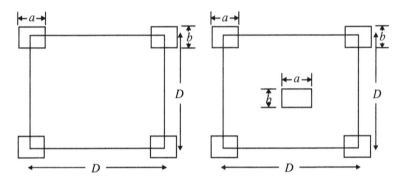

4.3 For a given composite of $v_f = 0.6$, estimate the longitudinal modulus and transverse modulus by using strength of materials approach and Halpin-Tsai empirical equations. Given $E_f = 220$ GPa, $E_m = 14$ GPa.

4.4 If the composite given in Problem 4.3 is subjected to a longitudinal loading, compute the stresses in the fiber and the matrix and also the energy stored in the fibers and the matrix.

4.5 A rod for carrying tension load has to be designed using graphite/epoxy composite as a replacement for aluminum alloy. Calculate the volume fraction of graphite fibers required to match the longitudinal modulus of aluminum alloy. The material properties are as follows: the elastic modulus of epoxy is 3.5 GPa and density 1.2 g/cm^3, the tensile modulus of 240 GPa and density 1.90 g/cm^3 and that of aluminum 70 GPa and density 2.7 g/cm^3.

4.6 For an e-glass/epoxy lamina with $E_f = 72.4$ GPa, $E_m = 2.0$ GPa, Poisson's ratio $= 0.37$, obtain the critical buckling stress expression for extension and shear as a function of fiber volume fraction. Plot fiber volume fraction vs. the critical buckling stress.

4.7 If the fiber volume fraction is varied from a small value to a large value, how does the mode of critical buckling change for the data given in Problem 4.6?

4.8 A cross-ply laminate [0°/90°] of unidirectional carbon/epoxy has the following axial strains and curvatures at the mid-plane of the laminate:

$$\varepsilon_x^\circ = 500 \times 10^{-6},\ \varepsilon_y^\circ = 0,\ \gamma_{xy}^\circ = 0,\ \kappa_x = \kappa_y = \kappa_{xy} = 0$$

The change in temperature after curing is $- 125°C$. The temperature coefficients of expansion for the material are $\alpha_L = - 0.9 \times 10^{-6}/°C$, $\alpha_T = 27 \times 10^{-6}/°C$. Calculate the total stresses in the $0°$ and $90°$ plies.

References

1. Jones, R.M., 2010, *Mechanics of Composite Materials*, Special Indian Edition, Taylor and Francis.
2. Paul, B., 1960, "Predictions of Elastic Constants of Multiphase Materials", *Trans. Met. Soc. AIME*, 36.
3. Hashin, Z. and Rosen, W.B., 1964, "The Elastic Moduli of Fiber Reinforced Materials", *J. Appl. Mec., Trans. ASME*, 31, p. 233.
4. Adams, D.F. and Doner, D.R., 1967, "Transverse Normal Loading of a Unidirectional Composite", *J. Comp. Mat.*, 1, p. 52.
5. Chen, P.E. and Lin, J.M., 1969, "Transverse Properties of Fibrous Composites", *Mat. Res. Stand., MTRSA*, 9, p. 29.
6. Hill, R., 1965, "Theory of Mechanical Properties of Fibre Strengthened Material: III Self-consistent Model", *J. Mech. Phys. Solids*, 13, p. 189.
7. Chamis, C.C. and Sendeckyj, G.P., 1968, "Critique on Theories Predicting Thermoelastic Properties of Fibrous Composites", *J. Comp. Mat.*, 2, p. 332.
8. Christensen, R.M., 1979, *Mechanics of Composite Materials*, John Wiley and Sons, New York.
9. Hashin, Z., 1983, "Analysis of Composite Materials: A Survey", *J. Appl. Mech.*, 50, p. 481.
10. Halpin, J.C., 1984, *Primer on Composite Materials: Analysis*, Technomic Pub. Co., Lancaster, Pennsylvania.
11. Gibson, R.F., 1994, *Principles of Composite Material Mechanics*, Int. Edition, McGraw Hill Inc.
12. Whitney, J.M. and Riley, M.B., 1966, "Elastic Properties of Fiber Reinforced Composite materials", *JAIAA*, 4, p. 1537.
13. Hopkins, D.A. and Chamis, C.C., 1988, "A Unique Set of Micromechanics Equations for High Temperature Metal Matrix Composites", *ASTM*, STP, 964, p. 159.
14. Halpin, J.C. and Tsai, S.W., 1969, "Effects of Environmental Factors on Composite Materials", *AFML-TR67-423*, Ohio.
15. Schapery, R.A., 1968, "Thermal Expansion Coefficients of Composite Materials Based on Energy Principles", *J. Comp. Mat.*, 2, p. 380.
16. Rosen, B.W., 1987, "Composite Materials Analysis and Design", in *Engineered Materials Handbook*, Vol. 1, Composite Sec. 4, Reinhart, T.J. (Ed.), ASM Int., Ohio.
17. Ishikawa, T., Koyama, K. and Kobayashi, S., 1978, "Thermal Expansion Coefficients of Unidirectional Composites", *J. Comp. Mat.*, 12, p. 153.
18. Agarwal, B.D., Broutman, L.J. and Chandrashekhara, K., 2006, *Analysis and Performance of Fiber Composites*, 3rd Edition, John Wiley and Sons.
19. Rosen, B.W., 1965, "Mechanics of Composite Strengthening", *Fiber Composite Materials*, Am. Soc. Metals, Metals Park, Ohio.
20. Greszczuk, L.B., 1974, "Microbuckling of Lamina Reinforced Composites", in *Composite Materials: Testing and Design*, Berg, C.A. et al. (Eds.), ASTM, STP 546.

5. Strength of Fiber Reinforced Composite Lamina

In this chapter, we will discuss:

- The failure of a fiber reinforced lamina under various combinations of in-plane loadings.
- The lamina could be subjected to a combination of tensile and compressive in-plane loads.

5.1 Introduction

As stated earlier, the lamina is the building block for the laminate. Therefore, the strength and the failure of the laminate are governed by the strength and failure characteristics of the lamina. Structures are subjected to various kinds of loads and environment. An optimal and efficient design is one which can carry the designed load safely without the failure of any of the components. Continuous fibers provide the necessary strength to the lamina. However, these are direction dependent. The maximum strength is obtained when the fibers are oriented along the direction of the load. The longitudinal tensile strength (σ_{LU}) of the lamina is much higher than the longitudinal compressive strength (σ'_{LU}). Further, transverse tensile strength (σ_{TU}) is lower than the tensile strength. The transverse compressive strength (σ'_{TU}) is also different. The in-plane shear strength (τ_{LTU}) is also different. These five independent strength values of the lamina define the failure of the lamina. It is not possible to obtain the strength characteristics of the lamina in all possible orientations in terms of strengths in the principal material directions. In the case of isotropic materials, there is no problem since the direction of stress or strain has no significance. Thus, for isotropic materials, the failure criterion is quite simple. The maximum normal, shear or effective stresses are compared with the ultimate strength values. If any of the strength values are violated, failure is likely to occur. However, for a fiber reinforced lamina which is orthotropic, the direction of highest stress may not coincide with the direction of highest strength. Furthermore, the axes of principal stress do not coincide with the direction of principal strain.

Hence, to determine failure, the actual stress field has to be compared with the allowable stress field.

Laminates are generally subjected to various kinds of loadings, both in-plane and out-of-plane. The failure of the lamina initiates the failure of the laminate. The failure theories are based on strengths defined with respect to the L-T axes. Therefore, the stresses acting on the lamina are transformed from the reference axes to the material axes.

Failure of a lamina does not necessarily imply that the stresses in all the laminae exceed the ultimate or designed value. It is possible that some of them would have exceeded the value in a particular direction. The lamina, and therefore, the laminate is still capable of taking the load in other directions. If the design is carried out on the basis of the failure of a lamina, it would lead to a conservative design.

We shall discuss the various failure theories for uniaxial and biaxial in-plane loadings in the following sections.

5.2 Maximum Stress Failure Theory

The stresses acting in a lamina in the X-Y coordinate system are transformed to the stresses in the material axes (L-T). The transformed stresses are then compared with the allowable stress in that direction. If any one of the stress components is equal or exceeds the corresponding ultimate strengths of the unidirectional lamina, then the lamina has failed in the failure mode associated with that allowable stress. This implies that there are no interactions between the stresses. Thus, there are five sub-criteria for failure. These can be stated as

$$\sigma_L \geq \sigma_{LU}, \ \sigma_T \geq \sigma_{TU}, \ \sigma_L \geq \sigma'_{LU}, \ \sigma_{TU} \geq \sigma'_{TU} \text{ and } \tau_{LT} \geq \tau_{LTU} \qquad (5.1)$$

where σ' corresponds to the compressive strength. Further, whenever the stress is compressive, the stresses as well as the strength values are treated to be negative.

Example 5.1

An angle-ply lamina of glass/epoxy has the following properties in the L-T coordinate system

$$\sigma_{LU} = 1280 \text{ MPa}, \ \sigma_{TU} = 49 \text{ MPa}, \ \sigma'_{LU} = 622 \text{ MPa},$$
$$\sigma'_{TU} = 245 \text{ MPa}, \ \tau_{LTU} = 69 \text{ MPa}$$

The fibers are oriented at an angle of 60° to the X-axis. A tensile stress $\sigma_x = 10$ MPa is applied along the X direction (Fig. 5.1). Check whether the lamina is safe according to the maximum stress theory.

Solution

The applied stress σ_x is transformed to the L-T coordinate system by using Eq. (3.29).

$$\sigma_L = \sigma_x \cos^2\theta = 10 \times \cos^2 60° = 10 \times 0.25 = 2.5 \text{ MPa} < 1280 \text{ MPa}$$

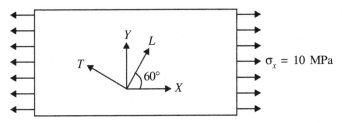

Fig. 5.1 Lamina subjected to a uniaxial σ_x.

$\sigma_T = \sigma_x \sin^2\theta = 10 \times \sin^2 60° = 10 \times 0.75 = 7.5$ MPa < 49 MPa

$\tau_{LT} = -\sigma_x \sin\theta\cos\theta = 10 \times \sin 60° \times \cos 60° = 10 \times 0.866 \times 0.5$
$= 4.33$ MPa < 69 MPa

The lamina is safe according to maximum stress theory. It is to be noted that, as far as the shear strength τ_{LTU} is concerned, it does not depend on the sign. This is similar to isotropic materials.

Example 5.2
In an orthotropic lamina, the fibers are oriented at an angle of 45° to the X-Y axis. It is subjected to a uniaxial state of stress $\sigma_x = 50$ MPa. The strength properties are

$$\sigma_{LU} = 500 \text{ MPa}, \ \sigma_{TU} = 10 \text{ MPa}, \ \sigma'_{LU} = 350 \text{ MPa},$$
$$\sigma'_{TU} = 75 \text{ MPa}, \ \tau_{LTU} = 35 \text{ MPa}$$

Check whether the lamina is safe.

Solution
Using Eq. (3.29), we get

$$\sigma_L = \sigma_x \cos^2 45° = 50 \times 0.5 = 25 \text{ MPa} < 500 \text{ MPa}$$
$$\sigma_T = \sigma_x \sin^2 45° = 50 \times 0.5 = 25 \text{ MPa} > 10 \text{ MPa}$$
$$\tau_{LT} = -\sigma_x \sin 45° \cos 45° = 50 \times 0.5 = 25 \text{ MPa} < 35 \text{ MPa}$$

It is observed that σ_T exceeds the ultimate stress prescribed in that direction. Thus, the lamina fails in the transverse mode.

5.3 Maximum Strain Failure Theory
The stresses acting in a lamina in the X-Y direction are first transformed to the stresses in the L-T coordinate system. These are then converted into strains by using the constitutive equations in the L-T coordinate system. The strains are then compared with the ultimate strains in the L-T coordinate system. The lamina has failed if any one of the strains exceeds the allowable strains. For failure to occur, any one of the following inequalities has to be satisfied:

$$\varepsilon_L \geq \varepsilon_{LU}, \ \varepsilon_T \geq \varepsilon_{TU}, \ \varepsilon_L \geq \varepsilon'_{LU}, \ \varepsilon_T \geq \varepsilon'_{TU} \text{ and } \tau_{LT} \geq \tau_{LTU} \qquad (5.2)$$

where ε' corresponds to the compressive strain. Thus, the maximum strain theory is similar to the maximum stress theory. There are five independent maximum strain values which define the failure of the lamina. Here again the theory assumes that there is no interaction between various strains.

The maximum stress failure theory and maximum strain failure theory give different results as the strain failure theory depends on Poisson's ratio.

Example 5.3
Using the strength properties given in Example 5.1 and the loading, check whether the lamina fails by applying the maximum strain theory for failure. The material properties are $E_L = 35$ GPa, $E_T = 7$ GPa, $G_{LT} = 3$ GPa, $v_{LT} = 0.12$.

Solution
Assuming the material to be elastic up to failure

$$\varepsilon_{LU} = \frac{1280 \times 10^6}{35 \times 10^9} = 0.0366, \varepsilon_{TU} = \frac{49 \times 10^6}{7 \times 10^9} = 0.007, \tau_{LTU} = \frac{69 \times 10^6}{3 \times 10^9} = 0.023$$

Rewriting Eq. (3.47), we get

$$\varepsilon_L = \frac{\sigma_x}{E_L}(\cos^2 60° - v_{LT} \sin^2 60°)$$

$$= \frac{10 \times 10^6}{35 \times 10^9}(0.25 - 0.3 \times 0.75)$$

$$= 7.143 \times 10^{-6} < 0.0366$$

$$\varepsilon_T = \sigma_x \left(\frac{\sin^2 60}{E_T} - \frac{\cos^2 60}{E_L} \right)$$

$$= 10 \times 10^6 \left(\frac{0.75}{7 \times 10^9} \frac{0.25 \times 0.3}{35 \times 10^9} \right)$$

$$= 1.05 \times 10^{-3} < 0.007$$

$$\gamma_{LT} = -\sigma_x \frac{\sin 60 \cos 60}{G_{LT}}$$

$$= -10 \times 10^6 \times \left(\sqrt{3}/4 \right)/3 \times 10^9$$

$$= 1.443 \times 10^{-3} < 0.023$$

As the strains in the lamina are much lower than the allowable strains, there is no failure of the lamina due to the applied load.

5.4 Interactive Failure Theories

The failure theories mentioned in Sections 5.2 and 5.3 do not take into account the interaction between the stresses. In fact, at any given point in a lamina, all the three stresses σ_x, σ_y and τ_{xy} exist. The effective stress is due to interaction between these stresses. A large number of interactive failure criteria have been suggested in literature. The initial effort to formulate the interactive failure criterion is attributed to Hill [1]. Hill's yield criterion is for a three-dimensional isotropic ideally plastic material. Tsai [2] extended this failure criterion for a unidirectional fiber composite, by assuming that it has the same mathematical form. Instead of yield criterion for the composite material, he termed it as a failure criterion. For a 3D material, Hill's yield criterion can be expressed as

$$A(\sigma_{11} - \sigma_{22})^2 + B(\sigma_{22} - \sigma_{33})^2 + C(\sigma_{33} - \sigma_{11})^2$$
$$+ 2D\tau_{12}^2 + 2E\tau_{23}^2 + 2F\tau_{31}^2 = 1 \tag{5.3}$$

where A, B, C, D, E and F are Hill's yield strength parameters. The coefficients can easily be obtained from one-dimensional uniaxial and shear failure tests. Hoffmann [3] modified Eq. (5.3) by adding linear stress terms in order to account for unequal failure stress in tension and compression with a single quadratic equation. A large number of failure theories have been proposed, which are essentially modifications of Eq. (5.3). Some of the theories proposed are quite involved, and in some cases, the efforts required in order to estimate parameters offset the advantage of using it. Most of these theories can be put into one of two categories, each of which has a different form. These can be represented as

$$F_{ij}\sigma_i\sigma_j = 1 \qquad\qquad i, j = 1, 2 \text{ (for a lamina)} \tag{5.4}$$

$$F_{ij}\sigma_i\sigma_j + F_i\sigma_i = 1 \qquad i, j = 1, 2 \text{ (for a lamina)} \tag{5.5}$$

5.4.1 Tsai-Hill Failure Criterion (Maximum Work Theory)

The Tsai-Hill theory [2] for failure of 3D composites is the extension of Hill's criterion. The tensor form of this criterion is given by Eq. (5.4). If this equation is expanded and the F_{ij} terms are replaced, the failure criterion is

$$(G + H)\sigma_L^2 + (F + H)\sigma_T^2 + (F + G)\sigma_{T'}^2 - 2(H\,\sigma_L\sigma_T + F\,\sigma_T\sigma_{T'} + G\,\sigma_L\sigma_{T'})$$
$$+ 2(L\tau_{TT'}^2 + M\tau_{LT'}^2 + N\tau_{LT}^2) = 1 \tag{5.6}$$

where F, G, H, L, M and N are anisotropic material strength parameters. These parameters can be obtained by applying uniaxial tensile stress along L, T and T' directions. For the case when only σ_L is acting and all other stresses are zero, we get

$$G + H = \frac{1}{\sigma_{LU}^2}, \, F + H = \frac{1}{\sigma_{TU}^2}, \, G + F = \frac{1}{\sigma_{TU}^2} \tag{5.7}$$

Solving for F, G and H, we obtain

$$2F = \frac{1}{\sigma_{TU}^2} + \frac{1}{\sigma_{T'U}^2} - \frac{1}{\sigma_{LU}^2}, \quad 2G = \frac{1}{\sigma_{LU}^2} + \frac{1}{\sigma_{T'U}^2} - \frac{1}{\sigma_{TU}^2},$$

$$2H = \frac{1}{\sigma_{LU}^2} + \frac{1}{\sigma_{TU}^2} - \frac{1}{\sigma_{T'U}^2} \tag{5.8}$$

Applying pure shear stress τ_{LT}, and keeping all other applied stresses as zero, we get

$$2N = \frac{1}{\tau_{LTU}^2} \tag{5.9}$$

Substituting the values for F, G, H and N in Eq. (5.6), the plane stress failure criterion yields

$$\left(\frac{\sigma_L}{\sigma_{LU}}\right)^2 - \left(\frac{1}{\sigma_{LU}^2} + \frac{1}{\sigma_{TU}^2} - \frac{1}{\sigma_{T'U}^2}\right)\sigma_L \sigma_T + \left(\frac{\sigma_T}{\sigma_{TU}}\right)^2 + \left(\frac{\tau_{LT}}{\tau_{LTU}}\right)^2 = 1 \tag{5.10}$$

In the case of a lamina, the load-resisting constituent in the T and T' directions is the matrix. Therefore, $\sigma_{TU} = \sigma_{T'U}$. In view of this equality, Eq. (5.10) simplifies to

$$\frac{\sigma_L^2}{\sigma_{LU}^2} - \left(\frac{\sigma_L}{\sigma_{LU}}\right)\left(\frac{\sigma_T}{\sigma_{LU}}\right) + \frac{\sigma_T^2}{\sigma_{TU}^2} + \frac{\tau_{LT}^2}{\tau_{LTU}^2} = 1 \tag{5.11}$$

Equation (5.11) is valid if all the stresses are positive. However, if any of the stresses become compressive, the corresponding compressive ultimate strength needs to be considered.

Example 5.4
An orthotropic angle-ply lamina has the following properties: $\sigma_{LU} = 1800$ MPa, $\sigma_{TU} = 40$ MPa, $\sigma'_{LU} = 1400$ MPa, $\sigma'_{TU} = 230$ MPa and $\tau_{LTU} = 100$ MPa.

The fiber orientation is 30° measured anticlockwise with respect to the reference axis (Fig. 5.2). It is subjected to the following state of stress: $\sigma_x = 100$ MPa, $\sigma_y = -50$ MPa and $\tau_{xy} = 50$ MPa.

Check whether the lamina is safe for this loading by employing the Tsai-Hill criterion.

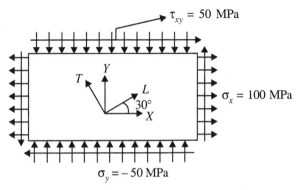

Fig. 5.2 Orthotropic lamina subjected to a multiaxial stress.

Solution

The applied stresses are transformed into the *L-T* coordinate system by making use of Eq. (3.29)

$$\begin{Bmatrix} \sigma_L \\ \sigma_T \\ \tau_{LT} \end{Bmatrix} = \begin{bmatrix} \cos^2\theta & \sin^2\theta & 2\sin\theta\cos\theta \\ \sin^2\theta & \cos^2\theta & -2\sin\theta\cos\theta \\ -\sin\theta\cos\theta & \sin\theta\cos\theta & \cos^2\theta-\sin^2\theta \end{bmatrix} \begin{Bmatrix} \sigma_x \\ \sigma_y \\ \tau_{xy} \end{Bmatrix} \quad (3.29)$$

Hence

$$\begin{Bmatrix} \sigma_L \\ \sigma_T \\ \tau_{LT} \end{Bmatrix} = \begin{bmatrix} \cos^2 30 & \sin^2 30 & 2\sin 30\cos 30 \\ \sin^2 30 & \cos^2 30 & -2\sin 30\cos 30 \\ -\sin 30\cos 30 & \sin 30\cos 30 & \cos^2 30-\sin^2 30 \end{bmatrix} \begin{Bmatrix} 100 \\ -50 \\ 50 \end{Bmatrix}$$

$$= \begin{bmatrix} 0.745 & 0.25 & 0.866 \\ 0.25 & 0.745 & -0.866 \\ -0.433 & 0.433 & 0.495 \end{bmatrix} \begin{Bmatrix} 100 \\ -50 \\ 50 \end{Bmatrix}$$

$$= \begin{Bmatrix} 105.3 \\ -55.55 \\ -40.20 \end{Bmatrix} \text{MPa}$$

Substituting the values in Eq. (5.11), we get

$$\left(\frac{105.3}{1800}\right)^2 - \left(\frac{105.3}{1800}\right)\left(\frac{-55.55}{-1400}\right) + \left(\frac{-55.55}{-230}\right)^2 + \left(\frac{-40.20}{100}\right)^2 = 1 \text{ (for failure)}$$

$$= 0.0034 - 0.0023 + 0.058 + 0.01616$$

$$= 0.075 < 1, \text{ hence the lamina is safe.}$$

Example 5.5

In an angle-ply lamina of carbon/epoxy, the fibers are oriented at an angle of 45° to the reference axis. It is subjected to the following state of stress: $\sigma_x = -100$ N/mm^2, $\sigma_y = 50$ N/mm^2 and $\tau_{xy} = 10$ N/mm^2. The strength values are: $\sigma_{LU} = 1550$ N/mm^2, $\sigma_{TU} = 60$ N/mm^2, $\sigma'_{LU} = 1150$ N/mm^2, $\sigma'_{TU} = 240$ N/mm^2 and $\tau_{LTU} = 75$ N/mm^2. Determine if the lamina failure has occurred for this state of stress.

Solution

Stresses in the L-T coordinate system are given by

$$\begin{Bmatrix} \sigma_L \\ \sigma_T \\ \tau_{LT} \end{Bmatrix} = \begin{bmatrix} 0.5 & 0.5 & 1.0 \\ 0.5 & 0.5 & -1.0 \\ -0.5 & 0.5 & 0 \end{bmatrix} \begin{Bmatrix} -100 \\ 50 \\ 10 \end{Bmatrix}$$

$$= \begin{Bmatrix} -15 \\ -35 \\ 75 \end{Bmatrix} \text{N/mm}^2$$

Substituting the corresponding values in the Tsai-Hill criterion, we get

$$\left(\frac{-15}{-1150} \right)^2 - \left(\frac{-15}{-1150} \right)\left(\frac{-35}{1150} \right) + \left(\frac{-35}{240} \right)^2 + \left(\frac{75}{75} \right)^2$$

$$= 0.00017 - 0.00039 + 0.0212 + 1.0$$

$$= 1.021 > 1.0, \text{ hence the lamina has failed.}$$

The Tsai-Hill criterion does not give the possible mode of failure as it takes into consideration the interaction between the stresses. To ascertain the possible mode of failure, one has to look into the individual stresses. A comparison of the stresses shows that the shear stress $\tau_{LT} = \tau_{LTU}$. Hence, the mode of failure for the lamina will be the shear mode of failure.

Example 5.6

Fibers are inclined at an angle of 45° to the reference axis. The properties of the lamina are: $E_L = 145$ GPa, $E_T = 12$ GPa, $G_{LT} = 6$ GPa and $\nu_{LT} = 0.3$. The strength properties are: $\sigma_{LU} = 1500$ MPa, $\sigma_{TU} = 50$ MPa, $\sigma'_{LU} = 1200$ MPa, $\sigma'_{TU} = 250$ MPa and $\tau_{LTU} = 70$ MPa. The lamina is subjected only to a positive shear stress as shown in Fig. 5.3. Find (a) the maximum positive shear stress that can be applied without causing failure of the lamina and (b) what will be the strength of the lamina if the direction of the loading is reversed.

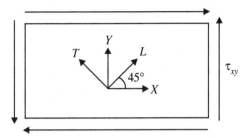

Fig. 5.3 Orthotropic lamina subjected to a pure shear stress.

Solution

$$\begin{Bmatrix} \sigma_L \\ \sigma_T \\ \tau_{LT} \end{Bmatrix} = \begin{bmatrix} 0.5 & 0.5 & 1.0 \\ 0.5 & 0.5 & -1.0 \\ -0.5 & 0.5 & 0 \end{bmatrix} \begin{Bmatrix} 0 \\ 0 \\ \tau \end{Bmatrix}$$

Applying the Tsai-Hill criterion, we get for failure

$$\left(\frac{\tau}{1500}\right)^2 - \left(\frac{\tau}{1500}\right)\left(\frac{-\tau}{-1200}\right) + \left(\frac{-\tau}{-250}\right)^2 = 1$$

$$\tau = 249.3 \text{ MPa}$$

If the shear stress applied is negative, then

$$\begin{Bmatrix} \sigma_L \\ \sigma_T \\ \tau_{LT} \end{Bmatrix} = \begin{bmatrix} 0.5 & 0.5 & 1.0 \\ 0.5 & 0.5 & -1.0 \\ -0.5 & 0.5 & 0 \end{bmatrix} \begin{Bmatrix} 0 \\ 0 \\ -\tau \end{Bmatrix}$$

Once again, applying the Tsai-Hill criterion, we get

$$\left(\frac{-\tau}{-1200}\right)^2 - \left(\frac{-\tau}{-1200}\right)\left(\frac{\tau}{1500}\right) + \left(\frac{\tau}{50}\right)^2 = 1$$

$$\tau = 49.9 \text{ MPa}$$

5.4.2 Tsai-Wu Failure Criterion [4]

This is an improvement over the Tsai-Hill theory [2]. It is based on the premise that the correlation between theory and experiment can be improved by increasing the number of terms in the failure prediction equation. For an orthotropic lamina subjected to plane stress, the failure equation is

$$F_{11}\sigma_L^2 + F_{22}\sigma_T^2 + 2F_{12}\sigma_L\sigma_T + F_{66}\tau_{LT}^2 + 2F_{16}\sigma_L\tau_{LT} + 2F_{26}\sigma_T\tau_{LT}$$
$$+ F_1\sigma_L + F_2\sigma_T + F_6\tau_{LT} = 1 \tag{5.12}$$

The coefficients are obtained by applying stresses one at a time.

Application of pure shear stress results in

$$F_{66}\tau_{LTU}^2 + F_6\tau_{LTU} = 1 \ (\text{for} + \tau)$$

$$F_{66}\tau_{LTU}^2 - F_6\tau_{LTU} = 1 \ (\text{for} - \tau)$$

(5.13)

The only way in which Eq. (5.13) is satisfied is $F_6 = 0$ and $F_{66} = \dfrac{1}{\tau_{LTU}^2}$.
In a similar manner, $F_{16} = F_{26} = 0$.

Applying a uniaxial stress (tensile and compressive) along the longitudinal direction, we get for failure

$$F_{11}\sigma_{LU}^2 + F_1\sigma_{LU} = 1 \ (\text{for tension})$$

$$F_{11}\sigma_{LU}'^2 + F_1\sigma_{LU}' = 1 \ (\text{for compression})$$

(5.14)

Since positive and negative shear do not have different values for failure, then

$$F_{11} = \frac{1}{\sigma_{LU}\sigma_{LU}'}, \ F_1 = \frac{1}{\sigma_{LU}} - \frac{1}{\sigma_{LU}'}$$

(5.15)

Similarly, applying the uniaxial stress along the T direction, we get

$$F_{22}\sigma_{TU}^2 + F_2\sigma_{TU} = 1 \ (\text{for tension})$$

$$F_{22}\sigma_{TU}'^2 - F_1\sigma_{TU}' = 1 \ (\text{for compression})$$

(5.16)

Solving Eq. (5.16), we obtain

$$F_{22} = \frac{1}{\sigma_{TU}\sigma_{TU}'}, \ F_2 = \frac{1}{\sigma_{TU}} - \frac{1}{\sigma_{TU}'}$$

(5.17)

The Tsai-Wu failure criterion reduces to

$$F_{11}\sigma_L^2 + F_{22}\sigma_T^2 + 2F_{12}\sigma_L\sigma_T + F_{66}\tau_{LT}^2 + F_1\sigma_L + F_2\sigma_T = 1$$

(5.18)

F_{12} is still to be determined. We have made use of all the strength values obtained by using tensile and compressive axial stresses and in-plane shear stress. Wu [5, 6] suggested that this constant can be obtained by subjecting the specimen to a biaxial state of stress such that $\sigma_L = \sigma_T = \sigma$. This state of stress causes the failure of the lamina. Substituting these values in Eq. (5.18), we get

$$F_1 = \frac{1 - (F_1 + F_2)\sigma - (F_{11} + F_{22})\sigma^2}{2\sigma^2}$$

(5.19)

It can be seen from Eq. (5.19), that the value of F_{12} depends on different strength values and the biaxial failure stress. The accuracy with which the Tsai-Wu theory predicts failure depends on how F_{12} has been obtained.

Hashin [7] has pointed out that there could be four different values of F_{12} as there are four different pairs of σ_1 and σ_2 that will cause failure. Tsai and Hahn [8] have proposed the relation

$$F_{12} = \frac{\sqrt{(F_{11}F_{22})}}{2} \tag{5.20}$$

This reduces Eq. (5.18) to the generalized form of the von Mises yield criterion for isotropic materials. Further, Eq. (5.18) reduces to the Tsai-Hill criterion, when tensile and compressive strengths are the same and $F_{12} = 1/(2\sigma_{LU}^2)$.

Example 5.7

An orthotropic lamina has the following strength properties: σ_{LU} = 1800 MPa, σ_{TU} = 40 MPa, σ'_{TU} = 230 MPa, σ'_{LU} = 1400 MPa, τ_{LTU} = 100 MPa. The lamina is subjected to a biaxial tensile stress, $\sigma_L = \sigma_T$ = 30 MPa. Determine the Tsai-Hill interaction parameter F_{12}. Also determine using the Tsai-Hill criterion, whether the lamina fails for the loading σ_L = 150 MPa, σ_T = – 25 MPa and τ_{LT} = 95 MPa.

Solution

We need to calculate F_1, F_2, F_{11} and F_{22} in order to obtain F_{12}.

$$F_{11} = \frac{1}{\sigma_{LU}\sigma'_{LU}}, \ F_1 = \frac{1}{\sigma_{LU}} - \frac{1}{\sigma'_{LU}}, \ F_{22} = \frac{1}{\sigma_{TU}\sigma'_{TU}}, \ F_2 = \frac{1}{\sigma_{TU}} - \frac{1}{\sigma'_{TU}}, \ F_{66} = \frac{1}{\tau_{LTU}^2}$$

$$F_{11} = \frac{1}{1800 \times 1400} = 3.97 \times 10^{-7}, \ F_1 = \frac{1}{1800} - \frac{1}{1400} = -1.587 \times 10^{-4}$$

$$F_{22} = \frac{1}{40 \times 230} = 1.087 \times 10^{-4}, \ F_2 = \frac{1}{40} - \frac{1}{230} = 2.065 \times 10^{-2}, \ F_{66} = \frac{1}{100^2} = 1 \times 10^{-4}$$

$$F_{12} = \frac{1 - (-1.587 \times 10^{-4} + 206.5 \times 10^{-4}) \times 35 - (3.97 \times 10^{-7} + 1087 \times 10^{-7}) \times 35^2}{2 \times 35^2}$$

$$= 6.088 \times 10^{-5}$$

Substituting the values in Eq. (5.18), we get

$$3.97 \times 10^{-7} \times (150)^2 + 2 \times 6.088 \times 10^{-5} \times 150 \times (-25) + (1.087 \times 10^{-4}) \times (-25)^2$$
$$1 \times 10^{-4} \times (95)^2 + (-1.587 \times 10^{-4}) \times 150 + (2.065 \times 10^{-2}) \times (-25)$$
$$= 1.01466 > 1$$

Therefore, failure is predicted for the given loading.

Example 5.8

Find the maximum value of $P > 0$, if the following stresses are applied on a 30°-graphite/epoxy lamina. σ_x = 2P, σ_y = – 3P and τ_{xy} = 3P. The strength properties are: σ_{LU} = 1500 MPa, σ_{TU} = 40 MPa, σ'_{LU} = 1500 MPa,

σ'_{TU} = 246 MPa and τ_{LTU} = 68 MPa. The material properties are E_L = 181 GPa, E_T = 10.30 GPa, G_{LT} = 7.17 GPa and ν_{LT} = 0.28.

Solution

Making use of Eq. (3.29), transform the applied stresses into the *L-T* coordinate system.

$$
\begin{Bmatrix} \sigma_L \\ \sigma_T \\ \tau_{LT} \end{Bmatrix} = \begin{bmatrix} \cos^2 30° & \sin^2 30° & 2\sin 30°\cos 30° \\ \sin^2 30° & \cos^2 30° & -2\sin 30°\cos 30° \\ -\sin 30°\cos 30° & \sin 30°\cos 30° & \cos^2 30° - \sin^2 30° \end{bmatrix} \begin{Bmatrix} 2P \\ -3P \\ 3P \end{Bmatrix}
$$

$$
= \begin{bmatrix} 0.745 & 0.25 & 0.866 \\ 0.25 & 0.745 & -0.866 \\ -0.433 & 0.433 & 0.495 \end{bmatrix} \begin{Bmatrix} 2P \\ -3P \\ 3P \end{Bmatrix} = \begin{Bmatrix} 3.338P \\ -3.175P \\ -0.68P \end{Bmatrix}
$$

(a) Maximum stress failure theory

$$\sigma_{LU} = 1500 = 3.338P, \text{ hence for failure } P = 449.3 \text{ MPa}$$
$$\sigma'_{TU} = 246 = 3.175P, \text{ hence for failure } P = 77.48 \text{ MPa}$$
$$\tau_{LTU} = 68 = 0.68P, \text{ hence for failure } P = 100 \text{ MPa}$$

The maximum value of P = 77.48 MPa and the failure mode will be the transverse compressive mode.

(b) Maximum strain failure theory

$$\varepsilon_{LU} = \frac{\sigma_{LU}}{E_L} = \frac{1500}{181} \times 10^{-3} = 8.287 \times 10^{-3}, \varepsilon_{TU} = \frac{\sigma_{TU}}{E_T} = \frac{40}{10.3} \times 10^{-3} = 3.883 \times 10^{-3}$$

$$\varepsilon'_{LU} = \frac{\sigma'_{LU}}{E_L} = \frac{1500}{181} \times 10^{-3} = 8.287 \times 10^{-3}, \varepsilon'_{TU} = \frac{\sigma'_{TU}}{E_T} = \frac{246}{10.3} \times 10^{-3} = 23.88 \times 10^{-3}$$

$$\gamma_{LTU} = \frac{\tau_{LTU}}{G_{LT}} = \frac{68}{7.17} \times 10^{-3} = 9.484 \times 10^{-3}$$

$$\varepsilon_L = \frac{\sigma_L}{E_L} - \nu_{LT}\frac{\sigma_T}{E_L} = \frac{3.338P}{E_L} + 0.28\frac{3.175P}{E_L} = \frac{4.227P}{181} \times 10^{-3} = 0.023P \times 10^{-3}$$

$$\varepsilon_{LU} = 8.287 \times 10^{-3} = 0.023P \times 10^{-3}, \text{ hence for failure } P = 360.3 \text{ MPa}$$

$$\varepsilon_T = \frac{\sigma_T}{E_T} - \nu_{LT}\frac{\sigma_L}{E_L} = \frac{-3.175P \times 10^6}{10.3 \times 10^9} - 0.28\frac{3.338P \times 10^6}{181 \times 10^9} = -0.313P \times 10^{-3}$$

$$\varepsilon'_{TU} = 23.88 \times 10^{-3} = 0.313P \times 10^{-3}, \text{ hence for failure } P = 76.293 \text{ MPa}$$

$$\gamma_{LT} = \frac{0.68P}{7.17} \times 10^{-3} = 0.095P \times 10^{-3}$$

$\gamma_{LTU} = 9.484 \times 10^{-3} = 0.095P \times 10^{-3}$, hence for failure $P = 99.83$ MPa

(c) Tsai-Hill failure theory
Substituting the values in Eq. (5.11), we get for failure

$$\left(\frac{3.338P}{1500}\right)^2 - \left(\frac{3.338P}{1500}\right)\left(\frac{3.175P}{1500}\right) + \left(\frac{3.175P}{246}\right)^2 + \left(\frac{0.68P}{68}\right)^2 = 1$$

$$P = 61.22 \text{ MPa}$$

Thus, the application of different theories gives different results.

Some of the other failure theories which, though complex, are also employed to assess the failure of the given lamina both under uniaxial and multiaxial in-plane loadings are given below.

Hashin Criterion [7]
In this criterion, four distinct modes – tensile matrix, tensile fiber, compressive matrix and compressive fiber – are modeled separately, resulting in a piece-wise smooth failure surface. Another unique feature of this failure criterion is that it avoids the necessity of predicting the multiaxial tensile (compressive) modes in terms of compressive (tensile) failure stresses. The four criteria for a 3D case corresponding to the different failure modes are:

(i)   Tensile fiber mode: $\sigma_L \geq 0$

$$\left(\frac{\sigma_L}{\sigma_{LU}}\right)^2 + \frac{1}{\tau_{LTU}^2}(\tau_{LT'}^2 + \tau_{TT'}^2) = 1$$

(ii)   Tensile matrix mode: $\sigma_T + \sigma_{T'} \geq 0$

$$\frac{1}{\sigma_{TU}^2}(\sigma_T + \sigma_{T'})^2 + \frac{1}{\tau_{LTU}^2}(\tau_{LT}^2 - \sigma_T\sigma_{T'}) + \frac{1}{\tau_{LTU}^2}(\tau_{LT'}^2 + \tau_{TT'}^2) = 1$$

(iii) Compressive fiber mode: $\sigma_L \leq 0$

$$\sigma_L = \sigma_{LU}'$$

(iv)   Compressive matrix mode: $\sigma_T + \sigma_{T'} \leq 0$

$$\frac{1}{\sigma_{TU}'}\left[\left(\frac{\sigma_{TU}'}{2\tau_{LT'U}}\right)^2 - 1\right](\sigma_T + \sigma_{T'}) + \frac{1}{4\tau_{LT'U}^2}(\sigma_T + \sigma_{T'})^2$$

$$+ \frac{1}{\tau_{LT'U}^2}(\tau_{LT}^2 - \sigma_T\sigma_{T'}) + \frac{1}{\tau_{LTU}^2}(\tau_{LT'}^2 + \tau_{TT'}^2) = 1$$

Hoffman Criterion

Hoffman [3] modified the failure criterion of Hill [1] by adding linear stress terms to account for the unequal failure stress in tension and in compression with a single quadratic expression. The tensor polynomial form of the criterion can be obtained from the most general polynomial failure criterion, as proposed by Tsai [2]. This criterion is written as

$$F_1\sigma_L + F_2\sigma_T + F_3\sigma_{T'} + 2F_{12}\sigma_L\sigma_T + 2F_{13}\sigma_L\sigma_{T'} + 2F_{23}\sigma_T\sigma_{T'}$$
$$+ F_{11}\sigma_L^2 + F_{22}\sigma_T^2 + F_{33}\sigma_{T'}^2 + F_{44}\tau_{LT'}^2 + F_{55}\tau_{TT'}^2 + F_{66}\tau_{LT}^2 + \ldots \geq 1$$

The Hoffman criterion is obtained by substituting the values for F_1, F_2, etc.

$$F_1 = \frac{1}{\sigma_{LU}} - \frac{1}{\sigma'_{LU}}; \quad F_2 = \frac{1}{\sigma_{TU}} - \frac{1}{\sigma'_{TU}}; \quad F_3 = \frac{1}{\sigma_{T'U}} - \frac{1}{\sigma'_{T'U}}$$

$$F_{11} = \frac{1}{\sigma_{LU}\sigma'_{LU}}; \quad F_{22} = \frac{1}{\sigma_{TU}\sigma'_{TU}}; \quad F_{33} = \frac{1}{\sigma_{T'U}\sigma'_{T'U}}$$

$$F_{44} = \frac{1}{\tau_{LT'U}^2}; \quad F_{55} = \frac{1}{\tau_{TT'U}^2}; \quad F_{66} = \frac{1}{\tau_{LTU}^2}$$

$$F_{12} = -\frac{1}{2}\left(\frac{1}{\sigma_{LU}\sigma'_{LU}} + \frac{1}{\sigma_{TU}\sigma'_{TU}} - \frac{1}{\sigma_{T'U}\sigma'_{T'U}} \right)$$

$$F_{13} = -\frac{1}{2}\left(\frac{1}{\sigma_{LU}\sigma'_{LU}} - \frac{1}{\sigma_{TU}\sigma'_{TU}} + \frac{1}{\sigma_{T'U}\sigma'_{T'U}} \right)$$

$$F_{23} = -\frac{1}{2}\left(-\frac{1}{\sigma_{LU}\sigma'_{LU}} + \frac{1}{\sigma_{TU}\sigma'_{TU}} + \frac{1}{\sigma_{T'U}\sigma'_{T'U}} \right)$$

The other strength terms are zero.

Summary

*   The various failure theories proposed by a number of researchers were discussed.
*   The applications to 2D lamina subjected to in-plane loads, both tensile and compressive were illustrated.
*   A number of examples have been worked out to illustrate the failure behavior of the lamina.

Problems

5.1 An E-glass/epoxy lamina is subjected to the state of stress as $\sigma_x = 100$ MPa, $\sigma_y = 70$ MPa, $\tau_{xy} = 35$ MPa. Determine if the failure will be according to the maximum stress, maximum strain, and Tsai-Hill theories, assuming a fiber orientation of (i) 30°, (ii) − 30° and (iii) 60°. The material and strength properties are: $E_L = 60.7$ GPa, $E_T = 24.8$ GPa, $G_{LT} = 11.99$ GPa, $\nu_{LT} = 0.23$,

σ_{LU} = 1288 MPa, σ'_{LU} = 820.5 MPa, σ_{TU} = 45.9 MPa, σ'_{TU} = 174.4 MPa, τ_{LTU} = 44.8 MPa.

5.2 For each of the following lamina, the fiber orientation θ is variable and can lie between ($-90° \le \theta \le 90°$). Plot the failure stress σ_0 as a function of θ using the Tsai-Hill failure theory. Assume the lamina is made up of E-glass/epoxy. The material and strength properties are given in Problem 5.1. (i) $\sigma_x = \sigma_0$, $\sigma_y = -2\sigma_0$, (ii) $\sigma_x = \sigma_0$, $\tau_{xy} = \sigma_0$ and (iii) $\sigma_x = 2\sigma_0$, $\sigma_y = -\sigma_0$.

5.3 For a filament wound pressure vessel, the required lamina thickness is a function of winding angle, $t = f(\theta)$. Determine lamina thickness at which failure does not take place. The pressure vessel is assumed to be subjected to an internal pressure P. Employing the classical thin-walled pressure theory, obtain the relation for thickness, t and winding angle θ. Use the material and strength properties given in Problem 5.1. Employ the Tsai-Hill failure criterion.

5.4 If in addition to the internal pressure P, a torque T is applied to the pressure vessel mentioned in Problem 5.3, obtain the required lamina thickness as a function of angle θ to ensure safe design using maximum stress theory.

5.5 Determine the Tsai-Wu interaction parameter F_{12}, when a lamina is subjected to a biaxial tensile stress and the biaxial failure stress is found to be $\sigma_1 = \sigma_2 = 35$ MPa. Determine whether the lamina is safe for the state of stress as $\sigma_L = 120$ MPa, $\sigma_T = -60$ MPa, $\tau_{LT} = 70$ MPa. The material and strength properties are: $E_L = 160$ GPa, $E_T = 10$ GPa, $G_{LT} = 7$ GPa, $v_{LT} = 0.3$ and $\sigma_{LU} = 1800$ MPa, $\sigma'_{LU} = 1400$ MPa, $\sigma_{TU} = 40$ MPa, $\sigma'_{TU} = 230$ MPa, $\tau_{LTU} = 100$ MPa.

5.6 Find the Tsai-Hill failure criterion for a lamina subjected to pure shear loading at various fiber orientation angles.

5.7 A balanced orthotropic lamina is subjected to the following state of stress: $\sigma_x = 100$ MPa, $\sigma_y = -60$ MPa, $\tau_{xy} = 50$ MPa. The fiber orientation is 30°. The strength and material properties of the lamina are as follows: $E_L = E_T = 70$ GPa, $G_{LT} = 5$ GPa, $v_{LT} = 0.25$. $\sigma_{LU} = \sigma'_{LU} = \sigma_{TU} = \sigma'_{TU} = 560$ MPa. $\tau_{LTU} = 25$ MPa. Use maximum stress theory and maximum strain theory to check whether the lamina fails or not for the given loading.

References

1. Hill, R., 1948, "A Theory of Yielding and Plastic Flow of Anisotropic Materials", *Proc. of the Royal Society*, Series A, 193, p. 281.

2. Tsai, S.W., 1965, *Strength Characteristics of Composite Materials*, NASA, CR-224.

3. Hoffman, O., 1967, "The Brittle Strength of Orthotropic Materials", *J. of Composite Materials*, 1, p. 200.

4. Tsai, S.W. and Wu, E.M., 1971, "A General Theory of Strength for Anisotropic Materials", *J. Comp. Mat.*, 5, p. 58.

5. Wu, E.M., 1974, "Phenomenological Anisotropic Failure Criteria", in Sendeckyj, G.P. (Ed.), *Composite Materials: 2, Mechanics of Composite Materials*, p. 353, Academic Press, New York.

6. Wu, E.M., 1972, "Optimal Experimental Measurements of Anisotropic Failure Tensors", *J. Comp. Mat.*, 6, p. 472.

7. Hashin, Z., 1980, "Failure Criteria for Unidirectional Fiber Composites", *J. Appl. Mech.*, 47, p. 329.
8. Tsai, S.W. and Hahn, H.T., 1980, *Introduction to Composite Materials*, Technomic Publishing Co., Lancaster, Pennsylvania.

6. Laminate Analysis

In this chapter, we will:

- Discuss the construction of a laminate formed by stacking a number of laminae of different fiber orientations.
- Describe laminates supporting not only in-plane loads but out-of-plane loads as well.
- Use the constitutive equations for the lamina derived in earlier chapters and make certain assumptions regarding the behavior of the laminate
- Derive the constitutive equation for the laminate.
- Look at the hygrothermal behavior of the laminate.

6.1 Introduction

In Chapters 3 to 5, we discussed the behavior of lamina subjected to in-plane loading as these are not capable of resisting other kind of loading. Their failure behavior was also discussed. Lamina being very thin cannot resist loads along the z-directions. In order that the fiber reinforced composite resists loads in several directions, the lamina has to be stacked together so that the required thickness can be obtained. A *laminate* is a stacking of two or more laminae arranged in a specified manner. The laminae are bonded together and hence act as a single unit. The fiber orientation of the laminae may be the same or different. They may be of same or different materials. Even if they are of the same material, the fiber volume fraction may not be the same in all the lamina. In general, the material axis of the lamina is oriented with respect to the reference axes of the laminate. The reference axis, also called structural axes, passes through the geometric mid-surface of the lamina as shown in Fig. 6.1. This is chosen for the purpose of convenience of the analysis. The out-of-plane axis, z-axis, is taken positive downwards.

6.2 Laminate Codes

Since there are a number of ways in which the lamina can be arranged, a code is generally used to identify the laminate. The code provides for

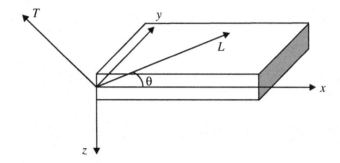

Fig. 6.1 Reference axes for the laminate.

accommodating many possible combinations of ply (lamina) orientations and stacking sequences. Several elements of the code are as follows:

(i)   Adjacent lamina are separated by a slash if the fiber orientation angles are different.

(ii)   The lamina is listed in a sequence starting from the top surface of the laminate.

(iii)   A number representing the angles in degrees between the fiber direction and *x*-axis designates each lamina. The fiber orientation θ is positive, if the direction vector is moved anticlockwise from *x*-axis to the material axis (Fig. 6.1).

(iv)   Adjacent laminae of the same orientation are denoted by a numerical subscript.

(v)   A square bracket indicates the beginning and the end of the code.

(vi)   A symmetric laminate can be described by listing only the upper half of the ply angles and by using the subscript *s* outside the bracket.

(vii)   Adjacent lamina oriented at angles equal in magnitude but opposite in sign, are indicated by plus (+) and minus (−) sign.

(viii) Symmetric laminates with odd number of plies are coded in the same manner as even number of plies, except that the center ply is over lined to indicate that half the thickness is considered for symmetry.

(ix)   Enclosing the set of angles in parenthesis identifies sets of ply angles, which are repeated in a laminate.

(x)   A laminate consisting of plies of different materials is called a hybrid laminate. The code remains the same, except that a subscript is placed on the ply angle to indicate the type of material used.

Figure 6.2 shows the laminate stacking sequence and the orientation codes.

6.3 Formulation for Laminated Plates

As stated in Section 6.1, a laminate consists of a number of plies or laminae stacked together in a desired manner. It is assumed that the laminae are perfectly bonded together in order to behave like a single unit. It is further

[90°/45°/0°/− 45°/30°]

+ 90°
+ 45°
0°
− 45°
+ 30°

[90°/45°/$\overline{0°}$]$_s$

90°
45°
0°
45°
90°

[0°/± 45°/$\overline{90°}$]$_s$

0°
+ 45°
− 45°
90°
− 45°
+ 45°
0°

[90°/45°/0°]$_2$

90°
45°
0°
90°
45°
0°

Fig. 6.2 Laminate stacking sequence and the orientation codes.

assumed that there is no slip between the lamina. As a consequence, the displacements are continuous across the thickness. Each lamina is thin and the overall laminate thickness is still very small compared to the other two dimensions. One can, therefore, treat the laminate to be a two-dimensional plane stress. Even though each lamina is orthotropic, the laminate exhibits anisotropy in general. It is assumed that the laminate is homogenous i.e. the material properties do not vary spatially. Figure 6.1 defines the coordinate system. The geometric middle surface lies in the *xy* plane. The displacements at any point along the *x*, *y*, *z* directions are *u*, υ and *w*, respectively.

The following assumptions are made regarding the behavior of the laminate, which are on the lines given by Whitney [1] and Halpin [2]:

(i) The laminate consists of homogenous orthotropic laminae adhesively bonded together, with the material axis of each lamina oriented differently with respect to the respective reference axis.

(ii) Transverse displacements are continuous across the thickness of the laminate.

(iii) The displacements *u*, υ and *w* are small compared to the thickness of the laminate.

(iv) The thickness *t* of the laminate is small when compared to the other two dimensions.

(v) Transverse shear strains γ_{xz} and γ_{yz} are negligible. This implies that a line originally perpendicular to the mid-plane remains straight and perpendicular to the deformed state. As will be shown later, this assumption is not strictly valid in the case of composites as shear rigidity is an independent property and could be very different from the elastic modulus.

(vi) The in-plane strains ε_x, ε_y and γ_{xy} are small compared to unity.

(vii) Strain-displacement and stress-strain relation are linear.
(viii) The transverse normal stress $\sigma_z = 0$. The state of stress is one of plane stress.

6.3.1 Strain-displacement Relations
Under the application of the load, a section undergoes both bending and rotation. As the deformation is linear, one can assume that the cross-section undergoes bending and then rotation. Figure 6.3 shows a cross-section of the laminate in the x-z plane before and after the deformation. The in-plane displacements of the mid-surface are defined as u^0 and v^0 along the x and y directions. The displacements u, v and w at any arbitrary point C (x, y, z) at a distance below the mid-surface is given by (assuming small deformation)

$$u(x, y) = u^0 - z\beta \qquad (6.1)$$

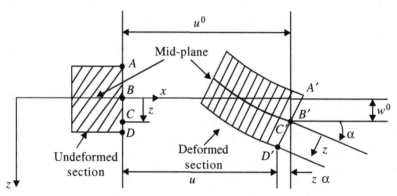

Fig. 6.3 Deformation of the laminate in the x-z plane.

This is true since there is no distortion of the cross-section. The slope β is the angle between the two normals before and after the deformation. It is defined as

$$\beta = \frac{\partial w^0}{\partial x} \qquad (6.2)$$

Therefore, the displacement u at any point z through the laminate thickness is

$$u(x, y) = u^0 - z\frac{\partial w^0}{\partial x} \qquad (6.3)$$

Similarly, taking a cross-section in the y-z plane, we have

$$v(x, y) = v^0 - z\frac{\partial w^0}{\partial y} \qquad (6.4)$$

Since there is no shortening of extension along the z direction, the mid-surface displacement w^0 at $(x, y, 0)$ is same at any z. Hence

$$w^0 (x, y, 0) = w (x, y, z) \qquad (6.5)$$

Therefore, for transverse displacement, we shall use w instead of w^0.

The laminate strains, ε_x, ε_y and γ_{xy} in the x-y plane as a result of small strains are

$$\varepsilon_x = \frac{\partial u}{\partial x} = \frac{\partial u^0}{\partial x} - z \frac{\partial^2 w}{\partial x^2} \qquad (6.6a)$$

$$\varepsilon_y = \frac{\partial v}{\partial y} = \frac{\partial v^0}{\partial y} - z \frac{\partial^2 w}{\partial y^2} \qquad (6.6b)$$

$$\gamma_{xy} = \frac{\partial u}{\partial y} + \frac{\partial v}{\partial x} = \frac{\partial u^0}{\partial y} + \frac{\partial v^0}{\partial x} - 2z \frac{\partial^2 w}{\partial x \partial y} \qquad (6.6c)$$

Equations (6.6a) to (6.6c) can be rewritten in the matrix form as

$$\left\{ \begin{array}{c} \varepsilon_x \\ \varepsilon_y \\ \gamma_{xy} \end{array} \right\} = \left\{ \begin{array}{c} \dfrac{\partial u^0}{\partial x} \\[2ex] \dfrac{\partial v^0}{\partial y} \\[2ex] \dfrac{\partial u^0}{\partial y} + \dfrac{\partial v^0}{\partial x} \end{array} \right\} + z \left\{ \begin{array}{c} -\dfrac{\partial^2 w}{\partial x^2} \\[2ex] -\dfrac{\partial^2 w}{\partial y^2} \\[2ex] -2\dfrac{\partial^2 w}{\partial x \partial y} \end{array} \right\} \qquad (6.7)$$

Equation (6.7) can also be defined in terms of mid-plane strains and curvatures. These are defined as

$$\left\{ \begin{array}{c} \varepsilon_x \\ \varepsilon_y \\ \gamma_{xy} \end{array} \right\} = \left\{ \begin{array}{c} \varepsilon_x^0 \\ \varepsilon_y^0 \\ \gamma_{xy}^0 \end{array} \right\} + z \left\{ \begin{array}{c} \kappa_x \\ \kappa_y \\ \kappa_{xy} \end{array} \right\} \qquad (6.8)$$

where the mid-plane strains are defined as

$$\left\{ \begin{array}{c} \varepsilon_x^0 \\ \varepsilon_y^0 \\ \gamma_{xy}^0 \end{array} \right\} = \left\{ \begin{array}{c} \dfrac{\partial u^0}{\partial x} \\[2ex] \dfrac{\partial v^0}{\partial y} \\[2ex] \dfrac{\partial u^0}{\partial y} + \dfrac{\partial v^0}{\partial x} \end{array} \right\} \qquad (6.9)$$

The curvature terms are defined as

$$\begin{Bmatrix} \kappa_x \\ \kappa_y \\ \kappa_{xy} \end{Bmatrix} = -\begin{Bmatrix} \dfrac{\partial^2 w}{\partial x^2} \\[2mm] \dfrac{\partial^2 w}{\partial y^2} \\[2mm] 2\dfrac{\partial^2 w}{\partial x \partial y} \end{Bmatrix} \tag{6.10}$$

The first two terms in Eq. (6.10) represent the bending curvature, while the last term corresponds to twisting curvature.

From Eq. (6.7), it is observed that the strains vary linearly with respect to the laminate thickness. Further, at any given z, the strains are functions of x and y only.

6.3.2 Stress-strain Relations
Equation (6.7) can be used to find the stresses in a given lamina. The stress-strain relations for the kth lamina in terms of the reference axis are given by Eq. (3.42)

$$\begin{Bmatrix} \sigma_x \\ \sigma_y \\ \tau_{xy} \end{Bmatrix}_k = \begin{bmatrix} \overline{Q}_{11} & \overline{Q}_{12} & \overline{Q}_{16} \\ \overline{Q}_{12} & \overline{Q}_{22} & \overline{Q}_{26} \\ \overline{Q}_{16} & \overline{Q}_{26} & \overline{Q}_{66} \end{bmatrix}_k \begin{Bmatrix} \varepsilon_x \\ \varepsilon_y \\ \gamma_{xy} \end{Bmatrix}_k \tag{6.11}$$

$[\overline{Q}]$ matrix corresponds to the transformed stiffness matrix of the kth lamina of a given laminate. Substituting Eq. (6.8) into Eq. (6.11) for strains, we obtain

$$\begin{Bmatrix} \sigma_x \\ \sigma_y \\ \tau_{xy} \end{Bmatrix}_k = \begin{bmatrix} \overline{Q}_{11} & \overline{Q}_{12} & \overline{Q}_{16} \\ \overline{Q}_{12} & \overline{Q}_{22} & \overline{Q}_{26} \\ \overline{Q}_{16} & \overline{Q}_{26} & \overline{Q}_{66} \end{bmatrix}_k \begin{Bmatrix} \varepsilon_x^0 \\ \varepsilon_y^0 \\ \gamma_{xy}^0 \end{Bmatrix}_k + z \begin{bmatrix} \overline{Q}_{11} & \overline{Q}_{12} & \overline{Q}_{16} \\ \overline{Q}_{12} & \overline{Q}_{22} & \overline{Q}_{26} \\ \overline{Q}_{16} & \overline{Q}_{26} & \overline{Q}_{66} \end{bmatrix}_k \begin{Bmatrix} \kappa_x \\ \kappa_y \\ \kappa_{xy} \end{Bmatrix}_k \tag{6.12}$$

Thus, the stresses vary linearly with respect to the thickness coordinate.

However, since the $[\overline{Q}]$ matrix is different for each ply (the fiber orientation may not be the same), there is a discontinuity across the lamina. The stresses and strains defined in the x-y coordinate system can be transformed to the material coordinates of the lamina by making use of Eqs. (3.29) and (3.33).

Example 6.1
A laminate of 0.4 mm thickness subjected to loads gives the following mid-plane strains and curvatures in the reference coordinate system:

$$
\begin{Bmatrix} \varepsilon_x^0 \\ \varepsilon_y^0 \\ \gamma_{xy}^0 \end{Bmatrix} = \begin{Bmatrix} 2\times10^{-6} \\ 3\times10^{-6} \\ 5\times10^{-6} \end{Bmatrix} \text{mm/mm} + z \begin{Bmatrix} 0.05 \\ 0.06 \\ 0.10 \end{Bmatrix} / \text{mm}
$$

Obtain the total strains in the *x-y* coordinate system at the top and middle surface of the laminate.

Solution
The global strain at any z is given by Eq. (6.7).

For the mid-surface, $z = 0$ mm. Substituting this value of z in Eq. (6.7), we get

$$
\begin{Bmatrix} \varepsilon_x^0 \\ \varepsilon_y^0 \\ \gamma_{xy}^0 \end{Bmatrix} = \begin{Bmatrix} 2\times10^{-6} \\ 3\times10^{-6} \\ 5\times10^{-6} \end{Bmatrix} \text{mm/mm}
$$

For the upper surface, $z = 0.2$, the strain is given by

$$
\begin{Bmatrix} \varepsilon_x^0 \\ \varepsilon_y^0 \\ \gamma_{xy}^0 \end{Bmatrix} = \begin{Bmatrix} 2\times10^{-6} \\ 3\times10^{-6} \\ 5\times10^{-6} \end{Bmatrix} + 0.2\times \begin{Bmatrix} 0.05\times10^{-6} \\ 0.06\times10^{-6} \\ 0.10\times10^{-6} \end{Bmatrix} = \begin{Bmatrix} 2.01\times10^{-6} \\ 3.012\times10^{-6} \\ 5.02\times10^{-6} \end{Bmatrix}
$$

6.3.3 Force and Moment Resultants
As observed from Eq. (6.12), the stresses in the laminate are functions of *x*, *y* and *z*. As a consequence, they vary from point to point within the lamina. It is, therefore, convenient to express them in terms of resultant force and moments per unit length. It is also more convenient to use forces and moments per unit length rather than forces and moments. These can be obtained by integrating the stresses over the thickness of the laminate and are defined as

$$
N_x = \int_{-t/2}^{t/2} \sigma_x dz, \quad N_y = \int_{-t/2}^{t/2} \sigma_y dz, \quad N_{xy} = \int_{-t/2}^{t/2} \tau_{xy} dz \qquad (6.13)
$$

$$
Q_x = \int_{-t/2}^{t/2} \tau_{xz} dz, \quad Q_y = \int_{-t/2}^{t/2} \tau_{yz} dz \qquad (6.14)
$$

$$
M_x = \int_{-t/2}^{t/2} \sigma_x dz, \quad M_y = \int_{-t/2}^{t/2} \sigma_y z dz, \quad M_{xy} = \int_{-t/2}^{t/2} \tau_{xy} z dz \qquad (6.15)
$$

where N_x, N_y, N_{xy} are in-plane forces per unit width, Q_x and Q_y are transverse shear forces per unit width, M_x, M_y and M_{xy} are bending and twisting moments per unit width. Figure 6.4 shows the positive directions of forces and moments. t is the total thickness of the laminate.

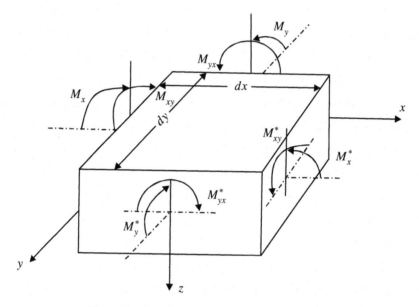

Fig. 6.4 Plate bending and twisting moments.

Equations (6.13) to (6.15) cannot be integrated as they are, since the stresses are discontinuous across the lamina interface. Therefore, these have to be integrated over each layer thickness and then summed up to obtain the total forces and moments. Since the kth layer occupies the region between t_{k-1} and t_k, Eqs. (6.13) to (6.15) can be expressed as

$$\begin{Bmatrix} N_x \\ N_y \\ N_{xy} \end{Bmatrix} = \sum_{i=1}^{NL} \int_{z_{i-1}}^{z_i} \begin{Bmatrix} \sigma_x \\ \sigma_y \\ \tau_{xy} \end{Bmatrix}_k dz \tag{6.16}$$

$$\begin{Bmatrix} Q_y \\ Q_x \end{Bmatrix} = \sum_{k=1}^{NL} \int_{z_{i-1}}^{z_i} \begin{Bmatrix} \tau_{yz} \\ \tau_{xz} \end{Bmatrix} dz \tag{6.17}$$

$$\begin{Bmatrix} M_x \\ M_y \\ M_{xy} \end{Bmatrix} = \sum_{k=1}^{NL} \int_{z_{i-1}}^{z_i} z \begin{Bmatrix} \sigma_x \\ \sigma_y \\ \tau_{xy} \end{Bmatrix} dz \tag{6.18}$$

where *NL* is the number of plies in a given laminate. Generally all the plies are of uniform thickness. The stresses in Eqs. (6.16) to (6.18) can be rewritten in terms of mid-plane strains and curvatures by substituting for stresses from Eq. (6.12). This leads to

$$
\begin{Bmatrix} N_x \\ N_y \\ N_{xy} \end{Bmatrix} = \sum_{k=1}^{NL} \left[\int_{z_{i-1}}^{z_i} \begin{bmatrix} \overline{Q}_{11} & \overline{Q}_{12} & \overline{Q}_{16} \\ \overline{Q}_{12} & \overline{Q}_{22} & \overline{Q}_{26} \\ \overline{Q}_{16} & \overline{Q}_{26} & \overline{Q}_{66} \end{bmatrix}_k \begin{Bmatrix} \varepsilon_x^0 \\ \varepsilon_y^0 \\ \gamma_{xy}^0 \end{Bmatrix} dz \right.
$$

$$
\left. + \int_{z_{i-1}}^{z_i} \begin{bmatrix} \overline{Q}_{11} & \overline{Q}_{12} & \overline{Q}_{16} \\ \overline{Q}_{12} & \overline{Q}_{22} & \overline{Q}_{26} \\ \overline{Q}_{16} & \overline{Q}_{26} & \overline{Q}_{66} \end{bmatrix}_k \begin{Bmatrix} \kappa_x \\ \kappa_y \\ \kappa_{xy} \end{Bmatrix} z \, dz \right] \tag{6.19}
$$

$$
\begin{Bmatrix} M_x \\ M_y \\ M_{xy} \end{Bmatrix} = \sum_{k=1}^{NL} \left[\int_{z_{i-1}}^{z_i} \begin{bmatrix} \overline{Q}_{11} & \overline{Q}_{12} & \overline{Q}_{16} \\ \overline{Q}_{12} & \overline{Q}_{22} & \overline{Q}_{26} \\ \overline{Q}_{16} & \overline{Q}_{26} & \overline{Q}_{66} \end{bmatrix}_k \begin{Bmatrix} \varepsilon_x^0 \\ \varepsilon_y^0 \\ \gamma_{xy}^0 \end{Bmatrix} z \, dz \right.
$$

$$
\left. + \int_{z_{i-1}}^{z_i} \begin{bmatrix} \overline{Q}_{11} & \overline{Q}_{12} & \overline{Q}_{16} \\ \overline{Q}_{12} & \overline{Q}_{22} & \overline{Q}_{26} \\ \overline{Q}_{16} & \overline{Q}_{26} & \overline{Q}_{66} \end{bmatrix}_k \begin{Bmatrix} \kappa_x \\ \kappa_y \\ \kappa_{xy} \end{Bmatrix} z^2 \, dz \right] \tag{6.20}
$$

For a laminate where shear effects are of consequence, the out-of-plane shear relationship can be written from Eq. (3.2) as

$$
\begin{Bmatrix} \sigma_4 \\ \sigma_5 \end{Bmatrix} = \begin{bmatrix} Q_{44} & 0 \\ 0 & Q_{55} \end{bmatrix} \begin{Bmatrix} \varepsilon_4 \\ \varepsilon_5 \end{Bmatrix} \tag{6.21}
$$

Equation (6.21) in terms of the *x-y* coordinate system is

$$
\begin{Bmatrix} \tau_{xz} \\ \tau_{yz} \end{Bmatrix} = \begin{bmatrix} \overline{Q}_{44} & \overline{Q}_{45} \\ \overline{Q}_{45} & \overline{Q}_{55} \end{bmatrix} \begin{Bmatrix} \gamma_{xz} \\ \gamma_{yz} \end{Bmatrix} \tag{6.22}
$$

Substituting for τ_{xz} and τ_{yz} in Eq. (6.17), we obtain

$$
\begin{Bmatrix} Q_x \\ Q_y \end{Bmatrix} = \sum_{k=1}^{NL} \int_{z_{i-1}}^{z_i} \begin{bmatrix} \overline{Q}_{44} & \overline{Q}_{45} \\ \overline{Q}_{45} & \overline{Q}_{55} \end{bmatrix} \begin{Bmatrix} \gamma_{xz} \\ \gamma_{yz} \end{Bmatrix} dz \tag{6.23}
$$

Equations (6.19) and (6.20) can be integrated as the mid-plane strains $\{\varepsilon^0\}$ and $\{\kappa\}$ are independent of the z coordinate. Further, within the lamina, the $[\overline{Q}]$ matrix is a constant. The integrals in Eqs. (6.19) and (6.20) become integrals of $(1, z, z^2)$. The in-plane forces and moments are expressed in the form, after integration, as

$$
\begin{Bmatrix} N_x \\ N_y \\ N_{xy} \\ M_x \\ M_y \\ M_{xy} \end{Bmatrix} = \begin{bmatrix} A_{11} & A_{12} & A_{16} & B_{11} & B_{12} & B_{16} \\ A_{12} & A_{22} & A_{26} & B_{12} & B_{22} & B_{26} \\ A_{16} & A_{26} & A_{66} & B_{16} & B_{26} & B_{66} \\ B_{11} & B_{12} & B_{16} & D_{11} & D_{12} & D_{16} \\ B_{12} & B_{22} & B_{26} & D_{12} & D_{22} & D_{26} \\ B_{16} & B_{26} & B_{66} & D_{16} & D_{26} & D_{66} \end{bmatrix} \begin{Bmatrix} \varepsilon_x^0 \\ \varepsilon_y^0 \\ \gamma_{xy}^0 \\ \kappa_x \\ \kappa_y \\ \kappa_{xy} \end{Bmatrix}
\tag{6.24}
$$

where

$$
A_{ij} = \sum_{k=1}^{NL} [\overline{Q}_{ij}]_k \, (z_k - z_{k-1})
\tag{6.25a}
$$

$$
B_{ij} = \frac{1}{2} \sum_{k=1}^{NL} [\overline{Q}_{ij}]_k \, (z_k^2 - z_{k-1}^2)
\tag{6.25b}
$$

$$
D_{ij} = \frac{1}{3} \sum_{k=1}^{NL} [\overline{Q}_{ij}]_k \, (z_k^3 - z_{k-1}^3)
\tag{6.25c}
$$

Equation (6.24) can be written in terms of forces and moments separately as

$$
\begin{Bmatrix} N_x \\ N_y \\ N_{xy} \end{Bmatrix} = \begin{bmatrix} A_{11} & A_{12} & A_{16} \\ A_{12} & A_{22} & A_{26} \\ A_{16} & A_{26} & A_{66} \end{bmatrix} \begin{Bmatrix} \varepsilon_x^0 \\ \varepsilon_y^0 \\ \gamma_{xy}^0 \end{Bmatrix} + \begin{bmatrix} B_{11} & B_{12} & B_{16} \\ B_{12} & B_{22} & B_{26} \\ B_{16} & B_{26} & B_{66} \end{bmatrix} \begin{Bmatrix} \kappa_x \\ \kappa_y \\ \kappa_{xy} \end{Bmatrix}
\tag{6.26}
$$

$$
\begin{Bmatrix} M_x \\ M_y \\ M_{xy} \end{Bmatrix} = \begin{bmatrix} B_{11} & B_{12} & B_{16} \\ B_{12} & B_{22} & B_{26} \\ B_{16} & B_{26} & B_{66} \end{bmatrix} \begin{Bmatrix} \varepsilon_x^0 \\ \varepsilon_y^0 \\ \gamma_{xy}^0 \end{Bmatrix} + \begin{bmatrix} D_{11} & D_{12} & D_{16} \\ D_{12} & D_{22} & D_{26} \\ D_{16} & D_{26} & D_{66} \end{bmatrix} \begin{Bmatrix} \kappa_x \\ \kappa_y \\ \kappa_{xy} \end{Bmatrix}
\tag{6.27}
$$

It is observed that Eqs. (6.26) and (6.27) are a coupled set of equations. This implies that, if some elements of $[B]$ matrix are present, the in-plane applied forces result in bending of the laminate. A pure bending moment applied to the laminate gives rise to in-plane forces in addition to bending. This type of coupling exists even in the case of laminates made of layers of two dissimilar isotropic materials.

The matrix $[A]$ is called the *extensional stiffness matrix* as it relates the normal forces to mid-plane strains. The elements of $[A]$ are independent of the stacking sequence of the laminate. They are affected by the thickness of the lamina. The matrix $[D]$ is called the *bending stiffness matrix* as it relates the bending and twisting moments to laminate curvatures. The coefficients of $[D]$ matrix depend strongly on the location and thickness of the lamina.

The matrix $[B]$ is called *coupling stiffness matrix* as it relates the in-plane forces to the curvatures and moments to in-plane strains. The normal and shear forces acting at the mid-plane of the laminate give rise not only to mid-plane strains but also bending and twisting curvatures. Ashton et al. [3] showed that in the presence of non-zero B_{ij}, normal force produces twisting of the specimen.

In addition to the coupling mentioned with regard to the $[B]$ matrix, there are weak couplings as well. The presence of stiffness coefficients like A_{16} and A_{26} couples the normal strains to shear strains. Similarly, the stiffness coefficients like D_{16} and D_{26} couple the bending moments to twisting moments. All these couplings can either be completely eliminated or partially eliminated by the proper choice of a stacking sequence of the lamina.

Coupling is not a desirable feature as it brings down the stiffness of the laminate. Coupling between bending and extension must be clearly understood. In many applications, the neglect of coupling can be catastrophic. For example, for a given load, the deflection will be more, the buckling load reduces and the natural frequency also reduces if coupling is taken into account. If the design is carried out neglecting coupling, the laminate behavior will be entirely different. Card and Jones [4] showed that if longitudinal stiffeners are placed on the outside of an axial loaded circular cylindrical shell, the buckling load is twice the value when the same stiffeners are placed on the inside wall of the shell. The best laminate design would be one in which there is no coupling. Iyengar [5] has shown that as the number of plies for a given thickness is increased, the effect of coupling decreases. However, this may not be feasible because of practical and other considerations. We shall discuss the possible stacking sequences to completely eliminate or partially eliminate the coupling in the laminate.

6.3.4 Special Cases of Laminates

Symmetric Laminates
Let us first consider the coupling matrix $[B]$. As stated earlier, this is the worst kind of coupling. By making the matrix $[B]$ zero, Eqs. (6.26) and (6.27) are decoupled. From Eq. (6.25b), we have

$$B_{ij} = \frac{1}{2} \sum_{k=1}^{NL} [\overline{Q}_{ij}]_k \ (z_k^2 - z_{k-1}^2) \qquad (6.25b)$$

The contribution of a lamina to the ij^{th} element of B_{ij} is given by the product of the corresponding term in the $[\overline{Q}]$ matrix and the difference of the squares of the z coordinates of the top and bottom of each ply. If two laminae identical in material property, thickness and fiber orientation are placed at equal distances from the reference surface, the contribution to B_{ij} from this pair will be zero. Thus, if we place pairs of lamina identical in all respects at an equal distance on either side of the reference surface, all elements of [B] matrix can be identically made zero. Such a laminate is termed a *symmetric laminate*. In order to have a symmetric laminate, there may be either an even or an odd number of plies. Such a laminate when subjected to only in-plane forces, will have zero mid-plane curvatures. Similarly, when subjected to pure moments, it will have no mid-plane strains. Therefore, the analysis becomes simpler. Such laminates will not twist when they are cured at high temperatures and cooled down to operating temperatures. Figure 6.5 shows examples of symmetric laminate stacking sequences and the notation according to the laminate code.

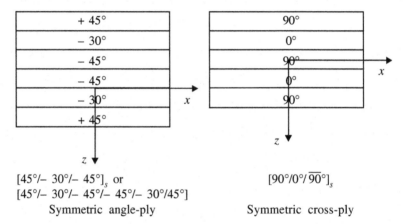

$[45°/- 30°/- 45°]_s$ or
$[45°/- 30°/- 45°/- 45°/- 30°/45°]$
Symmetric angle-ply

$[90°/0°/\overline{90°}]_s$

Symmetric cross-ply

Fig. 6.5 Examples of stacking sequence for symmetric laminates.

Symmetric Angle-ply Laminate
The symmetric angle-ply laminate has an odd number of plies. For such a laminate, [B] matrix is identically zero. The ply angles alternate between $+\theta$ and $-\theta$. From Eq. (3.43), it is observed that all $\overline{Q}_{16}$ and $\overline{Q}_{26}$ change sign for $+\theta$ and $-\theta$, while all other coefficients remain unchanged. For a laminate with NL layers of total thickness t, the coefficients of [A] and [D] matrices can be written as

$$[A_{11}, A_{12}, A_{22}, A_{66}] = t[\overline{Q}_{11}, \overline{Q}_{12}, \overline{Q}_{22}, \overline{Q}_{66}] \qquad (6.26a)$$

$$[A_{16}, A_{26}] = \frac{t}{NL}[\overline{Q}_{16}, \overline{Q}_{26}] \qquad (6.26b)$$

$$[D_{11}, D_{12}, D_{22}, D_{66}] = \frac{t^3}{12}[\overline{Q}_{11}, \overline{Q}_{12}, \overline{Q}_{22}, \overline{Q}_{66}] \qquad (6.26c)$$

$$[D_{16}, D_{26}] = \left(\frac{t^3 (3[NL]^2 - 2)}{12[NL]^3} \right) [\bar{Q}_{16}, \bar{Q}_{26}] \qquad (6.26d)$$

Symmetric Cross-ply Laminate

A cross-ply laminate with fiber oriented alternately at $0°$ and $90°$ along the material directions having odd number of plies is termed as symmetric cross-ply laminate. For such a laminate, $[B]$ matrix is identically zero. Further, A_{16}, A_{26}, D_{16} and D_{26} are zero as these are functions of $\bar{Q}_{ij}$. This means that the stacking sequence will not have any of the couplings. However, the laminate is weak in shear.

Anti-symmetric Laminates

There are many applications like, fan blades, aircraft wings, and helicopter blades that require pretwists. In such cases, one requires coupling between in-plane and out-of-plane quantities. This is achieved by placing plies in an anti-symmetric fashion with respect to the mid-surface. In general, the anti-symmetric laminates have an even number of layers.

Anti-symmetric Cross-ply Laminates

An anti-symmetric cross-ply laminate consists of even number of plies of alternating $0°$ and $90°$ layers oriented along and transverse to the material axes, respectively. The plies are of uniform thickness. For such a laminate, the non-zero elements of the coupling matrix $[B]$ are B_{11} and B_{22}. In addition, $B_{11} = - B_{22}$. Further, A_{16}, A_{26}, D_{16} and D_{26} are zero. Thus, an anti-symmetric cross-ply laminate differs from a symmetric cross-ply laminate in its behavior because of the presence of B_{11} and B_{12}. Since the thickness of the lamina is same, such a laminate is called a regular anti-symmetric cross-ply laminate. B_{11} has the maximum value when the number of plies is two. It reduces as the number of layers increase. It is generally observed that the magnitude of B_{11} is negligible when the number of plies increases beyond ten. Strictly speaking, it will be zero for infinite number of layers.

Anti-symmetric Angle-ply Laminates

The regular anti-symmetric laminate has an even number of equal thickness plies. It has the laminate oriented at $+ \theta°$ to the material axes on one side of the reference axes, while it has $- \theta°$ ply exactly at the same distance on the other side of the reference axes. The orientation θ could be same throughout the laminate or it could be different. For such a laminate, the non-zero elements of the $[B]$ matrix are B_{16} and B_{26}.

$$(B_{16}, B_{26}) = \left(\frac{t^2}{2NL} \right) (\bar{Q}_{16}, \bar{Q}_{26}) \qquad (6.26e)$$

As the number of plies, NL increases, B_{16} and B_{26} tend to zero.

The non-zero elements of $[A]$ and $[D]$ are A_{11}, A_{12}, A_{22}, A_{66}, D_{11}, D_{12}, D_{22} and D_{26}. These are given by Eqs. (6.26a) and (6.26c).

Balanced Laminate

A balanced angle-ply laminate consisting of pairs of plies with identical material properties and thickness are placed with fiber orientation of $+ \theta$ and $- \theta$ with respect to the reference axes. They need not be placed exactly at the same distance on either side of the reference axes. A balanced angle-ply laminate can be

Symmetric: $[\theta_1/- \theta_2/\theta_3/\theta_3/- \theta_2/\theta_1]$
Anti-symmetric: $[\theta_1/- \theta_2/\theta_3/- \theta_3/\theta_2/- \theta_1]$
Asymmetric: $[- \theta_1/- \theta_2/- \theta_3/\theta_1/\theta_2/\theta_3]$

A balanced cross-ply laminate has equal number of $0°$ and $90°$ plies. Depending on the ply location with respect to the reference axes, they could be symmetric or asymmetric. Such a laminate will have $A_{16} = A_{26} = D_{16} = D_{26} = 0$.

Quasi-isotropic Laminate

A laminate is called quasi-isotropic, if the coefficients of the extensional stiffness matrix $[A]$ correspond to the equivalent isotropic material. This implies that the coefficients of $[A]$ matrix are identical in all directions. Further, the coefficients are related in the following manner. $A_{11} = A_{22}$, $A_{16} = A_{26} = 0$ and $A_{66} = (A_{11} - A_{12})/2$. The coefficients of matrices $[B]$ and $[D]$ do not exhibit this behavior. For a laminate to be quasi-isotropic, the following conditions should be satisfied with regard to stacking sequence and fiber orientations:

(i) The number of plies should be three or more.
(ii) The fiber orientation between adjacent plies must be π/NL, where NL is the total number of plies.
(iii) The plies must be of the same thickness and material property.

Examples of quasi-isotropic laminates are

$$[- 60°/0°/60°], \quad [0°/45°/- 45°/90°]$$

Example 6.2

Show that a $[- 60°/0°/60°]$ laminate of graphite/epoxy is quasi-isotropic. Find $[A]$, $[B]$ and $[D]$ matrices for such a laminate. The thickness of the laminate is 6 mm.

Solution

The reduced stiffness matrix $[Q]$ is given as

$$[Q] = \begin{bmatrix} 181.8 & 2.897 & 0 \\ 2.897 & 10.35 & 0 \\ 0 & 0 & 7.17 \end{bmatrix} \text{GPa}$$

Making use of Eq. (3.43), the transformed reduced stiffness matrix for 60° and – 60° is given as

$$[\bar{Q}]_{\substack{+60° \\ -60°}} = \begin{bmatrix} 23.65 & 32.46 & \pm 20.05 \\ 32.46 & 109.4 & \pm 54.19 \\ \pm 20.05 & \pm 54.19 & 36.74 \end{bmatrix} \text{GPa}$$

The transformed stiffness matrix for 0° is $[\bar{Q}]_0 = [Q]$.

The stiffness matrices [A], [B] and [D] are given by

$$A_{ij} = \sum_{k=1}^{NL} [\bar{Q}_{ij}]_k \, (z_k - z_{k-1}) \qquad (6.25a)$$

$$B_{ij} = \frac{1}{2} \sum_{k=1}^{NL} [\bar{Q}_{ij}]_k \, (z_k^2 - z_{k-1}^2) \qquad (6.25b)$$

$$D_{ij} = \frac{1}{3} \sum_{k=1}^{NL} [\bar{Q}_{ij}]_k \, (z_k^3 - z_{k-1}^3) \qquad (6.25c)$$

For the given laminate, $t = 6$ mm. Therefore, $t_0 = -3$, $t_1 = -1$, $t_2 = 1$ and $t_3 = 3$.

$$A_{ij} = (\bar{Q}_{ij})_1(-1+3) + (\bar{Q}_{ij})_2(1+1) + (\bar{Q}_{ij})_3(3-1)$$
$$= \left[(\bar{Q}_{ij})_1 + (\bar{Q}_{ij})_2 + (\bar{Q}_{ij})_3 \right] \times 2$$

Substituting the values of $(\bar{Q}_{ij})_1, (\bar{Q}_{ij})_2$ and $(\bar{Q}_{ij})_3$, we get

$$[A] = \begin{bmatrix} 458.30 & 135.63 & 0 \\ 135.63 & 458.30 & 0 \\ 0 & 0 & 161.30 \end{bmatrix} \text{GPa-mm}$$

$$B_{ij} = \frac{1}{2} \left\{ (\bar{Q}_{ij})_1 \left[(-1)^2 - (-3)^2 \right] + (\bar{Q}_{ij})_2 \left[(1)^2 - (-1)^2 \right] + (\bar{Q}_{ij})_3 \left[(3)^2 - (1)^2 \right] \right\}$$
$$= \frac{1}{2} \left\{ -(\bar{Q}_{ij})_1 + (\bar{Q}_{ij})_3 \right\} \times 8$$

$$[B] = \begin{bmatrix} 0 & 0 & 160.4 \\ 0 & 0 & 433.52 \\ 160.4 & 433.52 & 0 \end{bmatrix} \text{GPa-mm}^2$$

$$D_{ij} = \frac{1}{3} \left\{ (\bar{Q}_{ij})_1 \left[(-1)^3 - (-3)^3 \right] + (\bar{Q}_{ij})_2 \left[(1)^3 - (-1)^3 \right] + (\bar{Q}_{ij})_3 \left[(3)^3 - (1)^3 \right] \right\}$$

$$= \frac{1}{3} \left\{ (\bar{Q}_{ij})_1 [26] + (\bar{Q}_{ij})_2 [2] + (\bar{Q}_{ij})_3 [26] \right\}$$

$$= \frac{1}{3} \left\{ (\bar{Q}_{ij})_2 \times 2 + \left[(\bar{Q}_{ij})_1 + (\bar{Q}_{ij})_3 \right] \times 26 \right\}$$

$$[D] = \begin{bmatrix} 530.73 & 564.57 & 0 \\ 564.57 & 1903.19 & 0 \\ 0 & 0 & 641.61 \end{bmatrix} \text{GPa-mm}^3$$

It can be seen that the coefficients of [A] matrix meet the conditions required to be classified as a quasi-isotropic laminate.

Example 6.3

For an anti-symmetric four ply laminate [– 45°/45°/– 45°/45°], obtain the stiffness matrices [A], [B] and [D]. Each ply is of thickness 0.3 mm. [Q] matrix is given as

$$[Q] = \begin{bmatrix} 138.82 & 2.72 & 0 \\ 2.72 & 9.05 & 0 \\ 0 & 0 & 6.9 \end{bmatrix} \text{GPa}$$

Solution

The first step is to obtain the $[\bar{Q}]$ matrix for 45° and – 45° lamina. This is obtained by using Eq. (3.43). For convenience, the equation is reproduced below

$$\bar{Q}_{11} = Q_{11} \cos^4 \theta + Q_{22} \sin^4 \theta + 2(Q_{12} + 2Q_{66}) \sin^2 \theta \cos^2 \theta$$

$$\bar{Q}_{22} = Q_{11} \sin^4 \theta + Q_{22} \cos^4 \theta + 2(Q_{12} + 2Q_{66}) \sin^2 \theta \cos^2 \theta$$

$$\bar{Q}_{12} = (Q_{11} + Q_{22} - 4Q_{66}) \cos^2 \theta \sin^2 \theta + Q_{12} (\cos^4 \theta + \sin^4 \theta)$$

$$\bar{Q}_{66} = (Q_{11} + Q_{22} - 2Q_{12} - 2Q_{66}) \cos^2 \theta \sin^2 \theta + Q_{66} (\cos^4 \theta + \sin^4 \theta)$$

$$\bar{Q}_{16} = (Q_{12} - Q_{22} - 2Q_{66}) \cos^3 \theta \sin \theta - (Q_{22} - Q_{12} - 2Q_{66}) \sin^3 \theta \cos \theta$$

$$\bar{Q}_{26} = (Q_{12} - Q_{22} - 2Q_{66}) \cos \theta \sin^3 \theta - (Q_{22} - Q_{12} - 2Q_{66}) \sin \theta \cos^3 \theta$$

Substituting the values of θ as 45° and – 45°, $[\bar{Q}]$ is obtained as

$$[\bar{Q}]_{\substack{45° \\ -45°}} = \begin{bmatrix} 45.23 & 31.43 & \pm 32.44 \\ 31.43 & 45.23 & \pm 32.44 \\ \pm 32.44 & \pm 32.44 & 35.62 \end{bmatrix} \text{GPa}$$

$$A_{ij} = 2 \times \left[(\bar{Q}_{ij})_{45} + (\bar{Q}_{ij})_{-45} \right] \times 0.3 \text{ GPa-mm}$$

$$[A] = \begin{bmatrix} 54.276 & 77.752 & 0 \\ 77.752 & 54.276 & 0 \\ 0 & 0 & 42.744 \end{bmatrix} \text{GPa-mm}$$

$$B_{ij} = \frac{1}{2} \left\{ (\bar{Q}_{ij})_{45}(z_1^2 - z_0^2) + (\bar{Q}_{ij})_{-45}(z_2^2 - z_1^2) + (\bar{Q}_{ij})_{45}(z_3^2 - z_2^2) + (\bar{Q}_{ij})_{-45}(z_4^2 - z_3^2) \right\}$$

$$= \frac{1}{2} \left\{ (\bar{Q}_{ij})_{45}(z_1^2 - z_0^2 + z_3^2 - z_2^2) + (\bar{Q}_{ij})_{-45}(z_2^2 - z_1^2 + z_4^2 - z_3^2) \right\}$$

$$= \frac{1}{2} \left\{ (\bar{Q}_{ij})_{45}[(-0.3)^2 - (-0.6)^2 + (0.3)^2 - 0] + (\bar{Q}_{ij})_{-45}[0 - (-0.3)^2 + (0.6)^2 - (0.3)^2] \right\}$$

$$= \frac{1}{2} \left\{ (\bar{Q}_{ij})_{45}[0.09 - 0.36 + 0.09] + (\bar{Q}_{ij})_{-45}[-0.09 + 0.36 - 0.09] \right\}$$

$$= \frac{1}{2} \left\{ (\bar{Q}_{ij})_{45}[-0.18] + (\bar{Q}_{ij})_{-45}[0.18] \right\}$$

$$= 0.09 \times \left[(\bar{Q}_{ij})_{-45} - (\bar{Q}_{ij})_{45} \right]$$

$$[B] = \begin{bmatrix} 0 & 0 & -5.839 \\ 0 & 0 & -5.839 \\ -5.839 & -5.839 & 0 \end{bmatrix} \text{GPa-mm}^2$$

$$D_{ij} = \frac{1}{3} \left\{ (\bar{Q}_{ij})_{45}(z_1^3 - z_0^3) + (\bar{Q}_{ij})_{-45}(z_2^3 - z_1^3) + (\bar{Q}_{ij})_{45}(z_3^3 - z_2^3) + (\bar{Q}_{ij})_{-45}(z_4^3 - z_3^3) \right\}$$

$$= \frac{1}{3} \left\{ (\bar{Q}_{ij})_{45}(z_1^3 - z_0^3 + z_3^3 - z_2^3) + (\bar{Q}_{ij})_{-45}(z_2^3 - z_1^3 + z_4^3 - z_3^3) \right\}$$

$$= \frac{1}{3} \left\{ (\bar{Q}_{ij})_{45}[(-0.3)^3 - (-0.6)^3 + (0.3)^3 - 0] + (\bar{Q}_{ij})_{-45}[0 - (-0.3)^3 + (0.6)^3 - (0.3)^3] \right\}$$

$$= \left[(\bar{Q}_{ij})_{45} + (\bar{Q}_{ij})_{-45} \right] \times 0.072$$

$$[D] = \begin{bmatrix} 6.513 & 4.526 & 0 \\ 4.526 & 6.513 & 0 \\ 0 & 0 & 5.129 \end{bmatrix} \text{GPa-mm}^3$$

Example 6.4

For a three ply laminate, [0°/30°/45°] of graphite/epoxy, obtain the [A], [B], and [D] matrices. Each ply has a thickness of 2 mm. The elastic properties of graphite/epoxy are as follows: E_L = 181 GPa, E_T = 10.3 GPa, G_{LT} = 7.17 GPa and ν_{LT} = 0.28.

Solution

The first step is to obtain the $[Q]$ matrix of the lamina. Making use of Eq. (3.24), we obtain

$$Q_{11} = \frac{181}{1-0.28\times0.0159} = 181.8 \text{ GPa}, \; Q_{12} = \frac{0.28\times10.3}{1-0.28\times0.0159} = 2.897 \text{ GPa}$$

$$Q_{22} = \frac{10.3}{1-0.28\times0.0159} = 10.35 \text{ GPa}, \; Q_{66} = 7.17 \text{ GPa}$$

Making use of Eq. (3.43), we can write the transformed stiffness matrices as

$$[\bar{Q}]_{0°} = \begin{bmatrix} 181.8 & 2.897 & 0 \\ 2.897 & 10.35 & 0 \\ 0 & 0 & 7.17 \end{bmatrix} \text{GPa}$$

$$[\bar{Q}]_{30°} = \begin{bmatrix} 109.4 & 32.46 & 54.19 \\ 32.46 & 23.65 & 20.05 \\ 54.19 & 20.05 & 36.74 \end{bmatrix} \text{GPa}$$

$$[\bar{Q}]_{45°} = \begin{bmatrix} 56.66 & 42.32 & 42.87 \\ 42.32 & 56.66 & 42.87 \\ 42.87 & 42.87 & 46.59 \end{bmatrix} \text{GPa}$$

The total thickness of the laminate, $t = 6$ mm, $z_0 = -3$, $z_1 = -1$, $z_2 = 1$ and $z_3 = 3$.

The stiffness matrices for the laminate are obtained by making use of Eqs. (6.25a, b, c)

$$A_{ij} = (\bar{Q}_{ij})_1(z_1 - z_0) + (\bar{Q}_{ij})_2(z_2 - z_1) + (\bar{Q}_{ij})_3(z_3 - z_2)$$

$$= (\bar{Q}_{ij})_1(-1+3) + (\bar{Q}_{ij})_2(1+1) + (\bar{Q}_{ij})_3(3-1)$$

$$= \left[(\bar{Q}_{ij})_1 + (\bar{Q}_{ij})_2 + (\bar{Q}_{ij})_3 \right] \times 2$$

$$[A] = \begin{bmatrix} 695.60 & 155.35 & 194.92 \\ 155.35 & 181.32 & 125.84 \\ 194.92 & 125.84 & 181.10 \end{bmatrix} \text{GPa-mm}$$

$$B_{ij} = \frac{1}{2} \left[(\bar{Q}_{ij})_1(z_1^2 - z_0^2) + (\bar{Q}_{ij})_2(z_2^2 - z_1^2) + (\bar{Q}_{ij})_3(z_3^2 - z_2^2) \right]$$

$$= \frac{1}{2} \left[(\bar{Q}_{ij})_1(1-9) + (\bar{Q}_{ij})_2(1-1) + (\bar{Q}_{ij})_3(9-1) \right]$$

$$= 4 \times \left[(\bar{Q}_{ij})_3 - (\bar{Q}_{ij})_1 \right]$$

$$[B] = \begin{bmatrix} -600.56 & 157.69 & 171.48 \\ 157.69 & 185.24 & 171.48 \\ 171.48 & 171.48 & 157.28 \end{bmatrix} \text{GPa-mm}^2$$

$$D_{ij} = \frac{1}{3}\left[(\bar{Q}_{ij})_1 (z_1^3 - z_0^3) + (\bar{Q}_{ij})_2 (z_2^3 - z_1^3) + (\bar{Q}_{ij})_3 (z_3^3 - z_2^3) \right]$$

$$= \frac{1}{3}\left[(\bar{Q}_{ij})_1 (z_1^3 - z_0^3) + (\bar{Q}_{ij})_2 (z_2^3 - z_1^3) + (\bar{Q}_{ij})_3 (z_3^3 - z_2^3) \right]$$

$$= \frac{1}{3}\left\{ \left[(\bar{Q}_{ij})_1 + (\bar{Q}_{ij})_3 \right] \times 26 \, (\bar{Q}_{ij})_2 \times 2 \right\}$$

$$[D] = \begin{bmatrix} 2130.92 & 428.48 & 407.66 \\ 428.48 & 596.52 & 384.91 \\ 407.66 & 384.91 & 490.41 \end{bmatrix} \text{GPa-mm}^3$$

Example 6.5

For a two ply laminate [45°/– 45°] of total thickness t, calculate [A], [B], and [D] matrices with the stiffness matrix [Q] defined with respect to the material axes as

$$[\bar{Q}]_{0°} = \begin{bmatrix} 181.8 & 2.897 & 0 \\ 2.897 & 10.35 & 0 \\ 0 & 0 & 7.17 \end{bmatrix} \text{GPa}$$

Solution

The transformed stiffness matrix with respect to the reference axes is given by

$$[\bar{Q}]_{\pm45°} = \begin{bmatrix} 56.66 & 42.32 & \pm 42.87 \\ 42.32 & 56.66 & \pm 42.87 \\ \pm 42.87 & \pm 42.87 & 46.59 \end{bmatrix} \text{GPa}$$

For this laminate, $z_0 = -\dfrac{t}{2}$, $z_1 = 0$ and $z_2 = \dfrac{t}{2}$

$$A_{ij} = (\bar{Q}_{ij})_{45} (z_1 - z_0) + (\bar{Q}_{ij})_{-45} (z_2 - z_1)$$

$$= (\bar{Q}_{ij})_{45} (0 + t/2) + (\bar{Q}_{ij})_{-45} (t/2 - 0)$$

$$[A] = \begin{bmatrix} 56.66 & 42.32 & 0 \\ 42.32 & 56.66 & 0 \\ 0 & 0 & 46.59 \end{bmatrix} t \text{ GPa-mm}$$

$$B_{ij} = \frac{1}{2}\left[(\overline{Q}_{ij})_{45}(z_1^2 - z_0^2) + (\overline{Q}_{ij})_{-45}(z_2^2 - z_1^2)\right]$$

$$= \frac{t^2}{8}\left[(\overline{Q}_{ij})_{-45} - (\overline{Q}_{ij})_{45}\right]$$

$$[B] = t^2 \begin{bmatrix} 0 & 0 & -685.92 \\ -685.92 & 0 & -685.92 \\ -685.92 & -685.92 & 0 \end{bmatrix} \text{GPa-mm}^2$$

$$D_{ij} = \frac{1}{3}\left[(\overline{Q}_{ij})_{45}(z_1^3 - z_0^3) + (\overline{Q}_{ij})_{-45}(z_2^3 - z_1^3)\right]$$

$$= \frac{t^3}{24}\left[(\overline{Q}_{ij})_{45} + (\overline{Q}_{ij})_{-45}\right]$$

$$[D] = t^3 \begin{bmatrix} 2719.68 & 2031.36 & 0 \\ 2031.36 & 2719.68 & 0 \\ 0 & 0 & 2236.32 \end{bmatrix} \text{GPa-mm}^3$$

6.4 Lamina Stresses and Strains

A laminate is generally subjected to in-plane forces or moments or a combination of both. In some cases, conditions on some of the strains and curvatures may be imposed. Under these conditions, one may be required to find the lamina stresses and strains and to ascertain whether the lamina fails for the given loads. Equation (6.24) relates the mid-plane strains, curvatures to the applied in-plane loads and moments. It can be seen that these are functions of stacking sequence, material properties and fiber orientation. The strains and curvatures obtained are with respect to the reference axes (or structural axes). Equation (6.12) is used to obtain the lamina stresses and Eq. (6.11) is employed to obtain the lamina stresses and strains with respect to the reference axes. Equations (3.29) and (3.33) are then used to obtain the stresses and strains with respect to the L-T axes. Knowing these, one can then employ (i) maximum stress failure theory, (ii) maximum strain failure theory, (iii) Tsai-Hill theory, or (iv) Tsai-Wu theory to determine failure of a lamina. However, this does not imply failure of the laminate. The failure of the laminate initiates with the failure of the lamina, but the failure takes place when all the lamina fail or the laminate satisfies some other failure criterion.

It is seen from Eq. (6.24) for a general laminate, that the coupling matrix $[B]$ is non-zero. The solution of Eq. (6.24) to obtain the strains and curvatures requires inverting the 6×6 matrix. This takes quite a bit of time. It is possible to reduce the time and complexity by means of a transformation. This requires only the solution of a 3×3 matrix even though $[B]$ is non-zero.

Rewriting Eqs. (6.26) and (6.27) in a concise form as

$$\{N\} = [A]\{\varepsilon^0\} + [B]\{\kappa\} \tag{6.28}$$

$$\{M\} = [B]\{\varepsilon^0\} + [D]\{\kappa\} \tag{6.29}$$

Premultiplying both sides of Eq. (6.28) by $[A]^{-1}$ and solving for $\{\varepsilon^0\}$, we get

$$\{\varepsilon^0\} = [A]^{-1}\{N\} - [A]^{-1}[B]\{\kappa\} \tag{6.30}$$

Substituting for $\{\varepsilon^0\}$ from Eq. (6.30) into Eq. (6.29), we can rewrite Eq. (6.30) as

$$\{M\} = [B][A]^{-1}\{N\} + [[D] - [B][A]^{-1}[B]]\{\kappa\} \tag{6.31}$$

Equations (6.30) and (6.31) can be used if in-plane forces and curvatures are known and we are interested in solving for mid-plane strains and moments. As can be seen, this requires inversion of only a 3 × 3 matrix. If in-plane forces and moments are prescribed, mid-plane strains and curvatures can be obtained by using Eq. (6.30). Equations (6.30) and (6.31) can be grouped together as

$$\left\{ \begin{matrix} \varepsilon^0 \\ M \end{matrix} \right\} = \left[\begin{matrix} [A^*] & [B^*] \\ [C^*] & [D^*] \end{matrix} \right] \left\{ \begin{matrix} N \\ \kappa \end{matrix} \right\} \tag{6.32}$$

where $[A^*] = [A]^{-1}$, $[B^*] = - [A]^{-1}[B]$, $[C^*] = [B][A]^{-1}$, $[D^*] = [D] - [B][A]^{-1}[B]$.

From the second relation of Eq. (6.32), we have

$$\{\kappa\} = [D^*]^{-1}\{M\} - [D^*]^{-1}[C]\{N\} \tag{6.33}$$

Substituting for $\{\kappa\}$ in Eq. (6.30) and rearranging the terms, we get

$$\{\varepsilon^0\} = ([A^*] - [B^*][D^*]^{-1}[C^*])\{N\} + [B^*][D^*]^{-1}\{M\} \tag{6.34}$$

Combining Eqs. (6.33) and (6.34), we obtain

$$\left\{ \begin{matrix} \{\varepsilon^0\} \\ \{\kappa\} \end{matrix} \right\} = \left[\begin{matrix} [a'] & [b'] \\ [c'] & [d'] \end{matrix} \right] \left\{ \begin{matrix} \{N\} \\ \{M\} \end{matrix} \right\} \tag{6.35}$$

where

$$[a'] = [A^*] - [B^*][D^*]^{-1}[C^*]$$
$$[b'] = [B^*][D^*]^{-1}$$
$$[c'] = -[D^*]^{-1}[C^*] = [b']^T \tag{6.36}$$
$$[d'] = [D^*]^{-1}$$

It can be seen from Eq. (6.36), that by this process, the complete inversion of Eq. (6.24) requires inversion of only a 3 × 3 matrix. This method involves multiplication of many matrices. The time taken for these steps is very much less compared to inversion of a 6×6 matrix.

Example 6.6

Consider the laminate given in Example 6.2. The laminate is subjected to uniform in-plane forces $N_x = 1000$ N/mm, $N_y = 0$ and $N_{xy} = 0$. Calculate the stresses and strains associated with the reference axes for each of the lamina. The material properties of the lamina are the same as in Example 6.2.

Solution

For the laminate, the $[A]$, $[B]$, $[D]$ matrices are

$$[A] = \begin{bmatrix} 458.3 & 135.63 & 0 \\ 135.63 & 458.3 & 0 \\ 0 & 0 & 161.3 \end{bmatrix} \text{GPa-mm}$$

$$[B] = \begin{bmatrix} 0 & 0 & 160.4 \\ 0 & 0 & 433.52 \\ 160.4 & 433.52 & 0 \end{bmatrix} \text{GPa-mm}^2$$

$$[D] = \begin{bmatrix} 530.73 & 564.57 & 0 \\ 564.57 & 1826.26 & 0 \\ 0 & 0 & 641.61 \end{bmatrix} \text{GPa-mm}^3$$

Since $[B]$ matrix is non-zero, the equations are coupled. Even though only N_x is applied, mid-plane strains and curvatures will develop. Equation (6.24) can be written as

$$\begin{Bmatrix} 1000 \\ 0 \\ 0 \\ 0 \\ 0 \\ 0 \end{Bmatrix} = \begin{bmatrix} 458.30 & 135.63 & 0 & 0 & 0 & 160.4 \\ 135.63 & 458.30 & 0 & 0 & 0 & 433.2 \\ 0 & 0 & 161.3 & 160.4 & 433.2 & 0 \\ 0 & 0 & 160.4 & 530.73 & 564.57 & 0 \\ 0 & 0 & 433.2 & 564.57 & 1903.19 & 0 \\ 160.4 & 433.2 & 0 & 0 & 0 & 641.61 \end{bmatrix} \begin{Bmatrix} \varepsilon_x^0 \\ \varepsilon_y^0 \\ \gamma_{xy}^0 \\ \kappa_x \\ \kappa_y \\ \kappa_{xy} \end{Bmatrix}$$

This equation can be solved as a system of simultaneous equations since there are a few terms which are non-zero. In general, one has to make use of a computer to obtain the result.

The strains obtained are:

$$\varepsilon_x^0 = 2.418, \quad \varepsilon_y^0 = -0.3985 \text{ and } \kappa_{xy} = -0.3354$$

and all other strains and curvatures are zero.

Example 6.7

A symmetric laminate of graphite/epoxy $[0°/-60°/60°]_s$ is subjected to a biaxial loading $N_x = 1000$ MPa/mm and $N_y = 1000$ MPa/mm. The total thickness of the laminate is 6 mm. Obtain the strains developed as a result of the biaxial loading.

Solution

Since the laminate is symmetric, $[B]$ matrix is identically zero. We need not calculate the $[D]$ matrix, since there is no applied moment. Only $[A]$ matrix needs to be calculated. The transformed reduced stiffness matrix is given as

$$[\bar{Q}]_{0°} = \begin{bmatrix} 181.8 & 2.897 & 0 \\ 2.897 & 10.35 & 0 \\ 0 & 0 & 7.17 \end{bmatrix} \text{GPa}$$

$$[\bar{Q}]_{\mp 60°} = \begin{bmatrix} 23.65 & 32.46 & \mp 20.05 \\ 32.46 & 109.4 & \mp 54.19 \\ \mp 20.05 & \mp 54.19 & 36.74 \end{bmatrix} \text{GPa}$$

For the laminate, $z_0 = -3$, $z_1 = -2$, $z_2 = -1$, $z_3 = 0$, $z_4 = 1$, $z_5 = 2$ and $z_6 = 3$ mm:

$$A_{ij} = (\bar{Q}_{ij})_1(z_1 - z_0) + (\bar{Q}_{ij})_2(z_2 - z_1) + (\bar{Q}_{ij})_3(z_3 - z_2) + (\bar{Q}_{ij})_3(z_4 - z_3)$$

$$+ (\bar{Q}_{ij})_2(z_5 - z_4) + (\bar{Q}_{ij})_1(z_6 - z_5)$$

$$= (\bar{Q}_{ij})_1(-2 + 3) + (\bar{Q}_{ij})_2(-1 + 2) + (\bar{Q}_{ij})_3(0 + 1) + (\bar{Q}_{ij})_3(1 - 0)$$

$$+ (\bar{Q}_{ij})_2(2 - 1) + (\bar{Q}_{ij})_1(3 - 2)$$

$$= 2\{(\bar{Q}_{ij})_1 + (\bar{Q}_{ij})_2 + (\bar{Q}_{ij})_3\}$$

$$[A] = \begin{bmatrix} 458.30 & 135.64 & 0 \\ 135.64 & 458.30 & 0 \\ 0 & 0 & 161.3 \end{bmatrix} \text{GPa-mm}$$

Therefore

$$\begin{Bmatrix} N_x \\ N_y \\ N_{xy} \end{Bmatrix} = \begin{bmatrix} 458.30 & 135.64 & 0 \\ 135.64 & 458.30 & 0 \\ 0 & 0 & 161.3 \end{bmatrix} \times 10^9 \begin{Bmatrix} \varepsilon_x \\ \varepsilon_y \\ \gamma_{xy} \end{Bmatrix}$$

Substituting the values of N_x and N_y and solving the simultaneous equation, we get

$$\varepsilon_x = 1.6817 \times 10^{-3}, \ \varepsilon_y = 1.6843 \times 10^{-3}, \ \gamma_{xy} = 0$$

Example 6.8
A cross-ply symmetric laminate $[0°/90°]_s$ fabricated from graphite/epoxy unidirectional plies is subject to a uniform tensile force of $N_x = 50$ N/m. Each ply is 0.2 mm thick. The material properties are as given in Example 6.6. Calculate the stresses along the material axes.

Solution
The stiffness matrix $[Q]$ for $0°$ lamina is given by

$$[\bar{Q}]_{0°} = \begin{bmatrix} 181.8 & 2.897 & 0 \\ 2.897 & 10.35 & 0 \\ 0 & 0 & 7.17 \end{bmatrix} \times 10^9 \text{ Pa}$$

The transformed stiffness matrix for $90°$ lamina is

$$[\bar{Q}]_{90°} = \begin{bmatrix} 10.35 & 2.897 & 0 \\ 2.897 & 181.8 & 0 \\ 0 & 0 & 7.17 \end{bmatrix} \times 10^9 \text{ Pa}$$

Since the laminate is symmetric, $[B]$ is identically zero. We need to calculate only matrix $[A]$ since only in-plane load is applied. For this problem

$$z_0 = -0.4 \text{ mm}, \ z_1 = -0.2 \text{ mm}, \ z_2 = 0, \ z_3 = 0.2 \text{ mm}, \ z_4 = 0.4 \text{ mm}$$

$$A_{ij} = (\bar{Q}_{ij})_1 (z_1 - z_0) + (\bar{Q}_{ij})_2 (z_2 - z_1) + (\bar{Q}_{ij})_3 (z_3 - z_2) + (\bar{Q}_{ij})_4 (z_4 - z_3)$$

$$= (\bar{Q}_{ij})_1 (-0.2 + 0.4) + (\bar{Q}_{ij})_2 (0 + 0.2) + (\bar{Q}_{ij})_3 (0.2 - 0) + (\bar{Q}_{ij})_4 (0.4 - 0.2)$$

$$= 2\{(\bar{Q}_{ij})_1 + (\bar{Q}_{ij})_2\} \times 0.2$$

$$[A] = \begin{bmatrix} 76.86 & 2.318 & 0 \\ 2.318 & 76.86 & 0 \\ 0 & 0 & 5.736 \end{bmatrix} \times 10^6 \text{ Pa-m}$$

$$\begin{Bmatrix} 50 \\ 0 \\ 0 \end{Bmatrix} = \begin{bmatrix} 76.82 & 2.318 & 0 \\ 2.318 & 76.82 & 0 \\ 0 & 0 & 5.736 \end{bmatrix} \times 10^6 \begin{Bmatrix} \varepsilon_x \\ \varepsilon_y \\ \gamma_{xy} \end{Bmatrix}$$

Solving, we get $\varepsilon_x = 0.6515 \times 10^{-6}$, $\varepsilon_y = -0.00197 \times 10^{-6}$

$$\begin{Bmatrix} \sigma_x \\ \sigma_y \\ \tau_{xy} \end{Bmatrix}_{0°} = \begin{bmatrix} 181.8 & 2.897 & 0 \\ 2.897 & 10.35 & 0 \\ 0 & 0 & 7.17 \end{bmatrix} \times 10^9 \times 10^{-6} \begin{Bmatrix} 0.6515 \\ -0.0197 \\ 0 \end{Bmatrix}$$

$$= \begin{Bmatrix} 118.3856 \\ 1.6835 \end{Bmatrix} \times 10^3 \text{ N/m}^2$$

$$\begin{Bmatrix} \sigma_x \\ \sigma_y \\ \tau_{xy} \end{Bmatrix}_{90°} = \begin{bmatrix} 10.35 & 2.897 & 0 \\ 2.897 & 181.8 & 0 \\ 0 & 0 & 7.17 \end{bmatrix} \times 10^9 \times 10^{-6} \begin{Bmatrix} 0.6515 \\ -0.0197 \\ 0 \end{Bmatrix}$$

$$= \begin{Bmatrix} 6.686 \\ -1.694 \end{Bmatrix} \times 10^3 \text{ N/m}^2$$

Example 6.9

Find the lamina stresses and strains along the material axes for a symmetric laminate $[0°/-60°/60°]_s$ of graphite/epoxy. The laminate is subjected to a bending moment, $M_x = 50$ N-m/m. The lamina thickness is 0.125 mm.

Solution

Since the laminate is symmetric, $[B]$ matrix is identically zero. Further, the in-plane stresses are zero as the laminate is subjected only to bending moment. Hence, we need to calculate $[A]$ matrix as it relates only in-plane forces and mid-plane strains. The transformed stiffness matrices for the lamina are

$$[\bar{Q}]_{0°} = \begin{bmatrix} 181.8 & 2.897 & 0 \\ 2.897 & 10.35 & 0 \\ 0 & 0 & 7.17 \end{bmatrix} \text{ GPa}$$

$$[\bar{Q}]_{\mp 60°} = \begin{bmatrix} 23.65 & 32.46 & \mp 20.05 \\ 32.46 & 109.4 & \mp 54.19 \\ \mp 20.05 & \mp 54.19 & 36.74 \end{bmatrix} \text{ GPa}$$

For the given laminate

$$z_0 = -0.375 \text{ mm}, \; z_1 = -0.25 \text{ mm}, \; z_2 = -0.125 \text{ mm}, \; z_3 = 0,$$
$$z_4 = 0.125 \text{ mm}, \; z_5 = 0.25 \text{ mm}, \; z_6 = 0.375 \text{ mm}$$

$$D_{ij} = \frac{1}{3} \Big[(\bar{Q}_{ij})_1 (z_1^3 - z_0^3) + (\bar{Q}_{ij})_2 (z_2^3 - z_1^3) + (\bar{Q}_{ij})_3 (z_3^3 - z_2^3)$$
$$+ (\bar{Q}_{ij})_3 (z_4^3 - z_3^3) + (\bar{Q}_{ij})_2 (z_5^3 - z_4^3) + (\bar{Q}_{ij})_1 (z_6^3 - z_5^3) \Big]$$

$$D_{ij} = \frac{1}{3} \Big[(\bar{Q}_{ij})_1 (z_1^3 - z_0^3 + z_6^3 - z_5^3) + (\bar{Q}_{ij})_2 (z_2^3 - z_1^3 + z_5^3 - z_4^3)$$
$$+ (\bar{Q}_{ij})_3 (z_3^3 - z_2^3 + z_4^3 - z_3^3) \Big]$$

$$= \frac{1}{3} \Big[(\bar{Q}_{ij})_1 \times (0.0742) + (\bar{Q}_{ij})_2 \times (0.0273) + (\bar{Q}_{ij})_3 \times (0.0039) \Big]$$

$$[D] = \begin{bmatrix} 4.7424 & 0.4092 & -0.1824 \\ 0.4092 & 1.3937 & -0.4218 \\ -0.1824 & -0.4218 & 0.5594 \end{bmatrix} \text{ N-m}^3$$

$$\begin{Bmatrix} 50 \\ 0 \\ 0 \end{Bmatrix} = \begin{bmatrix} 4.7424 & 0.4092 & -0.1824 \\ 0.4092 & 1.3937 & -0.4218 \\ -0.1824 & -0.4218 & 0.5594 \end{bmatrix} \begin{Bmatrix} \kappa_x \\ \kappa_y \\ \kappa_{xy} \end{Bmatrix}$$

Solving this equation, we get

$$\kappa_x = 10.8359, \ \kappa_y = -2.7371 \ \text{and} \ \kappa_{xy} = 1.4693$$

The stresses σ_x, σ_y and τ_{xy} for individual plies are given as

$$\begin{Bmatrix} \sigma_x \\ \sigma_y \\ \tau_{xy} \end{Bmatrix}_{0°} = z \begin{bmatrix} 181.8 & 2.897 & 0 \\ 2.897 & 10.35 & 0 \\ 0 & 0 & 7.17 \end{bmatrix} \begin{Bmatrix} 10.8359 \\ -2.7371 \\ 1.4693 \end{Bmatrix} \text{GPa}, \ -0.375 \le z \ge -0.25$$

$$\begin{Bmatrix} \sigma_x \\ \sigma_y \\ \tau_{xy} \end{Bmatrix}_{-60°} = z \begin{bmatrix} 23.65 & 32.46 & -20.05 \\ 32.46 & 109.4 & -54.19 \\ -20.05 & -54.19 & 36.74 \end{bmatrix} \begin{Bmatrix} 10.8359 \\ -2.7373 \\ 1.4693 \end{Bmatrix} \text{GPa}, \ -0.25 \le z \ge -0.125$$

$$\begin{Bmatrix} \sigma_x \\ \sigma_y \\ \tau_{xy} \end{Bmatrix}_{60°} = z \begin{bmatrix} 23.65 & 32.46 & 20.05 \\ 32.46 & 109.4 & 54.19 \\ 20.05 & 54.19 & 36.74 \end{bmatrix} \begin{Bmatrix} 10.8359 \\ -2.7373 \\ 1.4693 \end{Bmatrix} \text{GPa}, \ -0.125 \le z \ge 0$$

Similar relations are written for stresses in the bottom plies.

6.5 Hygrothermal Effects on Laminate Stresses

Stresses are generated within the lamina when they absorb moisture and/or are subjected to high temperature during curing. These stresses, which are locked-in stresses, affect the ultimate load carrying capacity of the laminate. Therefore, it is necessary to have a good estimate of these stresses. In addition to generating stresses in the laminate, they also degrade the properties of the laminate. Since the analysis is limited to linear elastic behavior, one can treat the thermal, hygroscopic and external loading separately, and then combine their effects on stresses and strains using the superposition principle.

6.5.1 Thermal Effects

The constitutive relations developed in Chapter 3 are based on constant environmental conditions. Generally, thermal effects influence strain only. In a given lamina, the fibers are surrounded by the matrix and do not allow free expansion of either the fiber or the matrix. This results in stresses being generated as the thermal coefficient of expansion of the fiber and matrix are not the same. In the case of a laminate, the laminae are not free to deform independent of each other. This also results in stresses. The magnitude of stresses generated depends on the coefficients of thermal expansion of fiber and the matrix, fiber orientation and stacking sequence. Since fibers and matrix materials have different coefficient of thermal expansion, the thermal strains in a lamina are different along the longitudinal (fiber) direction and transverse (normal) directions. For some materials, the longitudinal coefficients of thermal expansion are negative, whereas the transverse coefficients of thermal expansion are positive. For an orthotropic lamina, the thermal strains are expressed as

$$\begin{Bmatrix} \varepsilon_L^{Th} \\ \varepsilon_T^{Th} \end{Bmatrix} = \begin{Bmatrix} \alpha_L \\ \alpha_T \end{Bmatrix} \Delta T \tag{6.37}$$

The total strain along the material axes is obtained by adding the thermal strain to the strain developed because of the externally applied loads. Thus

$$\begin{Bmatrix} \varepsilon_L \\ \varepsilon_T \\ \gamma_{LT} \end{Bmatrix} = \begin{bmatrix} S_{11} & S_{12} & 0 \\ S_{12} & S_{22} & 0 \\ 0 & 0 & S_{66} \end{bmatrix} \begin{Bmatrix} \alpha_L \\ \alpha_T \\ \tau_{LT} \end{Bmatrix} + \begin{Bmatrix} \alpha_L \\ \alpha_T \\ 0 \end{Bmatrix} \Delta T \tag{6.38}$$

Premultiplying Eq. (6.38) by $[S]^{-1}$ on both sides, we get

$$\begin{Bmatrix} \alpha_L \\ \alpha_T \\ \tau_{LT} \end{Bmatrix} = \begin{bmatrix} Q_{11} & Q_{12} & 0 \\ Q_{12} & Q_{22} & 0 \\ 0 & 0 & Q_{66} \end{bmatrix} \left(\begin{Bmatrix} \varepsilon_L \\ \varepsilon_T \\ \gamma_{LT} \end{Bmatrix} - \begin{Bmatrix} \alpha_L \\ \alpha_T \\ 0 \end{Bmatrix} \Delta T \right) \tag{6.39}$$

If the lamina is restrained completely during thermal exposure, the total strain must be zero. The resulting thermal stresses are given by

$$\begin{Bmatrix} \sigma_L^{Th} \\ \sigma_T^{Th} \\ \tau_{LT}^{Th} \end{Bmatrix} = - \begin{bmatrix} Q_{11} & Q_{12} & 0 \\ Q_{12} & Q_{22} & 0 \\ 0 & 0 & Q_{66} \end{bmatrix} \begin{Bmatrix} \alpha_L \\ \alpha_T \\ 0 \end{Bmatrix} \Delta T \tag{6.40}$$

The thermal coefficients described in the above equations refer to the *L-T* coordinate system. It is observed that there are no thermal shear stresses and strains with respect to this coordinate system. However, this is not true when the material properties are referred to the reference axes or structural axes. The constitutive relations in the *X-Y* coordinate system in the presence of thermal effects can be written following the steps indicated in Chapter 3. The constitutive relations for a generally orthotropic lamina are

$$\begin{Bmatrix} \varepsilon_x \\ \varepsilon_y \\ \gamma_{xy} \end{Bmatrix} = \begin{bmatrix} \bar{S}_{11} & \bar{S}_{12} & \bar{S}_{16} \\ \bar{S}_{12} & \bar{S}_{22} & \bar{S}_{26} \\ \bar{S}_{16} & \bar{S}_{26} & \bar{S}_{66} \end{bmatrix} \begin{Bmatrix} \sigma_x \\ \sigma_y \\ \tau_{xy} \end{Bmatrix} + \begin{Bmatrix} \alpha_x \\ \alpha_y \\ \alpha_{xy} \end{Bmatrix} \Delta T \tag{6.41}$$

The apparent thermal coefficients of expansion are obtained by making use of Eq. (3.34). The relation is written as

$$\begin{Bmatrix} \alpha_x \\ \alpha_y \\ \alpha_{xy}/2 \end{Bmatrix} = [T]^{-1} \begin{Bmatrix} \alpha_L \\ \alpha_T \\ 0 \end{Bmatrix}$$ (6.42)

Thus, α_x, α_y and α_{xy} vary with lamina fiber orientation.

Laminates are fabricated by curing at high temperature. A set of lamina stacked in a particular sequence, are considered to be in a stress free state during curing. Initially, all the laminae are of the same length. If the laminae are not constrained, each of them will have its own deformation depending on the thermal coefficient of expansion. However, as the plies are constrained to deform with adjacent plies, residual stresses are generated. The residual stresses can be tensile or compressive. These depend on the material characteristics, lamina sequencing and temperature difference.

The stresses in the kth layer of the laminate are obtained by modifying Eq. (6.12) with the addition of a term due to thermal effect. These are

$$\begin{Bmatrix} \sigma_x \\ \sigma_y \\ \tau_{xy} \end{Bmatrix}_k = [\bar{Q}]_k \left(\begin{Bmatrix} \varepsilon_x^0 \\ \varepsilon_y^0 \\ \gamma_{xy}^0 \end{Bmatrix} + z \begin{Bmatrix} \kappa_x \\ \kappa_y \\ \kappa_{xy} \end{Bmatrix} - \begin{Bmatrix} \alpha_x \\ \alpha_y \\ \alpha_{xy} \end{Bmatrix} \Delta T \right)$$ (6.43)

This equation assumes that there is no variation of temperature across the thickness.

The thermal forces and moments are obtained by making use of Eqs. (6.16) and (6.18). If the laminate is not subjected to external load and is unconstrained, the thermal forces and moments will not produce any resultant forces and moments. In other words, the thermal forces and moments are self equilibrating. We can obtain the mid-plane strains and curvatures by employing the condition that these are not produced by externally applied loads. By substituting Eq. (6.43) into Eqs. (6.16) and (6.18), we get

$$\begin{Bmatrix} N_x \\ N_y \\ N_{xy} \end{Bmatrix} = \sum_{k=1}^{NL} \left[\int_{z_{i-1}}^{z_i} \begin{bmatrix} \bar{Q}_{11} & \bar{Q}_{12} & \bar{Q}_{16} \\ \bar{Q}_{12} & \bar{Q}_{22} & \bar{Q}_{26} \\ \bar{Q}_{16} & \bar{Q}_{26} & \bar{Q}_{66} \end{bmatrix}_k \begin{Bmatrix} \varepsilon_x^0 \\ \varepsilon_y^0 \\ \gamma_{xy}^0 \end{Bmatrix} dz + \int_{z_{i-1}}^{z_i} \begin{bmatrix} \bar{Q}_{11} & \bar{Q}_{12} & \bar{Q}_{16} \\ \bar{Q}_{12} & \bar{Q}_{22} & \bar{Q}_{26} \\ \bar{Q}_{16} & \bar{Q}_{26} & \bar{Q}_{66} \end{bmatrix} \begin{Bmatrix} \kappa_x \\ \kappa_y \\ \kappa_{xy} \end{Bmatrix} z \, dz \right.$$

$$\left. - \int_{z_{i-1}}^{z_i} [\bar{Q}]_k \begin{Bmatrix} \alpha_x \\ \alpha_y \\ \alpha_{xy} \end{Bmatrix} \Delta T \, dz \right] = \begin{Bmatrix} 0 \\ 0 \\ 0 \end{Bmatrix}$$ (6.44)

Equation (6.44) on integration and rearrangement leads to

$$
\begin{bmatrix} A_{11} & A_{12} & A_{16} \\ A_{12} & A_{22} & A_{26} \\ A_{16} & A_{26} & A_{66} \end{bmatrix} \begin{Bmatrix} \varepsilon_x^0 \\ \varepsilon_y^0 \\ \gamma_{xy}^0 \end{Bmatrix} + \begin{bmatrix} B_{11} & B_{12} & B_{16} \\ B_{12} & B_{22} & B_{26} \\ B_{16} & B_{26} & B_{66} \end{bmatrix} \begin{Bmatrix} \kappa_x \\ \kappa_y \\ \kappa_{xy} \end{Bmatrix} = \begin{Bmatrix} N_x^{\mathrm{Th}} \\ N_y^{\mathrm{Th}} \\ N_{xy}^{\mathrm{Th}} \end{Bmatrix}
$$ (6.45)

and

$$
\begin{Bmatrix} M_x \\ M_y \\ M_{xy} \end{Bmatrix} = \sum_{k=1}^{NL} \left[\int_{z_{i-1}}^{z_i} \begin{bmatrix} \bar{Q}_{11} & \bar{Q}_{12} & \bar{Q}_{16} \\ \bar{Q}_{12} & \bar{Q}_{22} & \bar{Q}_{26} \\ \bar{Q}_{16} & \bar{Q}_{26} & \bar{Q}_{66} \end{bmatrix}_k \begin{Bmatrix} \varepsilon_x^0 \\ \varepsilon_y^0 \\ \gamma_{xy}^0 \end{Bmatrix} z\, dz + \int_{z_{i-1}}^{z_i} \begin{bmatrix} \bar{Q}_{11} & \bar{Q}_{12} & \bar{Q}_{16} \\ \bar{Q}_{12} & \bar{Q}_{22} & \bar{Q}_{26} \\ \bar{Q}_{16} & \bar{Q}_{26} & \bar{Q}_{66} \end{bmatrix} \begin{Bmatrix} \kappa_x \\ \kappa_y \\ \kappa_{xy} \end{Bmatrix} z^2\, dz \right.
$$

$$
\left. - \int_{z_{i-1}}^{z_i} [\bar{Q}]_k \begin{Bmatrix} \alpha_x \\ \alpha_y \\ \alpha_{xy} \end{Bmatrix} \Delta T\, z\, dz \right] = \begin{Bmatrix} 0 \\ 0 \\ 0 \end{Bmatrix}
$$ (6.46)

Equation (6.46), on integration and rearrangement, leads to

$$
\begin{bmatrix} B_{11} & B_{12} & B_{16} \\ B_{12} & B_{22} & B_{26} \\ B_{16} & B_{26} & B_{66} \end{bmatrix} \begin{Bmatrix} \varepsilon_x^0 \\ \varepsilon_y^0 \\ \gamma_{xy}^0 \end{Bmatrix} + \begin{bmatrix} D_{11} & D_{12} & D_{16} \\ D_{12} & D_{22} & D_{26} \\ D_{16} & D_{26} & D_{66} \end{bmatrix} \begin{Bmatrix} \kappa_x \\ \kappa_y \\ \kappa_{xy} \end{Bmatrix} = \begin{Bmatrix} M_x^{\mathrm{Th}} \\ M_y^{\mathrm{Th}} \\ M_{xy}^{\mathrm{Th}} \end{Bmatrix}
$$ (6.47)

$$
\begin{Bmatrix} N_x^{\mathrm{Th}} \\ N_y^{\mathrm{Th}} \\ N_{xy}^{\mathrm{Th}} \end{Bmatrix} = \int_{-t/2}^{t/2} \begin{Bmatrix} \sigma_x \\ \sigma_y \\ \tau_{xy} \end{Bmatrix}_k^{\mathrm{Th}} dz = \int_{-t/2}^{t/2} [\bar{Q}]_k \begin{Bmatrix} \alpha_x \\ \alpha_y \\ \alpha_{xy} \end{Bmatrix}_k \Delta T\, dz
$$ (6.48)

$$
\begin{Bmatrix} M_x^{\mathrm{Th}} \\ M_y^{\mathrm{Th}} \\ M_{xy}^{\mathrm{Th}} \end{Bmatrix} = \int_{-t/2}^{t/2} \begin{Bmatrix} \sigma_x \\ \sigma_y \\ \tau_{xy} \end{Bmatrix}_k^{\mathrm{Th}} z\, dz = \int_{-t/2}^{t/2} [\bar{Q}]_k\, z \begin{Bmatrix} \alpha_x \\ \alpha_y \\ \alpha_{xy} \end{Bmatrix}_k \Delta T\, dz
$$ (6.49)

$$
\begin{Bmatrix} N_x^{\mathrm{Th}} \\ N_y^{\mathrm{Th}} \\ N_{xy}^{\mathrm{Th}} \end{Bmatrix} = \sum_{k=1}^{NL} [\bar{Q}]_k \Delta T \int_{z_{i-1}}^{z_i} \begin{Bmatrix} \alpha_x \\ \alpha_y \\ \alpha_{xy} \end{Bmatrix}_k dz = \sum_{k=1}^{NL} [\bar{Q}]_k \Delta T \begin{Bmatrix} \alpha_x \\ \alpha_y \\ \alpha_{xy} \end{Bmatrix} (z_k - z_{k-1})
$$ (6.50)

$$\begin{Bmatrix} M_x^{Th} \\ M_y^{Th} \\ M_{xy}^{Th} \end{Bmatrix} = \sum_{k=1}^{NL} [\bar{Q}]_k \Delta T \int_{z_{i-1}}^{z_i} \begin{Bmatrix} \alpha_x \\ \alpha_y \\ \alpha_{xy} \end{Bmatrix}_k z\,dz = \frac{1}{2}\sum_{k=1}^{NL} [\bar{Q}]_k \Delta T \begin{Bmatrix} \alpha_x \\ \alpha_y \\ \alpha_{xy} \end{Bmatrix} (z_k^2 - z_{k-1}^2) \quad (6.51)$$

Example 6.10

Determine the thermal forces N_x^{Th}, N_y^{Th}, N_{xy}^{Th} for a $[\pm\,45°]_s$ laminate. The material properties are $E_L = 181$ GPa, $E_T = 10.3$ GPa, $G_{LT} = 7.17$ GPa and $\nu_{LT} = 0.28$.

The thermal coefficients of expansion are $\alpha_L = 0.02 \times 10^{-6}/°C$ and $\alpha_T = 22.5 \times 10^{-6}/°C$.

Assume $\Delta T = 60°C$. The thickness of the laminate is t. Thus, the thickness of each lamina is $t/4$.

Solution

From Eq. (3.24), we have

$$Q_{11} = \frac{E_L}{1 - \nu_{TL}\nu_{LT}} = \frac{181}{1 - 0.28 \times 0.016} = 181.81\,\text{GPa}$$

$$Q_{22} = \frac{E_T}{1 - \nu_{TL}\nu_{LT}}\,\text{GPa}$$

$Q_{12} = Q_{11} \times \nu_{LT} = 181.81 \times 0.016 = 2.909$ GPa, $Q_{66} = G_{LT} = 7.17$ GPa

We make use of relations given by Eq. (3.43) to find the elements of $[\bar{Q}]$ matrix for $\pm\,45°$ lamina. For $45°$ lamina

$\bar{Q}_{11} = 181.81 \times 0.2498 + 10.35 \times 0.2498 + 2 \times (2.909 + 2 \times 7.17) \times 0.2498$
$\quad = 56.55$ GPa

$\bar{Q}_{22} = 56.55$ GPa

$\bar{Q}_{12} = (181.81 + 10.35 - 4 \times 7.17) \times 0.2498 + 2.909 \times (0.2498 + 0.2498)$
$\quad = 42.290$ GPa

$\bar{Q}_{66} = (181.81 + 10.35 - 2 \times 2.909 - 2 \times 7.17) \times 0.2498 + 7.17 \times (0.2498 + 0.2498)$
$\quad = 43.460$ GPa

$\bar{Q}_{16} = (2.909 - 10.35 - 2 \times 7.17) \times 0.2498 + (10.35 - 2.909 - 2 \times 7.17) \times 0.2498$
$\quad = -7.164$ GPa

$\bar{Q}_{26} = -7.164$ GPa

For $-45°$ lamina

$$\bar{Q}_{11} = \bar{Q}_{22} = 56.55\text{ GPa}$$
$$\bar{Q}_{12} = 42.29\text{ GPa}$$
$$\bar{Q}_{66} = 43.460\text{ GPa}$$
$$\bar{Q}_{16} = \bar{Q}_{26} = 7.164\text{ GPa}$$

Thermal strain, assuming free expansion of the lamina in the material direction is

$$\varepsilon_L^{Th} = \alpha_L \, \Delta T = 0.02 \times 10^{-6} \times 60 = 0.12 \times 10^{-6}$$

$$\varepsilon_T^{Th} = \alpha_T \, \Delta T = 22.5 \times 10^{-6} \times 60 = 1350 \times 10^{-6}$$

$$\begin{Bmatrix} \varepsilon_x \\ \varepsilon_y \\ \gamma_{xy}/2 \end{Bmatrix} = \begin{bmatrix} \cos^2\theta & \sin^2\theta & -2\sin\theta\cos\theta \\ \sin^2\theta & \cos^2\theta & 2\sin\theta\cos\theta \\ \sin\theta\cos\theta & -\sin\theta\cos\theta & \cos^2\theta - \sin^2\theta \end{bmatrix} \begin{Bmatrix} \alpha_L \\ \alpha_T \\ 0 \end{Bmatrix} \Delta T$$

$$= \begin{bmatrix} 0.4998 & 0.4998 & -0.9997 \\ 0.4998 & 0.4998 & 0.9997 \\ 0.4998 & -0.4998 & 0 \end{bmatrix} \begin{Bmatrix} 0.12 \times 10^{-6} \\ 1350 \times 10^{-6} \\ 0 \end{Bmatrix}$$

$$= \begin{Bmatrix} 674.7899 \times 10^{-6} \\ 674.7899 \times 10^{-6} \\ -674.6700 \times 10^{-6} \end{Bmatrix}$$

Therefore, for 45° lamina

$$\begin{Bmatrix} \varepsilon_x^{Th} \\ \varepsilon_y^{Th} \\ \gamma_{xy}^{Th} \end{Bmatrix} = \begin{Bmatrix} 674.7899 \times 10^{-6} \\ 674.7899 \times 10^{-6} \\ -1349.3400 \times 10^{-6} \end{Bmatrix}$$

and for − 45° lamina

$$\begin{Bmatrix} \varepsilon_x^{Th} \\ \varepsilon_y^{Th} \\ \gamma_{xy}^{Th} \end{Bmatrix} = \begin{Bmatrix} 674.7899 \times 10^{-6} \\ 674.7899 \times 10^{-6} \\ 1349.3400 \times 10^{-6} \end{Bmatrix}$$

Making use of Eq. (6.50), we obtain

$$\begin{Bmatrix} N_x^{Th} \\ N_y^{Th} \\ N_{xy}^{Th} \end{Bmatrix} = 2 \times \begin{bmatrix} 56.55 & 42.29 & -7.164 \\ 42.29 & 56.55 & -7.164 \\ -7.164 & -7.164 & 43.46 \end{bmatrix} \begin{Bmatrix} 674.7899 \\ 674.7899 \\ -1349.34 \end{Bmatrix} \frac{t}{4} \times 10^3$$

$$+ 2 \times \begin{bmatrix} 56.55 & 42.29 & 7.164 \\ 42.29 & 56.55 & 7.164 \\ 7.164 & 7.164 & 43.46 \end{bmatrix} \begin{Bmatrix} 674.7899 \\ 674.7899 \\ 1349.34 \end{Bmatrix} \frac{t}{4} \times 10^3$$

$$= \begin{Bmatrix} 0.0763 \\ 0.0763 \end{Bmatrix} t \text{ GPa}$$

Example 6.11

Calculate the residual stresses for the two ply laminate. The plies are of equal thickness of 5 mm. The top lamina is of 0° and the bottom lamina is of 45°. The stiffness matrix $[Q]$ for the material is given as

$$\begin{bmatrix} 20 & 0.7 & 0 \\ 0.7 & 2.0 & 0 \\ 0 & 0 & 0.7 \end{bmatrix} \text{GPa}$$

The laminate is cured at 125°C and cooled to a temperature of 25°C. $\alpha_L = 7.0 \times 10^{-6}/°C$ and $\alpha_T = 23 \times 10^{-6}/°C$.

Solution

The coefficients of thermal expansion in the x-y coordinate system by using transformation are

$$\begin{Bmatrix} \alpha_x \\ \alpha_y \\ \alpha_{xy} \end{Bmatrix}_{0°} = \begin{Bmatrix} 7 \\ 23 \\ 0 \end{Bmatrix} \times 10^{-6}, \quad \begin{Bmatrix} \alpha_x \\ \alpha_y \\ \alpha_{xy} \end{Bmatrix}_{45°} = \begin{Bmatrix} 15 \\ 15 \\ -16 \end{Bmatrix}_{45°} \times 10^{-6}$$

$$\Delta_T = -100°C$$

Since the laminate is not symmetric, we have to evaluate $[A]$, $[B]$ and $[D]$ matrices for the laminate. This requires the determination of $[\bar{Q}]_{0°}$ and $[\bar{Q}]_{45°}$. For the 0° lamina, $\bar{Q}$ and Q are identical. $[\bar{Q}]_{45°}$ is obtained by making use of Eq. (3.43). Thus

$$[\bar{Q}]_{0°} = \begin{bmatrix} 20 & 0.7 & 0 \\ 0.7 & 2.0 & 0 \\ 0 & 0 & 0.7 \end{bmatrix} \text{GPa and } [\bar{Q}]_{45°} = \begin{bmatrix} 6.55 & 5.15 & 4.50 \\ 5.15 & 6.55 & 4.50 \\ 4.50 & 4.50 & 5.15 \end{bmatrix} \text{GPa}$$

For the given laminate: $z_0 = -5$ mm, $z_1 = 0$ mm and $z_2 = 5$ mm.

$$A_{ij} = (\bar{Q}_{ij})_1(z_1 - z_0) + (\bar{Q}_{ij})_2(z_2 - z_1)$$

$$= (\bar{Q}_{ij})_{0°}[0 - (-5)] + (\bar{Q}_{ij})_{45°}[5 - 0]$$

$$= 5 \times [(\bar{Q}_{ij})_{0°} + (\bar{Q}_{ij})_{45°}]$$

$$[A] = 5 \times \left(\begin{bmatrix} 20 & 0.7 & 0 \\ 0.7 & 2.0 & 0 \\ 0 & 0 & 0.7 \end{bmatrix} + \begin{bmatrix} 6.55 & 5.15 & 4.50 \\ 5.15 & 6.55 & 4.50 \\ 4.50 & 4.50 & 5.15 \end{bmatrix} \right)$$

$$[A] = \begin{bmatrix} 132.75 & 29.25 & 22.50 \\ 29.25 & 42.75 & 22.50 \\ 22.50 & 22.50 & 29.25 \end{bmatrix} \text{GPa-mm}$$

$$B_{ij} = \frac{1}{2}\Big[(\bar{Q}_{ij})_{0°}(z_1^2 - z_0^2) + (\bar{Q}_{ij})_{45°}(z_2^2 - z_1^2)\Big]$$

$$= \frac{1}{2}\Big[(\bar{Q}_{ij})_{0°}(0 - 25) + (\bar{Q}_{ij})_{45°}(25 - 0)\Big]$$

$$= \frac{1}{2} \times 25 \times \Big[(\bar{Q}_{ij})_{45°} - (\bar{Q}_{ij})_{0°}\Big]$$

$$[B] = \begin{bmatrix} -170.625 & 55.625 & 56.250 \\ 55.625 & 56.875 & 56.250 \\ 56.250 & 56.250 & 55.625 \end{bmatrix} \text{GPa-mm}^2$$

$$D_{ij} = \frac{1}{3}\Big[(\bar{Q}_{ij})_{0°}(z_1^3 - z_0^3) + (\bar{Q}_{ij})_{45°}(z_2^3 - z_1^3)\Big]$$

$$= \frac{1}{3}\Big[(\bar{Q}_{ij})_{0°}(0 + 125) + (\bar{Q}_{ij})_{45°}(125 - 0)\Big]$$

$$= \frac{125}{3}\Big[(\bar{Q}_{ij})_{0°} + (\bar{Q}_{ij})_{45°}\Big]$$

$$[D] = \begin{bmatrix} 1106.25 & 243.75 & 187.50 \\ 243.75 & 356.25 & 187.50 \\ 187.50 & 187.50 & 243.75 \end{bmatrix} \text{GPa-mm}^3$$

Thermal forces and moments are calculated by employing Eqs. (6.50) and (6.51).

$$\begin{Bmatrix} N_x^{Th} \\ N_y^{Th} \\ N_{xy}^{Th} \end{Bmatrix} = \sum_{k=1}^{NL}[\bar{Q}]_k \Delta T \int_{z_{i-1}}^{z_i} \begin{Bmatrix} \alpha_x \\ \alpha_y \\ \alpha_{xy} \end{Bmatrix}_k dz = \sum_{k=1}^{NL}[\bar{Q}]_k \Delta T \begin{Bmatrix} \alpha_x \\ \alpha_y \\ \alpha_{xy} \end{Bmatrix}(z_k - z_{k-1})$$

$$\begin{Bmatrix} M_x^{Th} \\ M_y^{Th} \\ M_{xy}^{Th} \end{Bmatrix} = \sum_{k=1}^{NL}[\bar{Q}]_k \Delta T \int_{z_{i-1}}^{z_i} \begin{Bmatrix} \alpha_x \\ \alpha_y \\ \alpha_{xy} \end{Bmatrix}_k z\, dz = \frac{1}{2}\sum_{k=1}^{NL}[\bar{Q}]_k \Delta T \begin{Bmatrix} \alpha_x \\ \alpha_y \\ \alpha_{xy} \end{Bmatrix}(z_k^2 - z_{k-1}^2)$$

$$\begin{Bmatrix} N_x^{Th} \\ N_y^{Th} \\ N_{xy}^{Th} \end{Bmatrix} = -100 \times \left\{ \begin{bmatrix} 20 & 0.7 & 0 \\ 0.7 & 2.0 & 0 \\ 0 & 0 & 0.7 \end{bmatrix} \begin{Bmatrix} 7 \\ 23 \\ 0 \end{Bmatrix}(5) + \begin{bmatrix} 6.55 & 5.15 & 4.50 \\ 5.15 & 6.55 & 4.50 \\ 4.50 & 4.50 & 6.55 \end{bmatrix} \begin{Bmatrix} 15 \\ 15 \\ -16 \end{Bmatrix}(5) \right\} \times 10^{-6}$$

$$= 10^{-4} \begin{Bmatrix} -1298 \\ -772 \\ -151 \end{Bmatrix} \text{GPa-mm}$$

$$\begin{Bmatrix} M_x^{Th} \\ M_y^{Th} \\ M_{xy}^{Th} \end{Bmatrix} = -100 \times \frac{1}{2} \left\{ \begin{bmatrix} 20 & 0.7 & 0 \\ 0.7 & 2.0 & 0 \\ 0 & 0 & 0.7 \end{bmatrix} \begin{Bmatrix} 7 \\ 23 \\ 0 \end{Bmatrix} (-25) + \begin{bmatrix} 6.55 & 5.15 & 4.50 \\ 5.15 & 6.55 & 4.50 \\ 4.50 & 4.50 & 6.55 \end{bmatrix} \begin{Bmatrix} 15 \\ 15 \\ -16 \end{Bmatrix} (25) \right\} \times 10^{-6}$$

$$= \begin{Bmatrix} 657.5 \\ -657.5 \\ -377.5 \end{Bmatrix} \times 10^{-4} \text{ GPa-mm}$$

One can then write the following equation to evaluate the mid-plane strains and curvatures as

$$\begin{Bmatrix} N^{Th} \\ M^{Th} \end{Bmatrix} = \begin{bmatrix} A & B \\ B & D \end{bmatrix} \begin{Bmatrix} \varepsilon^0 \\ \kappa \end{Bmatrix}$$

The residual stresses can then be obtained.

6.5.2 Moisture Stresses

Composite laminates are susceptible to moisture absorption. Polymer based composites absorb moisture. This results in swelling of the laminate which in turn reduces the strength of the laminate. Moisture absorption causes changes in the dimension. It introduces hygroscopic strains in the laminates. In a given laminate, the lamina is constrained against free deformation due to the presence of adjacent lamina. This results in hygroscopic stresses. Thus, the overall load carrying ability of the laminate is reduced. Therefore, it is necessary to evaluate the stresses caused by the moisture absorption.

The hygroscopic strain in an isotropic material is defined as the product of the coefficient of moisture expansion β of the lamina and the change in the moisture content ΔC

$$\varepsilon^H = \beta \, \Delta C \tag{6.52}$$

For an orthotropic lamina, the coefficient of moisture changes with direction. For a unidirectional lamina, the moisture strains along and transverse to the direction of the fiber are given by the following relations

$$\varepsilon_L^H = \beta_L \, \Delta C \tag{6.53a}$$

$$\varepsilon_T^H = \beta_T \, \Delta C \tag{6.53b}$$

The total strain developed along the material axes is obtained by adding the moisture strain to the strain developed by the externally applied load as

$$\begin{Bmatrix} \varepsilon_L \\ \varepsilon_T \\ \gamma_{LT} \end{Bmatrix} = \begin{bmatrix} S_{11} & S_{12} & 0 \\ S_{12} & S_{22} & 0 \\ 0 & 0 & S_{66} \end{bmatrix} \begin{Bmatrix} \sigma_L \\ \sigma_T \\ \tau_{LT} \end{Bmatrix} + \begin{Bmatrix} \beta_L \\ \beta_T \\ 0 \end{Bmatrix} \Delta C \tag{6.54}$$

Premultiplying Eq. (6.54) by $[S]^{-1}$ on both sides, we get

$$\begin{Bmatrix} \sigma_L \\ \sigma_T \\ \tau_{LT} \end{Bmatrix} = \begin{bmatrix} Q_{11} & Q_{12} & 0 \\ Q_{12} & Q_{22} & 0 \\ 0 & 0 & Q_{66} \end{bmatrix} \begin{Bmatrix} \varepsilon_L \\ \varepsilon_T \\ \gamma_{LT} \end{Bmatrix} - \begin{Bmatrix} \beta_L \\ \beta_T \\ 0 \end{Bmatrix} \Delta C \tag{6.55}$$

The apparent coefficients of moisture expansion are obtained by making use of Eq. (3.34). The relation is written as

$$\begin{Bmatrix} \beta_x \\ \beta_y \\ \beta_{xy}/2 \end{Bmatrix} = [T]^{-1} \begin{Bmatrix} \beta_L \\ \beta_T \\ 0 \end{Bmatrix} \tag{6.56}$$

It can be observed from Eq. (6.56) that there exists an apparent coefficient of moisture shear in the x-y coordinate system. These give rise to hygroscopic strains in the lamina. If there are no constraints on the lamina, these do not generate any resultant force or moment. However, in a laminate, an individual lamina is constrained against deformations. The adjacent lamina prevents its deformation. As a result, stresses are induced in the lamina. The free expansion strain due to moisture is given by the following relation

$$\begin{Bmatrix} \varepsilon_x^H \\ \varepsilon_y^H \\ \gamma_{xy}^H \end{Bmatrix} = \begin{Bmatrix} \beta_x \\ \beta_y \\ \beta_{xy} \end{Bmatrix} \Delta C \tag{6.57}$$

The lamina hygroscopic stresses are then obtained as

$$\begin{Bmatrix} \sigma_x^H \\ \sigma_y^H \\ \tau_{xy}^H \end{Bmatrix} = \begin{bmatrix} \overline{Q}_{11} & \overline{Q}_{12} & \overline{Q}_{16} \\ \overline{Q}_{12} & \overline{Q}_{22} & \overline{Q}_{26} \\ \overline{Q}_{16} & \overline{Q}_{26} & \overline{Q}_{26} \end{bmatrix} \begin{Bmatrix} \varepsilon_x^0 + z\kappa_x - \beta_x \Delta C \\ \varepsilon_y^0 + z\kappa_y - \beta_y \Delta C \\ \gamma_{xy}^0 + z\kappa_{xy} - \beta_{xy} \Delta C \end{Bmatrix} \tag{6.58}$$

If there is no external load acting on the laminate and the laminate is not constrained against deformation, no resultant forces and moments are generated. Even if there is an external loading, the mid-plane strains and curvatures can still be evaluated by assuming that there is no applied external force or moment. The stresses and the moments generated internally due to moisture can be added to the external load by employing the superposition principle as the system considered is linear. Therefore, substitution of Eq. (6.58) into Eqs. (6.16) and (6.18), and subsequent integration and rearrangement leads to

$$
\begin{bmatrix} A_{11} & A_{12} & A_{16} \\ A_{12} & A_{22} & A_{26} \\ A_{16} & A_{26} & A_{66} \end{bmatrix} \begin{Bmatrix} \varepsilon_x^0 \\ \varepsilon_y^0 \\ \gamma_{xy}^0 \end{Bmatrix} + \begin{bmatrix} B_{11} & B_{12} & B_{16} \\ B_{12} & B_{22} & B_{26} \\ B_{16} & B_{26} & B_{66} \end{bmatrix} \begin{Bmatrix} \kappa_x \\ \kappa_y \\ \kappa_{xy} \end{Bmatrix} = \begin{Bmatrix} N_x^H \\ N_y^H \\ N_{xy}^H \end{Bmatrix} \quad (6.59)
$$

$$
\begin{bmatrix} B_{11} & B_{12} & B_{16} \\ B_{12} & B_{22} & B_{26} \\ B_{16} & B_{26} & B_{66} \end{bmatrix} \begin{Bmatrix} \varepsilon_x^0 \\ \varepsilon_y^0 \\ \gamma_{xy}^0 \end{Bmatrix} + \begin{bmatrix} D_{11} & D_{12} & D_{16} \\ D_{12} & D_{22} & D_{26} \\ D_{16} & D_{26} & D_{66} \end{bmatrix} \begin{Bmatrix} \kappa_x \\ \kappa_y \\ \kappa_{xy} \end{Bmatrix} = \begin{Bmatrix} M_x^H \\ M_y^H \\ M_{xy}^H \end{Bmatrix} \quad (6.60)
$$

$$
\begin{Bmatrix} N_x^H \\ N_y^H \\ N_{xy}^H \end{Bmatrix} = \sum_{k=1}^{NL} [\bar{Q}]_k \Delta C \int_{z_{i-1}}^{z_i} \begin{Bmatrix} \beta_x \\ \beta_y \\ \beta_{xy} \end{Bmatrix}_k dz = \sum_{k=1}^{NL} [\bar{Q}]_k \Delta C \begin{Bmatrix} \beta_x \\ \beta_y \\ \beta_{xy} \end{Bmatrix} (z_k - z_{k-1}) \quad (6.61a)
$$

$$
\begin{Bmatrix} M_x^H \\ M_y^H \\ M_{xy}^H \end{Bmatrix} = \sum_{k=1}^{NL} [\bar{Q}]_k \Delta C \int_{z_{i-1}}^{z_i} \begin{Bmatrix} \beta_x \\ \beta_y \\ \beta_{xy} \end{Bmatrix}_k z\,dz = \frac{1}{2}\sum_{k=1}^{NL} [\bar{Q}]_k \Delta C \begin{Bmatrix} \beta_x \\ \beta_y \\ \beta_{xy} \end{Bmatrix} (z_k^2 - z_{k-1}^2) \quad (6.61b)
$$

If in addition to the external forces and moments, moisture absorption and thermal effects exist, the net forces and moments are obtained by summing up all the three components as

$$
\begin{Bmatrix} \bar{N}_x \\ \bar{N}_y \\ \bar{N}_{xy} \end{Bmatrix} = \begin{Bmatrix} N_x \\ N_y \\ N_{xy} \end{Bmatrix} + \begin{Bmatrix} N_x^{Th} \\ N_y^{Th} \\ N_{xy}^{Th} \end{Bmatrix} + \begin{Bmatrix} N_x^H \\ N_y^H \\ N_{xy}^H \end{Bmatrix} \quad (6.62a)
$$

$$
\begin{Bmatrix} \bar{M}_x \\ \bar{M}_y \\ \bar{M}_{xy} \end{Bmatrix} = \begin{Bmatrix} M_x \\ M_y \\ M_{xy} \end{Bmatrix} + \begin{Bmatrix} M_x^{Th} \\ M_y^{Th} \\ M_{xy}^{Th} \end{Bmatrix} + \begin{Bmatrix} M_x^H \\ M_y^H \\ M_{xy}^H \end{Bmatrix} \quad (6.62b)
$$

One can then make use of Eqs. (6.26) and (6.27) by replacing N_x, N_y, N_{xy}, M_x, M_y and M_{xy} by $\bar{N}_x, \bar{N}_y, \bar{N}_{xy}, \bar{M}_x, \bar{M}_y, \bar{M}_{xy}$ to obtain the mid-plane strains and the curvatures. This can be used to obtain the lamina stresses and strains.

Example 6.12

A symmetric cross ply graphite/epoxy laminate of four layers of equal thickness is cured at a temperature of 125°C. After curing, the assembly is cooled from 125°C to 25°C. Cooling occurs in a humid atmosphere. The laminate absorbs 0.5% of its weight in moisture. The thickness of each lamina is 0.25 mm. The laminate has the following properties

E_L = 140 GPa, E_T = 10 GPa, G_{LT} = 7 GPa, ν_{LT} = 0.3,
α_L = − 0.3 × 10^{-6}, α_T = 28 × 10^{-6}, β_L = 0.0 and β_T = 0.44

Determine the thermal and moisture absorption forces.

Solution

$$\nu_{TL} = \frac{E_T}{E_L} \times \nu_{LT} = \frac{10}{140} \times 0.3 = 0.021$$

$$Q_{11} = \frac{E_L}{1 - \nu_{TL}\nu_{LT}} = \frac{140}{1 - 0.3 \times 0.021} = 140.89 \text{ GPa}$$

$$Q_{12} = \frac{\nu_{TL}E_T}{1 - \nu_{LT}\nu_{TL}} = \frac{0.3 \times 10.0}{1 - 0.3 \times 0.021} = 3.019 \text{ GPa}$$

$$Q_{22} = \frac{E_T}{1 - \nu_{TL}\nu_{TL}} = \frac{10}{1 - 0.3 \times 0.021} = 10.063 \text{ GPa}$$

$$Q_{66} = G_{LT} = 7 \text{ GPa}$$

The next step is to obtain the $[\bar{Q}]$ matrix for both 0° and 90° lamina.

$$[\bar{Q}]_{0°} = [Q]_{0°} = \begin{bmatrix} 140.89 & 3.019 & 0 \\ 3.019 & 10.063 & 0 \\ 0 & 0 & 7.0 \end{bmatrix} \text{GPa}$$

$$[\bar{Q}]_{90°} = \begin{bmatrix} 10.063 & 3.019 & 0 \\ 3.019 & 140.89 & 0 \\ 0 & 0 & 7.0 \end{bmatrix} \text{GPa}$$

As the laminate is symmetric and there is no external loading, we need to calculate only the [A] matrix. The transformed thermal and moisture coefficients are

$$\begin{Bmatrix} \alpha_x \\ \alpha_y \\ \alpha_{xy} \end{Bmatrix}_{0°} = \begin{Bmatrix} -0.3 \\ 28.0 \\ 0 \end{Bmatrix} \times 10^{-6}, \begin{Bmatrix} \alpha_x \\ \alpha_y \\ \alpha_{xy} \end{Bmatrix}_{90°} = \begin{Bmatrix} 28.0 \\ -0.3 \\ 0 \end{Bmatrix} \times 10^{-6}$$

$$\begin{Bmatrix} \beta_x \\ \beta_y \\ \beta_{xy} \end{Bmatrix}_{0°} = \begin{Bmatrix} 0.0 \\ 0.44 \\ 0 \end{Bmatrix}, \begin{Bmatrix} \beta_x \\ \beta_y \\ \beta_{xy} \end{Bmatrix}_{90°} = \begin{Bmatrix} 0.44 \\ 0 \\ 0 \end{Bmatrix}$$

$$A_{ij} = (\bar{Q}_{ij})_0(z_1 - z_0) + (\bar{Q}_{ij})_{90}(z_2 - z_1) + (\bar{Q}_{ij})_{90}(z_3 - z_2) + (\bar{Q}_{ij})_0(z_4 - z_3)$$

$$= (\bar{Q}_{ij})_0(-0.25 + 0.5) + (\bar{Q}_{ij})_{90}(0 + 0.25) + (\bar{Q}_{ij})_{90}(0.25 - 0) + (\bar{Q}_{ij})_0(0.5 - 0.25)$$

$$= 0.5 \times \left[(\bar{Q}_{ij})_0 + (\bar{Q}_{ij})_{90} \right]$$

$$[A] = \begin{bmatrix} 75.476 & 3.019 & 0 \\ 3.019 & 75.476 & 0 \\ 0 & 0 & 7 \end{bmatrix} \text{GPa-mm}$$

$$\begin{Bmatrix} N_x^{Th} \\ N_y^{Th} \\ N_{xy}^{Th} \end{Bmatrix} = -100 \times 0.5 \times \left[\begin{bmatrix} 140.89 & 3.019 & 0 \\ 3.019 & 10.063 & 0 \\ 0 & 0 & 7 \end{bmatrix} \begin{Bmatrix} -0.3 \\ 28 \\ 0 \end{Bmatrix} + \begin{bmatrix} 10.063 & 3.019 & 0 \\ 3.019 & 140.89 & 0 \\ 0 & 0 & 7 \end{bmatrix} \begin{Bmatrix} 28 \\ -0.3 \\ 0 \end{Bmatrix} \right] \times 10^{-6}$$

$$= \begin{Bmatrix} -0.01656 \\ -0.01656 \\ 0 \end{Bmatrix} \text{GPa-mm}$$

$$\begin{Bmatrix} N_x^{H} \\ N_y^{H} \\ N_{xy}^{H} \end{Bmatrix} = 0.005 \times 0.5 \times \left[\begin{bmatrix} 140.89 & 3.019 & 0 \\ 3.019 & 10.063 & 0 \\ 0 & 0 & 7 \end{bmatrix} \begin{Bmatrix} 0 \\ 0.44 \\ 0 \end{Bmatrix} + \begin{bmatrix} 10.063 & 3.019 & 0 \\ 3.019 & 140.89 & 0 \\ 0 & 0 & 7 \end{bmatrix} \begin{Bmatrix} 0.44 \\ 0 \\ 0 \end{Bmatrix} \right]$$

$$= \begin{Bmatrix} 0.0144 \\ 0.0144 \\ 0 \end{Bmatrix} \text{GPa-mm}$$

Thermal stresses are produced when the laminate is cured at higher temperature and cooled to normal temperatures. This is invariably carried out in practice as the part is required at short notice. In many situations, the curing stresses induced may be sufficiently large to cause failure of the laminate. Laminates when exposed to the environment absorb moisture which induces stresses in the laminate. Residual stresses also result because of mismatch in the coefficient of thermal expansion between the fiber and the matrix. All these result in residual stresses. Some of the observations made above have also been discussed in detail elsewhere [6-10].

Summary

- We have derived the constitutive equations for the laminate under certain assumptions.
- We have also looked at the hygrothermal behavior of the laminate.
- A number of examples have been worked out for different loading conditions.

Problems

6.1 The stacking sequence of a laminate is $[0/90_5]_s$ and each lamina has a thickness of 0.13 mm. The total laminate thickness, $t = 1.56$ cm. The material properties are as follows

$$E_L = 40 \text{ GPa}, E_T = 10 \text{ GPa}, G_{LT} = 4 \text{ GPa and } v_{LT} = 0.25$$

Obtain the extensional matrix [A] and bending stiffness matrix [D]. Show that the coupling matrix [B] is a null matrix.

6.2 If in Problem 6.1, the stacking sequence is changed to $[0/90_5]_2$, keeping the material and other parameters same, what will be the new [A], [B] and [D] matrices?

6.3 The reduced stiffness matrix [Q] is given by

$$\begin{bmatrix} 181.8 & 2.897 & 0 \\ 0 & 10.34 & 0 \\ 0 & 0 & 7.17 \end{bmatrix} \text{GPa}$$

Determine E_L, E_T, G_{LT} and v_{LT}. Also determine elements of [S].

6.4 A symmetric cross-ply laminate $[90°/0°/0°/90°]$ is subjected to $N_y = 50$ MPa. Determine the stresses in each ply. Assume the ply to be of uniform thickness. The material properties are given in Problem 6.1.

6.5 A quasi-isotropic laminate $[0°/60°/- 60°]_s$ is subjected to a biaxial loading as follows

$$N_x = N_0, N_y = - N_0, N_{xy} = 0$$

Determine the strains in the plies in terms of N_0.

6.6 How does the stress change if the laminate sequencing is changed in Problem 6.4 to $[0°/90°/90°/0°]$.

6.7 For the stacking sequence shown for the laminates, plot the variation of σ_x, σ_y and τ_{xy} through the thickness of an AS/3501 laminate subjected to N_x only. Assume each ply has a thickness of 0.15 mm. The material properties of AS/3501 are

$$E_L = 138 \text{ GPa}, E_T = 9.0 \text{ GPa}, G_{LT} = 6.9 \text{ GPa}, v_{LT} = 0.30$$

1. $[0°/45°/- 45°]_s$, 2. $[45°/0°/- 45°]_s$, 3. $[30°/45°/- 45°/0°]_s$.

6.8 Determine the thermal forces and moments for an anti-symmetric laminate $[30°/- 45°/45°/- 30°]$ whose properties are given as

$$E_L = 142 \text{ GPa}, E_T = 10.3 \text{ GPa}, G_{LT} = 7.2 \text{ GPa}, v_{12} = 0.27, \Delta T = - 100°C$$

6.9 A 2-ply anti-symmetric angle-ply laminate $[\theta/- \theta]$ is made of graphite/epoxy and designed to have the laminate coefficient of thermal expansion, α_x to be zero. Determine the ply orientation θ required to meet this condition. The lamina thickness is 0.1 mm. The lamina properties are

$$E_L = 138 \text{ GPa}, E_T = 9.0 \text{ GPa}, G_{LT} = 7.0 \text{ GPa}, v_{LT} = 0.3$$

$$\alpha_L = - 0.3 \times 10^{-6} \text{ m/m/K}, \alpha_T = 28.0 \times 10^{-6} \text{ m/m/K}$$

6.10 Repeat Problem 6.9 for a Kevlar/epoxy laminate, with the properties

$$E_L = 76 \text{ GPa}, E_T = 5.5 \text{ GPa}, G_{LT} = 2.5 \text{ GPa}, v_{LT} = 0.35$$

$$\alpha_L = - 4.0 \times 10^{-6}, \alpha_T = 79.0 \times 10^{-6}$$

References

1. Whitney, J.M., 1987, *Structural Analysis of Laminated Plates*, Technomic Publication Co., Lancaster, Pennsylvania.
2. Halpin, J.C., 1987, *Primer on Composite Materials: Analysis*, Technomic Publication Co., Lancaster, Pennsylvania.
3. Ashton, J.E., Halpin, J.C. and Petit, P.H., 1969, *Primer on Composite Materials: Analysis*, Technomic Publishing Company, West Port, Connecticut.
4. Card, M.F. and Jones, R.M., 1966, *Experimental and Theoretical Results for Buckling of Eccentrically Stiffened Cylinders*, NASA TN D-3639.
5. Iyengar, N.G.R., 2007, *Elastic Stability of Structural Elements*, Macmillan India Ltd.
6. Hahn, H.T. and Pagano, N.J., 1975, "Curing Stresses in Composite Laminates", *J. Composite Materials*, 9, p. 91.
7. Hahn, H.T., 1976, "Residual Stresses in Ploymer Matrix Composite Laminates", *J. Composite Materials*, 10, p. 266.
8. Pipes, R.B., Vinson, J.R. and Chou, T.W., 1976, "On the Hygrothermal Response of Laminated Composite Systems", *J. Composite Materials*, 10, p. 129.
9. Ishai, O. and Maza, A., 1974, "The Effect of Environmental Loading History on Longitudinal Strength of Glass Fiber Reinforced Plastics", *Rheological Acta*, 13, p. 381.
10. Zerai, I.G., Daniel, I.M. and Gotro, J.T., 1987, "Residual Stresses and Wrapage in Woven-Glass Epoxy Laminates", *Experimental Mechanics*, 27, p. 44.

7. Analysis for Strength of the Laminate

In this chapter, we will discuss:

- The failure of the laminate.
- Its ultimate load carrying capacity assuming the material behavior to be elastic.
- Small deformation theory.

7.1 Introduction

In Chapter 5, we discussed the failure of the lamina. In the case of a laminate, which in general consists of plies with different fiber orientation, the failure of a lamina does not result in the failure of the laminate. There will usually be a sequence of lamina failures at different loads resulting in the ultimate failure of the laminate when all the plies fail. Thus, the ultimate load carrying ability of the laminate will generally be higher than the first ply failure. In view of this, the prediction of the laminate failure load as the first ply failure load will be too conservative. This will result in higher structural weight. In situations where the weight is a penalty, one needs to consider the ultimate failure load of the laminate. As stated in earlier chapters, the laminate will fail under increasing mechanical, thermal and hygroscopic loads. A failed lamina still contributes to the stiffness and strength of the laminate. A laminate may carry stresses in all possible directions. However, a lamina can take stresses in its plane only and is assumed to be in a state of plane stress. One can think of replacing the stiffness and strength of the failed lamina by any one of the possibilities:

(i) When a ply or lamina fails, it may generate cracks parallel to the fiber direction. These cracks are due to the failure of the matrix. The lamina is still capable of taking loads along the direction of the fibers. The lamina can now be thought of being replaced by a fictitious ply with no transverse stiffness, no transverse strength. However, the longitudinal stiffness and strength remain unchanged.

(ii) The second possibility is to completely eliminate the stiffness and strength of the failed lamina. The lamina, though, contributes to the overall thickness of the laminate.

(iii) A third possibility is to reduce the strength and stiffness of the failed lamina in proportion to the load carried by the lamina.

The analysis is generally carried out till all the lamina fail. The sum of all the incremental loads gives the total load carrying capacity of the laminate. Chou [1, 2] gives details of experiments carried out to show the influence on lamina properties after the failure of the lamina in one direction.

7.2 Load-deformation Curve for the Laminate

Figure 7.1 shows the piece-wise linear load-deformation curve for a laminate subjected to a uniaxial loading. It is assumed that the laminate consists of a number of plies with different orientation. The "kinks" or "knees" in the curve indicate the failure of a lamina. The "kinks" or "knees" may not appear explicitly if the fiber orientations of the laminae are close to each other. In many situations, it is possible that the complete failure of the laminate may result in a large deformation, which may be beyond the scope of a linear theory. At each loading leading to the failure of the lamina, the new stiffness of the laminate has to be calculated to arrive at the changed state of stress in the lamina for that load; it has to be then checked whether any other lamina has failed at that loading. If it has failed, further readjustment of the lamina stresses needs to be calculated. Depending on lamina sequencing, it is possible that some of the lamina may fail after readjustment of stiffness due to the initial failure of a lamina. It is to be pointed out that first ply failure does not necessarily mean the failure of the first lamina. First ply failure is defined as the failure of the first lamina irrespective of its location in the laminate.

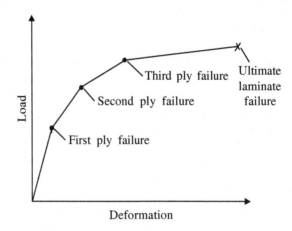

Fig. 7.1 Load-deformation characteristics of a uniaxially loaded lamina.

A similar statement is valid for the second ply failure, third ply failure and subsequent ply failures.

7.3 Stress/Load Analysis at First Ply Failure

Employing one of the failure theories discussed in Chapter 5, it is possible to find out whether for a given load, any of the plies of the laminate has failed. Since the strength of a ply is a function of fiber orientation, it is expected that all the plies will not fail at the same load. The first ply failure (FPF) generally occurs at a very small load, and there is no possibility of complete failure of the laminate except in very special cases. Once any of the plies fail, the procedure described earlier to estimate the in-plane loads and moments and hence the stresses and strains needs to be modified. The calculation of modified laminate stiffness requires the knowledge of modified ply stiffness coefficients, and hence, the type of ply failure. Halpin [3] analyzed the response of glass/epoxy laminate using a similar procedure. The predicted response curve shows good agreement with the experimental data. The kinks observed in the theoretical curve are not seen in the experimental work. This may be due to the fact that in theoretical work, ply failure is assumed to occur at a single strain level. A similar observation is made by Hahn and Tsai [4]. As shown in Fig. 7.1, the load-deformation characteristics are piece-wise linear in view of the assumption of linear behavior up to failure. The laminate constitutive equations for each increment of loading beyond the first ply failure can be written as

$$\left\{ \begin{matrix} \Delta N \\ \Delta M \end{matrix} \right\}_i = \begin{bmatrix} \bar{A} & \bar{B} \\ \bar{B} & \bar{D} \end{bmatrix}_i \left\{ \begin{matrix} \Delta \varepsilon^0 \\ \Delta \kappa \end{matrix} \right\}_i \tag{7.1}$$

where $[\bar{A}]$, $[\bar{B}]$ and $[\bar{D}]$ are the modified laminate stiffness matrices after incorporating the changes in the $[\bar{Q}]$ matrix of individual lamina due to the failure of the lamina or set of lamina. The changes are incorporated as indicated in Section 7.1. Once a lamina or a set of lamina fails, the loading has to be incremental. Further, it is difficult to know the exact value of the load at which the ply failure has taken place. This is achieved by halving the incremental loading at each stage, calculating the stresses and then applying the failure criterion. The total load, when the next failure occurs can be obtained as

$$\left\{ \begin{matrix} N_x \\ N_y \\ N_{xy} \end{matrix} \right\}_i = \left\{ \begin{matrix} N_x \\ N_y \\ N_{xy} \end{matrix} \right\}_{i-1} + \left\{ \begin{matrix} \Delta N_x \\ \Delta N_y \\ \Delta N_{xy} \end{matrix} \right\}_i \tag{7.2}$$

$$\begin{Bmatrix} M_x \\ M_y \\ M_{xy} \end{Bmatrix}_i = \begin{Bmatrix} M_x \\ M_y \\ M_{xy} \end{Bmatrix}_{i-1} + \begin{Bmatrix} \Delta M_x \\ \Delta M_y \\ \Delta M_{xy} \end{Bmatrix}_i \tag{7.3}$$

The total mid-plane strains and curvatures at any given stage of loading are obtained as

$$\begin{Bmatrix} \varepsilon_x^0 \\ \varepsilon_y^0 \\ \gamma_{xy}^0 \end{Bmatrix}_i = \begin{Bmatrix} \varepsilon_x^0 \\ \varepsilon_y^0 \\ \gamma_{xy}^0 \end{Bmatrix}_{i-1} + \begin{Bmatrix} \Delta\varepsilon_x^0 \\ \Delta\varepsilon_y^0 \\ \Delta\gamma_{xy}^0 \end{Bmatrix}_i \tag{7.4}$$

$$\begin{Bmatrix} \kappa_x \\ \kappa_y \\ \kappa_{xy} \end{Bmatrix}_i = \begin{Bmatrix} \kappa_x \\ \kappa_y \\ \kappa_{xy} \end{Bmatrix}_{i-1} + \begin{Bmatrix} \Delta\kappa_x \\ \Delta\kappa_y \\ \Delta\kappa_{xy} \end{Bmatrix}_i \tag{7.5}$$

Knowing the mid-plane strains and curvatures for the laminate, the stresses in individual lamina can be obtained by making use of Eq. (6.12). The strains in the lamina can be obtained by making use of the constitutive equations. These give stresses and strains in the X-Y coordinate system. The stresses and strains have to be transformed to the L-T coordinate system for applying the failure criterion. The procedures described above have to be applied at each stage of the loading till the overall failure of the laminate occurs. For further details, the reader can refer to Agarwal et al. [5] who discuss this procedure in detail for a cross-ply laminate, for which these equations can be simplified. In what follows, we shall analyze some laminates for ultimate failure subjected to in-plane loads.

Example 7.1
Find the failure load for a $[0/9\bar{0}]_s$ graphite/epoxy laminate. The plies are of equal thickness of 5 mm each. $0°$ plies are along the loading direction. The laminate is subjected to uniaxial tensile load.

Solution
The laminate is symmetric and only normal load is applied. We need to compute only the extensional stiffness matrix $[A]$. The transformed stiffness matrix for $0°$ and $90°$ lamina is given as

$$[\bar{Q}]_{0°} = \begin{bmatrix} 181.8 & 2.897 & 0 \\ 2.897 & 10.35 & 0 \\ 0 & 0 & 7.17 \end{bmatrix} \text{GPa}, [\bar{Q}]_{90°} = \begin{bmatrix} 10.35 & 2.897 & 0 \\ 2.897 & 181.8 & 0 \\ 0 & 0 & 7.17 \end{bmatrix} \text{GPa}$$

For the given thickness of plies

$$t_0 = -7.5 \text{ mm}, \ t_1 = -2.5 \text{ mm}, \ t_2 = 0.25 \text{ mm}, \ t_3 = 0.75 \text{ mm}$$

The extensional stiffness matrix for the laminate is given as

$$A_{ij} = \sum_{k=1}^{3} [\bar{Q}_{ij}]_k (t_k - t_{k-1})$$

$$[A] = \begin{bmatrix} 181.8 & 2.897 & 0 \\ 2.897 & 10.35 & 0 \\ 0 & 0 & 7.17 \end{bmatrix} \times 0.005 \text{ GPa}$$

$$+ \begin{bmatrix} 10.35 & 2.897 & 0 \\ 2.897 & 181.8 & 0 \\ 0 & 0 & 7.17 \end{bmatrix} \times 0.005 \text{ GPa}$$

$$+ \begin{bmatrix} 181.8 & 2.897 & 0 \\ 2.897 & 10.35 & 0 \\ 0 & 0 & 7.17 \end{bmatrix} \times 0.005 \text{ GPa}$$

$$= \begin{bmatrix} 1.870 & 0.043 & 0 \\ 0.043 & 1.012 & 0 \\ 0 & 0 & 1.075 \end{bmatrix} \text{ GPa-m}$$

$$\begin{Bmatrix} N_x \\ N_y \\ N_{xy} \end{Bmatrix} = \begin{bmatrix} A_{11} & A_{12} & A_{16} \\ A_{12} & A_{22} & A_{26} \\ A_{16} & A_{26} & A_{66} \end{bmatrix} \begin{bmatrix} \varepsilon_x^0 \\ \varepsilon_y^0 \\ \gamma_{xy}^0 \end{bmatrix}$$

Substituting the values of A_{ij} and the in-plane loads, we get

$$\begin{Bmatrix} N_x \\ 0 \\ 0 \end{Bmatrix} = \begin{bmatrix} 1.870 & 0.043 & 0 \\ 0.043 & 1.012 & 0 \\ 0 & 0 & 1.075 \end{bmatrix} \begin{Bmatrix} \varepsilon_x^0 \\ \varepsilon_y^0 \\ \gamma_{xy}^0 \end{Bmatrix} \text{ GPa-m}$$

Solving the above equation, we get

$$\begin{Bmatrix} \varepsilon_x^0 \\ \varepsilon_y^0 \\ \gamma_{xy}^0 \end{Bmatrix} = \begin{Bmatrix} 0.535 \ N_x \\ -0.0227 \ N_x \\ 0 \end{Bmatrix} \times 10^{-9}$$

The in-plane strains for the 0° are given by (since curvature terms are zero)

$$\begin{Bmatrix} \varepsilon_x \\ \varepsilon_y \\ \gamma_{xy} \end{Bmatrix}_{0°} = \begin{Bmatrix} 0.535\,N_x \\ -0.0227\,N_x \\ 0 \end{Bmatrix} \times 10^{-9}$$

The ply stresses are given by

$$\begin{Bmatrix} \sigma_x \\ \sigma_y \\ \tau_{xy} \end{Bmatrix}_{0°} = \begin{bmatrix} 181.8 & 2.897 & 0 \\ 2.897 & 10.35 & 0 \\ 0 & 0 & 7.17 \end{bmatrix} \times 10^9 \begin{Bmatrix} 0.535\,N_x \\ -0.0227\,N_x \\ 0 \end{Bmatrix} \times 10^{-9}$$

$$= \begin{Bmatrix} 97.197\,N_x \\ 1.315\,N_x \\ 0 \end{Bmatrix} Pa$$

The stresses in the L-T coordinate system for the lamina are given as

$$\begin{Bmatrix} \sigma_L \\ \sigma_T \\ \tau_{LT} \end{Bmatrix}_{0°} = \begin{bmatrix} 1 & 0 & 0 \\ 0 & 1 & 0 \\ 0 & 0 & 1 \end{bmatrix} \begin{Bmatrix} 97.917\,N_x \\ 1.315\,N_x \\ 0 \end{Bmatrix} = \begin{Bmatrix} 97.917\,N_x \\ 1.315\,N_x \\ 0 \end{Bmatrix} Pa$$

The strains for the 90° lamina are given by

$$\begin{Bmatrix} \sigma_x \\ \sigma_y \\ \tau_{xy} \end{Bmatrix}_{90°} = \begin{bmatrix} 10.35 & 2.897 & 0 \\ 2.897 & 181.8 & 0 \\ 0 & 0 & 7.17 \end{bmatrix} \times 10^9 \begin{Bmatrix} 0.535\,N_x \\ -0.0227\,N_x \\ 0 \end{Bmatrix} \times 10^{-9}$$

$$= \begin{Bmatrix} 5.471\,N_x \\ -2.577\,N_x \\ 0 \end{Bmatrix} Pa$$

The stresses in the L-T coordinate system for the 90° lamina are given by

$$\begin{Bmatrix} \sigma_L \\ \sigma_T \\ \tau_{LT} \end{Bmatrix}_{90°} = \begin{bmatrix} 0 & 1 & 0 \\ 1 & 0 & 0 \\ 0 & 0 & -1 \end{bmatrix} \begin{Bmatrix} 5.471\,N_x \\ -2.577\,N_x \\ 0 \end{Bmatrix}$$

$$= \begin{Bmatrix} -2.577\,N_x \\ 5.471\,N_x \\ 0 \end{Bmatrix} Pa$$

The failure criterion is then to be applied to determine the maximum values of N_x that can be sustained without failure of any of the layers. The failure criterion is to be applied to each layer separately. For the cross-ply laminate, the Tsai-Hill failure criterion for each layer can be expressed as

$$\left(\frac{\sigma_L}{\sigma_{LU}}\right)^2 - \frac{\sigma_L \sigma_T}{\sigma_{LU}^2} + \left(\frac{\sigma_T}{\sigma_{TU}}\right)^2 + \left(\frac{\tau_{LT}}{\tau_{LTU}}\right)^2 = 1$$

For the 0° and 90° layers, τ_{LT} is zero, therefore, the criterion simplifies to

$$\left(\frac{\sigma_L}{\sigma_{LU}}\right)^2 - \frac{\sigma_L \sigma_T}{\sigma_{LU}^2} + \left(\frac{\sigma_T}{\sigma_{TU}}\right)^2 = 1$$

Substituting the values for the 0° lamina, we get

$$\left(\frac{97.917N_x}{1500}\right)^2 - \left(\frac{97.917N_x}{1500}\right)\left(\frac{1.315N_x}{1500}\right) + \left(\frac{1.315N_x}{40}\right)^2 = 1$$

$$N_x^2[4.26 - 0.0572 + 1.08] \times 10^{-3} = 1$$

$$N_x = 13.758 \text{ MN/m}$$

The material properties for the graphite/epoxy unidirectional lamina are

$$E_L = 181 \text{ GPa}, \ E_T = 10.3 \text{ GPa}, \ G_{LT} = 7.17 \text{ GPa and } \nu_{LT} = 0.28$$

The strength properties for the unidirectional lamina are

$$\sigma_{LU} = \sigma'_{LU} = 1500 \text{ MPa}, \ \sigma_{TU} = 40 \text{ MPa}, \ \sigma'_{TU} = 246 \text{ MPa and } \tau_{LTU} = 68 \text{ MPa}$$

Assuming that the material behaves in a linearly elastic manner up to failure, the failure strains are

$$\varepsilon_{LU} = 0.00829, \ \varepsilon_{TU} = 0.00388$$

Substituting the values for the 90° lamina, we get for failure

$$\left(\frac{2.577N_x}{1500}\right)^2 - \left(\frac{2.577N_x}{1500}\right)\left(\frac{5.471N_x}{1500}\right) + \left(\frac{5.471N_x}{40}\right)^2 = 1$$

$$N_x = 7.311 \text{ MN/m}$$

Hence, the 90° lamina will fail first, when the applied value of $N_x = 7.311$ MN/m. For this load, the strain in the 0° lamina is

$$\varepsilon_L = 0.535 \times 10^{-9} \times 7.311 \times 10^6 = 0.00391$$

Response After 90 Ply Fails

As stated earlier, the failure of a ply may not mean the failure of the laminate. The remaining plies of the laminate can still take the load. In the present case, the inner ply at 90° to the *x*-axis has failed. The failure results in the formation of cracks in the matrix which are normal to the loading direction or series of cracks *parallel* to the fibers. Therefore, the load can still be carried by the inner layer in the fiber direction.

Conservative Analysis

One can obtain the conservative estimate of the total load carried by the laminate, by neglecting the stiffnesses of the degraded laminate/laminates. In the present case, all the transformed stiffness coefficients of the inner lamina are taken to be zero. The modified extensional stiffness matrix is calculated by making all elements of $[\bar{Q}]_{90}$ equal to zero.

$$[\bar{A}] = 2 \begin{bmatrix} 181.8 & 2.897 & 0 \\ 2.897 & 10.35 & 0 \\ 0 & 0 & 7.17 \end{bmatrix} \times 0.005 \text{ GPa-m}$$

$$\begin{Bmatrix} \Delta N_x \\ 0 \\ 0 \end{Bmatrix} = \begin{bmatrix} 1.818 & 0.029 & 0 \\ 0.029 & 0.103 & 0 \\ 0 & 0 & 0.072 \end{bmatrix} \begin{Bmatrix} \Delta\varepsilon_x^0 \\ \Delta\varepsilon_y^0 \\ \Delta\gamma_{xy}^0 \end{Bmatrix}$$

Solving for the strains, we get

$$\begin{aligned} \Delta N_x &= 1.809 \ \varepsilon_x^0 = 1.809 \times (0.00829 - 0.00391) \\ &= 0.00792 \text{ GPa} \\ &= 7.92 \text{ MPa} \end{aligned}$$

Therefore, the total axial load is given by

$$\bar{N}_x = N_x + \Delta N_x = 7.311 + 7.92 = 15.231 \text{ MPa}$$

Example 7.2

A $[90°/0°/90°]_s$ laminate consisting of 0.25 mm thickness of graphite/epoxy is subjected to a tensile uniaxial loading along the X direction. The lamina mechanical properties are E_L = 138 GPa, E_T = 9 GPa, G_{LT} = 6.9 GPa and v_{LT} = 0.3.

The strength properties are σ_{LU} = 1448 MPa, σ_{TU} = 48.3 MPa, σ'_{LU} = 1172 MPa, σ'_{TU} = 248 MPa and τ_{LTU} = 62.1 MPa.

Using the maximum strain criterion, find the failure loads corresponding to first ply failure and subsequent ply failures. Find the ultimate load carrying capacity of the laminate. This example is also discussed by Gibson [6].

Solution

For the given lamina, assuming a linear elastic behavior, the failure strains are

$$\varepsilon_{LU} = \frac{\sigma_{LU}}{E_{LU}} = \frac{1448}{138 \times 10^3} = 0.0105$$

$$\varepsilon_{TU} = \frac{\sigma_{TU}}{E_{TU}} = \frac{48.3}{9 \times 10^3} = 0.0054$$

As the axial load is gradually increased, the strain along the X direction ε_x increases and reaches the failure strain $\varepsilon_x = \varepsilon_{TU} = 0.0054$ for the ply. At this strain, the first ply failure takes place.

For the given material properties

$$[\bar{Q}]_{0°} = \begin{bmatrix} 138.8 & 2.72 & 0 \\ 2.72 & 9.05 & 0 \\ 0 & 0 & 6.9 \end{bmatrix} \text{GPa}, \quad [\bar{Q}]_{90°} = \begin{bmatrix} 9.05 & 2.72 & 0 \\ 2.72 & 138.8 & 0 \\ 0 & 0 & 6.9 \end{bmatrix} \text{GPa}$$

The laminate consists of 2 plies of 0° and 4 plies of 0.25 mm each; [A] matrix for the undamaged laminate is given as

$$[A] = \begin{bmatrix} 78.45 & 4.08 & 0 \\ 4.05 & 143.3 & 0 \\ 0 & 0 & 10.35 \end{bmatrix} \text{GPa-mm}$$

At the first ply failure, the axial load-deformation equations can be written as

$$\begin{Bmatrix} N_x \\ 0 \\ 0 \end{Bmatrix} = \begin{bmatrix} 78.45 & 4.08 & 0 \\ 4.05 & 143.3 & 0 \\ 0 & 0 & 10.35 \end{bmatrix} \begin{Bmatrix} 0.0054 \\ \varepsilon_y \\ \gamma_{xy} \end{Bmatrix}$$

Solving, we get

$$N_x = 0.423 \text{ GPa-mm}; \quad \varepsilon_y = -0.000154; \quad \gamma_{xy} = 0$$

The 90° lamina, which has no fibers along the x-direction is weak in that direction and hence it fails when the load reaches $N_x = 0.423$ GPa-mm. The failure is observed as cracks parallel to the fiber in the 90° plies.

Conservative Approach

In this approach, all ply stiffnesses of 90° plies are set equal to zero after the first ply failure. Thus, for further loading, only 0° plies contribute. The modified [$\bar{A}$] matrix is given by

$$[\bar{A}] = 0.5[\bar{Q}]_{0°} = \begin{bmatrix} 69.4 & 1.36 & 0 \\ 1.36 & 4.52 & 0 \\ 0 & 0 & 3.45 \end{bmatrix} \text{GPa-mm}$$

The 0° ply already has a strain equal to 0.0045 along the X-direction. The failure strain for the 0° ply is 0.0105. This implies that the ply can still take an additional amount of strain before it fails. The additional strain is given by

$$\Delta \varepsilon_x = 0.0105 - 0.0054 = 0.0051$$

The load deformation for the additional strain is given as

$$\begin{Bmatrix} N_x \\ 0 \\ 0 \end{Bmatrix} = \begin{bmatrix} 69.4 & 1.36 & 0 \\ 1.36 & 4.52 & 0 \\ 0 & 0 & 3.45 \end{bmatrix} \begin{Bmatrix} 0.0051 \\ \Delta\varepsilon_y \\ \Delta\gamma_{xy} \end{Bmatrix}$$

The solution of this equation leads to

$$\Delta N_x = 0.352 \text{ GPa-mm}, \quad \Delta\varepsilon_y = -0.00153, \quad \Delta\gamma_{xy} = 0$$

The total laminate failure is then

$$\bar{N}_x = N_x + \Delta N_x = 0.423 + 0.352 = 0.775 \text{ GPa-mm}$$

Modified Approach

Here it is assumed that not all stiffness coefficients of 90° are zero. Only coefficients associated with the following material constants, namely E_T, G_{LT} and v_{LT} are set equal to zero. The only non-zero material property is E_L corresponding to the 90° lamina. Thus, for the 90° lamina, for the first ply failure

$$[\bar{Q}_{22}] = 138 \text{ GPa}$$

and all other stiffness coefficients are zero.

The modified $[\bar{A}]$ for the laminate is

$$[\bar{A}] = \begin{bmatrix} 69.4 & 1.36 & 0 \\ 1.36 & 143.3 & 0 \\ 0 & 0 & 3.45 \end{bmatrix} \text{GPa-mm}$$

The modified load deformation equation is written as

$$\begin{Bmatrix} \Delta N_x \\ 0 \\ 0 \end{Bmatrix} = \begin{bmatrix} 69.4 & 1.36 & 0 \\ 1.36 & 143.3 & 0 \\ 0 & 0 & 3.45 \end{bmatrix} \begin{Bmatrix} 0.0051 \\ \Delta\varepsilon_y \\ \Delta\gamma_{xy} \end{Bmatrix}$$

The solution of this equation leads to

$$\Delta N_x = 0.354 \text{ GPa-mm}, \quad \Delta\varepsilon_y = -0.000048, \quad \Delta\gamma_{xy} = 0$$

The total load at laminate failure is

$$\bar{N}_x = 0.423 + 0.354 = 0.777 \text{ GPa-mm}$$

It is observed that the total load predicted by the two approaches is almost the same. This is just a coincidence. However, in general, the load prediction depends on ply material properties and stacking sequence.

Example 7.3

A symmetric laminate $[0°/45°/90°]_s$ is subjected to a uniaxial tensile load N_x. Each ply has a thickness of 0.125 mm. Obtain the total load of the laminate for ultimate failure. The properties of the lamina material are

E_L = 140 kN/m², E_T = 10 kN/m², G_{LT} = 5 kN/m², v_{LT} = 0.3
σ_{LU} = 1500 MN/m², σ_{TU} = 50 MN/m², σ'_{LU} = 1200 MN/m²,
σ'_{TU} = 250 MN/m², τ_{LTU} = 70 MN/m²

Solution

$[\bar{Q}]$ matrix for each of the lamina is obtained as

$$[\bar{Q}]_{0°} = \begin{bmatrix} 140.9 & 3 & 0 \\ 3 & 10.1 & 0 \\ 0 & 0 & 5 \end{bmatrix} \text{kN/m}^2$$

$$[\bar{Q}]_{45°} = \begin{bmatrix} 44.3 & 34.3 & 32.7 \\ 34.3 & 44.3 & 32.7 \\ 32.7 & 32.7 & 36.3 \end{bmatrix} \text{kN/m}^2$$

$$[\bar{Q}]_{90°} = \begin{bmatrix} 10.1 & 3 & 0 \\ 3 & 140.9 & 0 \\ 0 & 0 & 5 \end{bmatrix} \text{kN/m}^2$$

Making use of the $[\bar{Q}]$ matrix and the thickness of each of the lamina, we obtain the extensional stiffness matrix for the laminate as

$$[A] = \begin{bmatrix} 48.83 & 10.08 & 8.20 \\ 10.08 & 48.83 & 8.20 \\ 8.20 & 8.20 & 11.58 \end{bmatrix} \text{kN/mm}$$

Therefore

$$\begin{Bmatrix} N_x \\ 0 \\ 0 \end{Bmatrix} = \begin{bmatrix} 48.83 & 10.08 & 8.20 \\ 10.08 & 48.83 & 8.20 \\ 8.20 & 8.20 & 11.58 \end{bmatrix} \begin{Bmatrix} \varepsilon_x^0 \\ \varepsilon_y^0 \\ \gamma_{xy}^0 \end{Bmatrix}$$

Since there is only one non-zero load acting, one can solve the simultaneous equation to obtain the strains in terms of N_x. The in-plane strains are

$$\begin{Bmatrix} \varepsilon_x^0 \\ \varepsilon_y^0 \\ \gamma_{xy}^0 \end{Bmatrix} = \begin{Bmatrix} 23.474 \times 10^{-3} \\ -2.323 \times 10^{-3} \\ -15.974 \times 10^{-3} \end{Bmatrix} N_x$$

The stresses in the individual lamina along the material axes are obtained by employing the corresponding constitutive relations.

(i) For top and bottom plies oriented at 0°: Stresses in the *L-T* directions are

$$\begin{Bmatrix} \sigma_L \\ \sigma_T \\ \tau_{LT} \end{Bmatrix}_{0°} = \begin{bmatrix} 140.9 & 3 & 0 \\ 3 & 10.1 & 0 \\ 0 & 0 & 5 \end{bmatrix} \begin{Bmatrix} 23.474 \times 10^{-3} \\ -2.323 \times 10^{-3} \\ -15.974 \times 10^{-3} \end{Bmatrix} N_x \text{ N/mm}^2$$

$$= \begin{Bmatrix} 3.3005 \\ 0.0469 \\ -0.0799 \end{Bmatrix} \text{N/mm}^2$$

(ii) For plies in the middle of the laminate at 90°: The in-plane strains in the lamina are given as

$$\begin{Bmatrix} \varepsilon_x^0 \\ \varepsilon_y^0 \\ \gamma_{xy}^0 \end{Bmatrix} = \begin{Bmatrix} 23.474 \times 10^{-3} \\ -2.323 \times 10^{-3} \\ -15.974 \times 10^{-3} \end{Bmatrix} N_x$$

The stresses in the X-Y directions are

$$\begin{Bmatrix} \sigma_x \\ \sigma_y \\ \tau_{xy} \end{Bmatrix}_{90°} = \begin{bmatrix} 10.1 & 3 & 0 \\ 3 & 140.9 & 0 \\ 0 & 0 & 5 \end{bmatrix} \begin{Bmatrix} 23.474 \times 10^{-3} \\ -2.323 \times 10^{-3} \\ -15.974 \times 10^{-3} \end{Bmatrix} N_x$$

$$= \begin{Bmatrix} 0.230 \\ -0.257 \\ -0.079 \end{Bmatrix} N_x \text{ N/mm}^2$$

The stresses in the material directions are

$$\begin{Bmatrix} \sigma_L \\ \sigma_T \\ \tau_{LT} \end{Bmatrix}_{90°} = \begin{bmatrix} 0 & 1 & 0 \\ 1 & 0 & 0 \\ 0 & 0 & 1 \end{bmatrix} \begin{Bmatrix} 0.230 \\ -0.257 \\ -0.079 \end{Bmatrix} N_x = \begin{Bmatrix} -0.257 \\ 0.230 \\ -0.079 \end{Bmatrix} N_x \text{ N/mm}^2$$

(iii) The stresses in the 45° lamina are obtained as

$$\begin{Bmatrix} \sigma_x \\ \sigma_y \\ \tau_{xy} \end{Bmatrix}_{45°} = \begin{bmatrix} 44.3 & 34.3 & 32.7 \\ 34.3 & 44.3 & 32.7 \\ 32.7 & 32.7 & 36.3 \end{bmatrix} \begin{Bmatrix} 23.474 \times 10^{-3} \\ -2.323 \times 10^{-3} \\ -15.974 \times 10^{-3} \end{Bmatrix} N_x$$

$$= \begin{Bmatrix} 0.438 \\ 0.180 \\ 0.112 \end{Bmatrix} N_x$$

$$\left\{\begin{array}{c} \sigma_L \\ \sigma_T \\ \tau_{LT} \end{array}\right\}_{45°} = \left[\begin{array}{ccc} 0.5 & 0.5 & 1.0 \\ 0.5 & 0.5 & -1.0 \\ -0.5 & 0.5 & 0 \end{array}\right]\left\{\begin{array}{c} 0.438 \\ 0.180 \\ 0.112 \end{array}\right\}N_x$$

$$= \left\{\begin{array}{c} 0.421 \\ 0.197 \\ -0.129 \end{array}\right\}N_x$$

In order to cause failure of the 0° lamina according to the maximum stress theory, the possible N_x values are:

On the basis of σ_L

$$N_x = \frac{1500}{3.3005} = 454.476 \text{ N/mm}$$

On the basis of σ_T

$$N_x = \frac{50}{0.0469} = 1066.098 \text{ N/mm}$$

On the basis of τ_{LT}

$$N_x = \frac{70}{0.0799} = 876.095 \text{ N/mm}$$

In order to cause failure of the 90° lamina according to the maximum stress theory, the possible N_x values are:

On the basis of σ_L

$$N_x = \frac{1200}{0.257} = 4669.26 \text{ N/mm}$$

On the basis of σ_T

$$N_x = \frac{50}{0.23} = 217.391 \text{ N/mm}$$

On the basis of τ_{LT}

$$N_x = \frac{70}{0.079} = 886.076 \text{ N/mm}$$

In order to cause failure of the 45° lamina according to the maximum stress theory, the possible N_x values are:

On the basis of σ_L

$$N_x = \frac{1500}{0.421} = 3562.945 \text{ N/mm}$$

On the basis of σ_T

$$N_x = \frac{50}{0.192} = 260.416 \text{ N/mm}$$

On the basis of τ_{LT}

$$N_x = \frac{70}{0.129} = 542.636 \text{ N/mm}$$

On comparing the various values, it is observed that the minimum value of N_x that will cause first ply failure is 271.391 N/mm. The first ply failure takes place in the plies in which the fibers are oriented at an angle of 90° to the X-axis.

Second Ply Failure

Using the conservative approach, we assume that E_L, E_T and G_{LT} of the two 90° plies are set to zero. Though the lamina exists physically, it does not contribute to the overall stiffness of the laminate. We recalculate the extensional stiffness matrix $[A]$.

Therefore

$$[\bar{A}] = \begin{bmatrix} 46.3 & 9.316 & 8.175 \\ 9.316 & 13.60 & 8.175 \\ 8.175 & 8.175 & 10.325 \end{bmatrix} \text{kN/mm}$$

Before going further, it is worthwhile to check whether the laminate with reduced stiffness can withstand the load already applied on the laminate when all the plies were intact.

The strain in the laminate is given as

$$\begin{Bmatrix} \varepsilon_x^0 \\ \varepsilon_y^0 \\ \gamma_{xy}^0 \end{Bmatrix} = \begin{bmatrix} 46.3 & 9.316 & 8.175 \\ 9.316 & 13.60 & 8.175 \\ 8.175 & 8.175 & 10.325 \end{bmatrix}^{-1} \begin{Bmatrix} 217.391 \\ 0 \\ 0 \end{Bmatrix} \times 10^{-3}$$

$$= \begin{Bmatrix} 5.660 \\ -2.261 \\ -2.691 \end{Bmatrix} \times 10^{-3}$$

The stresses in the 0° lamina are

$$\begin{Bmatrix} \sigma_L \\ \sigma_T \\ \tau_{LT} \end{Bmatrix}_{0°} = \begin{bmatrix} 140.9 & 3 & 0 \\ 3 & 10.1 & 0 \\ 0 & 0 & 5 \end{bmatrix} \begin{Bmatrix} 5.660 \\ -2.261 \\ -2.691 \end{Bmatrix} \times 10^{-3} \times 10^3 \text{ N/mm}^2$$

$$= \begin{Bmatrix} 790.852 \\ -5.856 \\ -13.455 \end{Bmatrix} \text{N/mm}^2$$

For causing failure

- σ_L has to be increased by a factor $1500/790.852 = 1.896$
- σ_T has to be increased by a factor $250/5.856 = 42.691$
- τ_{LT} has to be increased by a factor $70/13.455 = 5.202$

Thus, the laminate can withstand the load even after the ply fails.

Plies at 45
The strains in the material direction of the plies are

$$\begin{Bmatrix} \varepsilon_L \\ \varepsilon_T \\ \gamma_{LT} \end{Bmatrix} = \begin{bmatrix} 0.5 & 0.5 & 1.0 \\ 0.5 & 0.5 & -1.0 \\ -0.5 & 0.5 & 0 \end{bmatrix} \begin{Bmatrix} 5.660 \\ -2.261 \\ -2.691 \end{Bmatrix} \times 10^{-3}$$

$$= \begin{Bmatrix} -.992 \\ 4.391 \\ -3.413 \end{Bmatrix} \times 10^{-3}$$

The stresses in the material directions are

$$\begin{Bmatrix} \sigma_L \\ \sigma_T \\ \tau_{LT} \end{Bmatrix} = \begin{bmatrix} 140.9 & 3 & 0 \\ 3 & 10.1 & 0 \\ 0 & 0 & 5 \end{bmatrix} \begin{Bmatrix} -.992 \\ 4.391 \\ -3.413 \end{Bmatrix} 10^3 \times 10^{-3} \text{ N/mm}^2$$

$$= \begin{Bmatrix} -126.66 \\ 41.371 \\ -17.065 \end{Bmatrix} \text{N/mm}^2$$

The factor by which σ_L has to be increased for failure is $1200/126.66 = 9.474$.
The factor by which σ_T has to be increased for failure is $50/41.371 = 1.208$.
The factor by which τ_{LT} has to be increased for failure is $70/17.065 = 4.101$.
    The minimum value of the different factors to cause failure is 1.208. This occurs in the 45° lamina. Therefore, the load to cause second ply failure is $1.208 \times 217 = 262.136$ N/mm. The failure occurs in the lamina in the transverse mode.

Third Ply Failure
Only two plies are left out for failure of the laminate. The stiffness of all other lamina is zero except these two.

Therefore

$$[A] = \begin{bmatrix} 35.23 & 0.75 & 0 \\ 0.75 & 2.53 & 0 \\ 0 & 0 & 1.25 \end{bmatrix} \text{kN/mm}$$

The mid-plane strains are obtained as

$$\begin{Bmatrix} 262.136 \\ 0 \\ 0 \end{Bmatrix} = \begin{bmatrix} 35.23 & 0.75 & 0 \\ 0.75 & 2.53 & 0 \\ 0 & 0 & 1.25 \end{bmatrix} \begin{Bmatrix} \varepsilon_x^0 \\ \varepsilon_y^0 \\ \gamma_{xy}^0 \end{Bmatrix} \times 10^{-3}$$

Solving, we get

$$\begin{Bmatrix} \varepsilon_x^0 \\ \varepsilon_y^0 \\ \gamma_{xy}^0 \end{Bmatrix} = \begin{Bmatrix} 7.487 \\ -2.216 \\ 0 \end{Bmatrix} \times 10^{-3}$$

The stresses in the material directions for this lamina are

$$\begin{Bmatrix} \sigma_L \\ \sigma_T \\ \tau_{LT} \end{Bmatrix} = \begin{bmatrix} 140.9 & 3 & 0 \\ 3 & 10.1 & 0 \\ 0 & 0 & 5 \end{bmatrix} \begin{Bmatrix} 7.487 \\ -2.216 \\ 0 \end{Bmatrix} \times 10^{-3} \times 10^3 \text{ N/mm}^2$$

$$= \begin{Bmatrix} 1048.2 \\ 0.072 \\ 0 \end{Bmatrix} \text{N/mm}^2$$

The factor by which σ_L has to be increased for failure is 1500/1048.2 = 1.431.

Therefore, applying the maximum stress theory, the ultimate failure load for the laminate is 262.136 × 1.431 = 375.11 N/mm.

Example 7.4

A four-ply [0/90]$_s$ symmetric graphite/epoxy laminate of uniform ply thickness of 0.25 cm is subjected to tensile loading along the X-direction. The thermal effects are considered with $\Delta T = -100°C$. Using the maximum strain criterion, find the loads corresponding to first ply failure and subsequent ply failures.

Solution

For this material, the material and strength properties are

$$E_L = 138 \text{ GPa}, \ E_T = 9.00 \text{ GPa}, \ G_{LT} = 6.9 \text{ GPa}, \ \nu_{LT} = 0.3$$
$$\sigma_{LU} = 1448 \text{ MPa}, \ \sigma'_{LU} = 1172 \text{ MPa}, \ \sigma_{TU} = 48.3 \text{ MPa}$$

$$\sigma'_{TU} = 248 \text{ MPa}, \ \tau_{LTU} = 62.1 \text{ MPa}$$
$$\alpha_L = 7.0 \times 10^{-6}/°C, \ \alpha_T = 23 \times 10^{-6}/°C$$

Assuming the stress-strain relation to be linear, the ultimate strains are

$$\varepsilon_{LU} = \frac{1448}{138 \times 10^3} = 0.0105, \ \varepsilon_{TU} = \frac{48.3}{9 \times 10^3} = 0.0054$$

$$\varepsilon'_{LU} = \frac{1172}{138 \times 10^3} = 0.0085, \ \varepsilon'_{TU} = \frac{248}{9 \times 10^3} = 0.0275$$

The elements of the stiffness matrix for 0° and 90° lamina are

$$[Q]_{0°} = \begin{bmatrix} 140.76 & 2.754 & 0 \\ 2.754 & 9.18 & 0 \\ 0 & 0 & 6.9 \end{bmatrix} \text{GPa}, \ [Q]_{90°} = \begin{bmatrix} 9.18 & 2.754 & 0 \\ 2.754 & 140.76 & 0 \\ 0 & 0 & 6.9 \end{bmatrix} \text{GPa}$$

$$\begin{Bmatrix} \alpha_x \\ \alpha_y \\ \alpha_{xy} \end{Bmatrix}_{0°} = \begin{Bmatrix} \alpha_L \\ \alpha_T \\ 0 \end{Bmatrix} = \begin{Bmatrix} 7 \\ 23 \\ 0 \end{Bmatrix} \times 10^{-6}, \ \begin{Bmatrix} \alpha_x \\ \alpha_y \\ \alpha_{xy} \end{Bmatrix}_{90°} = \begin{Bmatrix} 23 \\ 7 \\ 0 \end{Bmatrix} \times 10^{-6}$$

Since the laminate is symmetric and only axial load is applied, we need to calculate only the [A] matrix. The laminate extensional stiffness matrix is

$$[A] = 0.5 \times \begin{bmatrix} 140.76 & 2.754 & 0 \\ 2.754 & 9.18 & 0 \\ 0 & 0 & 6.9 \end{bmatrix} + 0.5 \begin{bmatrix} 9.18 & 2.754 & 0 \\ 2.754 & 140.76 & 0 \\ 0 & 0 & 6.9 \end{bmatrix} \text{GPa-mm}$$

$$= \begin{bmatrix} 74.97 & 2.754 & 0 \\ 2.754 & 74.97 & 0 \\ 0 & 0 & 6.9 \end{bmatrix} \text{GPa-mm}$$

$$\begin{Bmatrix} N_x^{Th} \\ N_y^{Th} \\ N_{xy}^{Th} \end{Bmatrix} = -100 \times \left[\begin{bmatrix} 140.76 & 2.754 & 0 \\ 2.754 & 9.18 & 0 \\ 0 & 0 & 6.9 \end{bmatrix} \begin{Bmatrix} 7 \\ 23 \\ 0 \end{Bmatrix} \times 0.5 \right.$$

$$\left. + \begin{bmatrix} 9.18 & 2.754 & 0 \\ 2.754 & 140.76 & 0 \\ 0 & 0 & 6.9 \end{bmatrix} \begin{Bmatrix} 23 \\ 7 \\ 0 \end{Bmatrix} \times 0.5 \right]$$

$$= \begin{Bmatrix} -0.063954 \\ -0.063954 \\ 0 \end{Bmatrix} \text{GPa-mm}$$

$$\begin{Bmatrix} N_x + N_x^{\text{Th}} \\ 0 + N_y^{\text{Th}} \\ 0 \end{Bmatrix} = \begin{bmatrix} 74.97 & 2.754 & 0 \\ 2.754 & 74.97 & 0 \\ 0 & 0 & 6.9 \end{bmatrix} \begin{Bmatrix} \varepsilon_x^0 \\ \varepsilon_y^0 \\ \gamma_{xy}^0 \end{Bmatrix}$$

Substituting for thermal forces developed, we get

$$N_x = 1.021 \text{ GPa-mm}$$

Thus, 90° ply will fail when $N_x = 1.021$ GPa-mm is applied to the laminate. After the failure of the 90° plies, the [A] matrix is modified by putting all elements of the [Q] matrix of 90° ply to zero. The modified [A] matrix is given by

$$[A] = \begin{bmatrix} 70.38 & 1.377 & 0 \\ 1.377 & 4.59 & 0 \\ 0 & 0 & 3.45 \end{bmatrix} \text{GPa-mm}$$

The thermal forces are to be recalculated, as the [A] matrix is modified.

$$\begin{Bmatrix} N_x^{\text{Th}} \\ N_y^{\text{Th}} \\ N_{xy}^{\text{Th}} \end{Bmatrix} = -100 \begin{bmatrix} 70.38 & 1.377 & 0 \\ 1.377 & 4.59 & 0 \\ 0 & 0 & 6.9 \end{bmatrix} \begin{Bmatrix} 7 \\ 23 \\ 0 \end{Bmatrix} \times 10^{-6} \times 0.5 \text{ GPa-mm}$$

$$= -\begin{Bmatrix} 0.02622 \\ 0.00576 \\ 0 \end{Bmatrix} \text{GPa-mm}$$

Hence

$$\begin{Bmatrix} \Delta N_x + N_x^{\text{Th}} \\ 0 + N_y^{\text{Th}} \\ 0 + N_{xy}^{\text{Th}} \end{Bmatrix} = \begin{bmatrix} 70.38 & 1.377 & 0 \\ 1.377 & 4.59 & 0 \\ 0 & 0 & 6.9 \end{bmatrix} \begin{Bmatrix} \Delta\varepsilon_x^0 \\ \Delta\varepsilon_y^0 \\ \Delta\gamma_{xy}^0 \end{Bmatrix}$$

The additional strain $\Delta\varepsilon_x^0$ that will cause failure of the 0° plies is

$$\Delta\varepsilon_x^0 = 0.0105 - 0.0054 = 0.0051$$

Substituting this value of strain and solving, we get

$$\Delta N_x = 0.3813 \text{ GPa-mm}$$

Thus, the total load that the laminate can carry = 1.021 + 0.3813 = 1.4023 GPa-mm.

Angle-ply laminates generally have full stiffness matrix as compared to cross-ply laminates in view of the change of coordinates. Unlike cross-ply laminates which have the deformation diagram as shown in Fig. 7.1, the angle-ply laminates have a smoother diagram with no kinks.

Example 7.5
Determine the first ply failure of the quasi-isotropic laminate $[0/\pm 45/90]_s$ subjected to a uniaxial loading. The laminate is fabricated at 125°C and cooled to room temperature of 25°C.

Assume that the laminae fail according to the maximum strain theory and all the stiffness coefficients of lamina become zero when the lamina fails. Thickness of the laminate is 2 mm.

Given the strength as

σ_{LU} = 1050 MPa, σ_{TU} = 28 MPa, σ'_{LU} = 650 MPa, σ'_{TU} = 140 MPa, σ_{LTU} = 65 MPa, α_L = 7.0 × 10^{-6}/°C, α_T = 23 × 10^{-6}/°C, ε_{LU} = 0.00153, ε_{TU} = 0.01914

Solution
The $[Q]$ matrix for the lamina is

$$[Q]_{0°} = \begin{bmatrix} 40.83 & 2.91 & 0 \\ 2.91 & 10.21 & 0 \\ 0 & 0 & 4 \end{bmatrix} \text{GPa}$$

Hence, the transformed stiffness matrices for different plies are

$$[\bar{Q}]_{0°} = \begin{bmatrix} 40.83 & 2.91 & 0 \\ 2.91 & 10.21 & 0 \\ 0 & 0 & 4 \end{bmatrix}$$

$$[\bar{Q}]_{90°} = \begin{bmatrix} 10.21 & 2.91 & 0 \\ 0 & 40.83 & 0 \\ 0 & 0 & 4 \end{bmatrix}$$

$$[\bar{Q}]_{\pm 45°} = \begin{bmatrix} 18.215 & 10.215 & \pm 7.655 \\ 10.215 & 18.215 & \pm 7.655 \\ \pm 7.655 & \pm 7.655 & 11.305 \end{bmatrix}$$

Since the applied load is only axial and the laminate is symmetric, we need to consider only the extensional stiffness matrix for the laminate.

$$[A] = \begin{bmatrix} 43.735 & 13.125 & 0 \\ 13.125 & 43.735 & 0 \\ 0 & 0 & 15.305 \end{bmatrix} \text{GPa-mm}$$

$$\begin{Bmatrix} \alpha_x \\ \alpha_y \\ \alpha_{xy} \end{Bmatrix}_{0°} = \begin{Bmatrix} 7 \\ 23 \\ 0 \end{Bmatrix} \times 10^{-6}, \quad \begin{Bmatrix} \alpha_x \\ \alpha_y \\ \alpha_{xy} \end{Bmatrix}_{90°} = \begin{Bmatrix} 23 \\ 7 \\ 0 \end{Bmatrix} \times 10^{-6}, \quad \begin{Bmatrix} \alpha_x \\ \alpha_y \\ \alpha_{xy} \end{Bmatrix}_{\pm45°} = \begin{Bmatrix} 15 \\ 15 \\ \mp16 \end{Bmatrix} \times 10^{-6}$$

Now the thermal forces can be calculated. The change in temperature, $\Delta T = 25 - 125 = -100°C$

$$-100 \begin{bmatrix} 40.83 & 2.91 & 0 \\ 2.91 & 10.21 & 0 \\ 0 & 0 & 4 \end{bmatrix}_{0°} \begin{Bmatrix} 7 \\ 23 \\ 0 \end{Bmatrix} \times 10^{-6} = \begin{Bmatrix} -35.27 \\ -25.52 \\ 0 \end{Bmatrix} \times 10^{-3}$$

$$-100 \begin{bmatrix} 10.21 & 2.91 & 0 \\ 2.91 & 40.83 & 0 \\ 0 & 0 & 4 \end{bmatrix}_{90°} \begin{Bmatrix} 23 \\ 7 \\ 0 \end{Bmatrix} \times 10^{-6} = \begin{Bmatrix} -25.52 \\ -35.27 \\ 0 \end{Bmatrix} \times 10^{-3}$$

$$-100 \begin{bmatrix} 18.215 & 10.215 & \pm7.655 \\ 10.215 & 18.215 & \pm7.655 \\ \pm7.655 & \pm7.655 & 11.305 \end{bmatrix}_{\pm45°} \begin{Bmatrix} 15 \\ 15 \\ \mp16 \end{Bmatrix} \times 10^{-6} = \begin{Bmatrix} -30.40 \\ -30.40 \\ \mp48.77 \end{Bmatrix} \times 10^{-3}$$

$$\begin{Bmatrix} N_x^{Th} \\ N_y^{Th} \\ N_{xy}^{Th} \end{Bmatrix} = 2[-0.75-(-1.0)]10^{-3} \begin{Bmatrix} -35.27 \\ -25.52 \\ 0 \end{Bmatrix} + 2[-0.5-(-0.75)]10^{-3} \begin{Bmatrix} -30.40 \\ -30.40 \\ -48.77 \end{Bmatrix}$$

$$+ 2[-0.25-(-0.5)]10^{-3} \begin{Bmatrix} -30.40 \\ -30.40 \\ +48.77 \end{Bmatrix} + 2[-0.0-(-0.25)]10^{-3} \begin{Bmatrix} -25.52 \\ -35.27 \\ 0 \end{Bmatrix}$$

$$= \begin{Bmatrix} -60.8 \\ -60.8 \\ 0 \end{Bmatrix} \times 10^{-3} \text{ GPa-mm}$$

$$\begin{Bmatrix} N_x + N_x^{Th} \\ N_y + N_y^{Th} \\ N_{xy} + N_{xy}^{Th} \end{Bmatrix} = \begin{Bmatrix} N_x - 0.0608 \\ 0 - 0.0608 \\ 0 + 0 \end{Bmatrix} = \begin{bmatrix} 43.735 & 13.125 & 0 \\ 13.125 & 43.735 & 0 \\ 0 & 0 & 15.305 \end{bmatrix} \begin{Bmatrix} \varepsilon_x^0 \\ \varepsilon_y^0 \\ \gamma_{xy}^0 \end{Bmatrix}$$

As the laminate is loaded axially, the first lamina will fail when the strain along the longitudinal direction reaches a value $\varepsilon_x = 0.00153$, which implies that the 90° lamina is going to fail. The load N_x is obtained by solving the simultaneous equation.

$$N_x = 103.3 \text{ MPa-mm}$$

Summary

- We discussed the failure and ultimate load carrying capacity of the laminate.
- Two different techniques have been discussed.
- The effect of the thermal stresses on the load carrying capacity is also discussed.

Problems

7.1 Determine the first ply failure (FPP) strength of a $[0°/90°]_s$ subjected to uniform tensile load along X-direction, by using: (i) Maximum stress theory, (ii) Tsai-Hill theory and (iii) Tsai-Wu theory. The thickness of each lamina is 0.15 mm. The material and strength properties are

E_L = 40 GPa, E_T = 8.5 GPa, G_{LT} = 3.5 GPa, v_{LT} = 0.28, σ_{LU} = 1080 MPa
σ'_{LU} = 620 MPa, σ_{TU} = 39 MPa, σ'_{TU} = 128 MPa, τ_{LT} = 89 MPa

7.2 Determine the first ply failure strength, if the laminate with the properties given in Problem 7.1 is subjected only to a uniform compressive load along the X-direction.

7.3 Determine the ultimate failure load of the laminate given in Problem 7.1 by employing all the three failure theories.

7.4 A quasi-isotropic laminate $[0°/\pm 45°/90°]_s$ of 250 × 20 × 4 (all dimensions in mm) is subjected to a uniaxial tensile load. The material and strength properties are as given in Problem 7.1. The fixtures in the uniaxial testing machine are 200 mm apart. Assuming that all the stiffness coefficients of the lamina are considered to be zero when it fails, calculate the lamina failure loads and the ultimate failure of the laminate. Use the Tsai-Hill theory to predict the failure of the lamina.

7.5 A composite laminate is fabricated from two isotropic materials, namely, aluminum and steel. The material is assumed to fail when the stress reaches the yield stress or ultimate strain of the material. Calculate the ultimate value of the moment that can be applied on the laminate. The material properties and thickness of the lamina are

Steel: E = 200 GPa, yield stress = 259 MPa, thickness = 3 mm
Aluminum: E = 78 GPa, yield stress = 98 MPa, thickness = 5 mm

7.6 A $[0°/45°/60°]_s$ is subjected to an in-plane shear load N_{xy}. Determine the ultimate failure load for the laminate. The thickness of each lamina is 0.125 mm. The material and strength properties are as given in Problem 7.1.

7.7 The laminate of Problem 7.6 is subjected to a biaxial load N_x and N_{xy}, where N_{xy} is 0.5 N_x. Obtain the ultimate failure for the laminate.

7.8 For a symmetric angle-ply laminate $[15°/-15°_4]_s$, the thickness of each ply is 0.125 mm and the total laminate thickness is 12.5 mm. The laminate is subjected to an axial load N_x. The material and strength properties are as given in Problem 7.1. Calculate the ply failure loads and the ultimate laminate load.

7.9 A symmetric cross-ply laminate $[0°/90°/0°]$ is subjected to a biaxial load N_x and N_y. Draw the failure ellipse in terms of σ_x and σ_y for the 0° lamina. Assume the total laminate thickness to be t. The material strength properties are as given in Problem 7.1.

7.10 The laminate $[0°/90°_4]_s$ is to be used as a cantilever beam. The width of the laminate is 5 cm and the length 40 cm. Thickness of each ply is 1.2 mm. A rigid attachment is fixed to the free end of the beam. The beam is the x-y plane. The rigid attachment is along the z-direction of length 15 cm. Find the load P that results in first ply failure on the x-plane.

References

1. Chou, S.C., Oringer, O. and Rainey, J.H., 1976, "Post-Failure Behavior of Laminates: I – No Stress Concentration", *J. Composite Materials*, 10, p. 371.
2. Chou, S.C., Oringer, O. and Rainey, J.H., 1977, "Post-Failure Behavior of Laminates: II – Stress Concentration", *J. Composite Materials*, 11, p. 71.
3. Halpin, J.C., 1984, *Primer on Composite Materials: Analysis*, Technomic Publishing Co., Lancaster, Pennsylvania.
4. Hahn, H.T. and Tsai, S.W., 1974, "On the Behavior of Composite Laminates after Initial Failures", *Journal of Composite Materials*, 8, p. 288.
5. Agarwal, B.D., Broutman, L.J. and Chandrashekhara, K., 2006, *Analysis and Performance of Fiber Composites*, John Wiley and Sons.
6. Gibson, R.F., 1994, *Principles of Composite Materials Mechanics*, McGraw-Hill, International Edition.

8. Analysis of Composite Beams

In this chapter, we will:

- Derive the governing equation for fiber reinforced laminated composite beams.
- Use the governing equations to study the buckling and vibration behavior.
- Modify the equations to include wide beams as well as sandwich panels.

8.1 Introduction

A beam is a one-dimensional structure. Depending on the type of loading, it is referred to as a beam or a column. For a one-dimensional member, the width and the height of the member is much smaller than the length. Depending on the cross-section and the loading, the classification of whether the member is one-dimensional or two-dimensional changes. In this chapter, we shall consider cases when the laminated beam is subjected to transverse loading, axial loading or is undergoing oscillations. A laminated composite beam may have a number of plies with different fiber orientations. Such a beam behaves differently from a solid beam.

8.2 Governing Equation for Laminated Beams

The theory described here is based on the analysis of Hoff [1] and Pagano [2]. The following assumptions are made for deriving the governing equations:

(i)   The assumptions made in deriving the Euler-Bernoulli equations for an elastic beam are still valid.
(ii) Plies of the beam are symmetrically arranged about the mid-surface.
(iii) A ply is linearly elastic in behavior.
(iv) There is no slip at the interfaces.

Figure 8.1 shows the laminated beam before and after the application of an external moment. The figure also shows the coordinate system employed. As a consequence, the only stress components are σ_x and τ_{xz}.

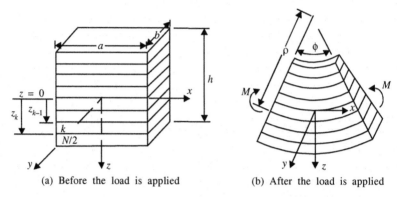

(a) Before the load is applied   (b) After the load is applied

Fig. 8.1 Laminated beam before and after application of bending moment.

The longitudinal normal strain at a distance z from the neutral surface is given as

$$\varepsilon_z = \frac{(\rho + z)\psi - \rho\psi}{\rho\psi} = \frac{z}{\rho} \tag{8.1}$$

where ρ is the radius of curvature of the neutral surface during bending and ψ is the angle of flexure.

The longitudinal stress $\sigma_x(x)$ in the kth ply is given by the constitutive equation for the ply as

$$(\sigma_x)_k = (E_x)_k \, (\varepsilon_x)_k \tag{8.2}$$

Making use of Eq. (8.1), Eq. (8.2) is written as

$$(\sigma_x)_k = (E_x)_k \frac{z}{\rho} \tag{8.3}$$

The stress σ_x results in internal bending moment, which for equilibrium must be equated with the externally applied bending moment M and is given as

$$M(x) = 2 \int_0^{NL/2} \sigma_x(x) \, z \, b \, dz \tag{8.4}$$

where $b \, dz$ is the elemental area on which the stress is acting. Substituting for σ_x from Eq. (8.3) and subsequent integration leads to

$$M(x) = \frac{2b}{3\rho} \sum_{k=1}^{NL/2} (E_x)_k (z_k^3 - z_{k-1}^3) \tag{8.5}$$

where NL is the total number of plies constituting the laminated beam and z_k is the distance of the kth ply from the neutral surface, positive in the downward direction. For an even number of plies of uniform thickness, $z_k = kh/NL$, Eq. (8.5) can be rewritten as

$$M(x) = \frac{2bh^3}{3\rho(NL)^3} \sum_{k=1}^{NL/2} (E_x)_k (3k^2 - 3k + 1) \tag{8.6}$$

Recalling that for a homogenous, isotropic elastic beam of same cross-section, the moment-curvature relation is given as

$$M(x) = \frac{E_{\text{eff}} bt^3}{12\rho} \tag{8.7}$$

Equating Eqs. (8.6) and (8.7), the E_{eff} of the laminated beam is expressed as

$$E_{\text{eff}} = \frac{8}{t^3} \sum_{k=1}^{NL/2} (E_x)_k (z_k^3 - z_{k-1}^3) \tag{8.8}$$

It can be seen that the modulus of the laminated beam depends on the stacking sequence and the modulus of each ply.

We shall illustrate the application of Eq. (8.8) for laminated beams through some examples.

Example 8.1

Obtain the maximum deflection and the longitudinal stress in a simply supported uniform beam subjected to a uniformly distributed load.

Solution

The bending moment at a distance x from the right-hand end is

$$M(x) = \frac{qL}{2} x - q \frac{x^2}{2}$$

For a rectangular cross-section beam, $I = bt^3/12$.

Assuming small deformation

$$\frac{1}{\rho} = \frac{d^2 \mathbf{w}}{dx^2}$$

Making use of these in Eq. (8.7), we can write

$$E_{\text{eff}} I \frac{d^2 \mathbf{w}}{dx^2} = M(x) = \frac{qL}{2} x - q \frac{x^2}{2}$$

or

$$\frac{d^2 \mathbf{w}}{dx^2} = \frac{1}{E_{\text{eff}} I} \left[\frac{qL}{2} x - q \frac{x^2}{2} \right]$$

Integrating twice, we get

$$\mathbf{w}(x) = \frac{1}{E_{\text{eff}} I} \left[\frac{qL}{12} x^3 - \frac{q}{24} x^4 + \frac{A}{2} x^2 + B \right]$$

The constants A, B are obtained from the boundary conditions of the beam. These are

$\mathbf{w} = 0$, at $x = 0$

$\dfrac{d\mathbf{w}}{dx} = 0$, at $x = L/2$ (because of symmetry of the problem)

Substituting the boundary conditions, $\mathbf{w}(x)$ is given as

$$\mathbf{w}(x) = \frac{qL^4}{24E_{\text{eff}} I} [2(x/L)^3 - (x/L)^4 - (x/L)^2]$$

The longitudinal stress can be determined by first obtaining strains from Eq. (8.1) and stresses from Eq. (8.2).

Example 8.2
Obtain the lamina stresses in a cantilever beam subjected to a tip load.

Solution
Figure 8.2 shows a laminated cantilever beam with a tip. From the knowledge of strength of materials, the tip deflection for an isotropic beam is given as

$$\mathbf{w}_{\text{max}} = \frac{PL^3}{3EI}$$

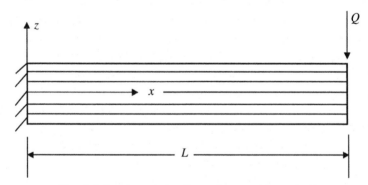

Fig. 8.2 Laminated cantilever beam with tip load.

For a corresponding laminated beam, $\mathbf{w}_{\text{max}}$ is given as

$$\mathbf{w}_{\text{max}} = \frac{PL^3}{3E_{\text{eff}} I} \text{ at the tip of the cantilever}$$

The stress in the kth ply can be obtained by combining Eqs. (8.3) and (8.7) as

$$(\sigma_x)_k = \frac{M}{E_{eff} I}(E_x)_k z = \frac{Mz}{I}\frac{(E_x)_k}{E_{eff}}$$

It can easily be seen that unlike the isotropic beam in which the stresses are maximum at the outer surface, in the case of a laminated composite beam, the maximum stress can occur in any ply depending on the value of the modulus.

8.3 Governing Equation for Laminated Composite Beams

In Section 8.2, we derived the governing equation for symmetric beams. There can be situations when the lay up is anisotropic. In such situations, the mid-surface may or may not coincide with the neutral surface. Therefore, coupling exists between in-plane and out-of-plane displacements, normal and shear strains, and bending moment and twisting moment. The governing equation developed in Section 8.2 is no longer valid.

For a general laminate, we write the laminate constitutive equation as

$$\begin{Bmatrix} N_x \\ N_y \\ N_{xy} \\ M_x \\ M_y \\ M_{xy} \end{Bmatrix} = \begin{bmatrix} A_{11} & A_{12} & A_{16} & B_{11} & B_{12} & B_{16} \\ A_{12} & A_{22} & A_{26} & B_{12} & B_{22} & B_{26} \\ A_{16} & A_{26} & A_{66} & B_{16} & B_{26} & B_{66} \\ B_{11} & B_{12} & B_{16} & D_{11} & D_{12} & D_{16} \\ B_{12} & B_{22} & B_{26} & D_{12} & D_{22} & D_{26} \\ B_{16} & B_{26} & B_{66} & D_{16} & D_{26} & D_{66} \end{bmatrix} \begin{Bmatrix} \varepsilon_x^0 \\ \varepsilon_y^0 \\ \gamma_y^0 \\ \kappa_x \\ \kappa_y \\ \kappa_{xy} \end{Bmatrix} \quad (8.9)$$

If the stacking is symmetric with respect to the mid-surface, $[B]$ matrix is a null matrix.

Therefore, the relationship related to $\{N\}$ and $\{M\}$ are uncoupled. This can be written as two independent equations.

If the beam is subjected to pure bending moment M_x, then $M_y = M_{xy} = 0$. We can then write the curvature in terms of the applied moment.

$$\begin{Bmatrix} \kappa_x \\ \kappa_y \\ \kappa_{xy} \end{Bmatrix} = \begin{bmatrix} d_{11} & d_{12} & d_{16} \\ d_{12} & d_{22} & d_{26} \\ d_{16} & d_{26} & d_{66} \end{bmatrix} \begin{Bmatrix} M_x \\ 0 \\ 0 \end{Bmatrix} \quad (8.10)$$

$$\kappa_x = -\frac{\partial^2 \mathbf{w}}{\partial x^2} = d_{11}M_x, \kappa_y = -\frac{\partial^2 \mathbf{w}}{\partial y^2} = d_{12}M_x, \kappa_{xy} = -2\frac{\partial^2 \mathbf{w}}{\partial x\partial y} = d_{16}M_x \quad (8.11)$$

Since d_{12} and d_{16} in general are not zero, **w** is also dependent on y. Hence one cannot assume that **w** is a function of x alone. If the length-to-width ratio is very large, d_{12} can be neglected. If d_{12} and d_{16} are neglected, which implies κ_y and κ_{xy} are zero, then

$$M_x = -D_{11}\frac{\partial^2 \mathbf{w}}{\partial x^2} \tag{8.12}$$

If it is assumed that $\mathbf{w} = \mathbf{w}(x)$

$$D_{11}b\frac{d^2 \mathbf{w}}{dx^2} = -M_x b \tag{8.13}$$

If we know M_x as a function of x, it is possible to write the governing equation for the beam.

8.3.1 Buckling of Composite Columns

Figure 8.3 shows the laminated column along with the coordinate system. Following the small deflection theory, we can write the strain-displacement relations for a uniaxial member as

$$\varepsilon_z = \frac{\partial \mathbf{w}}{\partial z} - y\frac{\partial^2 \mathbf{v}}{\partial z^2}, \varepsilon_x = 0, \gamma_{xz} = 0 \tag{8.14}$$

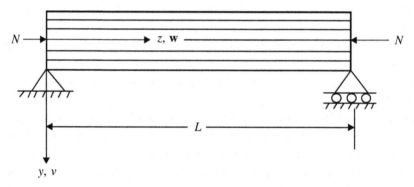

Fig. 8.3 Laminated beam under compressive loading.

where **w** is the displacement along the z direction and y is the distance of the surface measured from the mid-surface. The constitutive equations for the kth lamina can be written as

$$\sigma_z^k = Q_{11}^k \varepsilon_z$$

$$\sigma_x^k = Q_{12}^k \varepsilon_z = \frac{Q_{12}^k}{Q_{11}^k}\sigma_z^k \tag{8.15}$$

$$\tau_{xz}^k = Q_{16}^k \varepsilon_z^k = \frac{Q_{16}^k}{Q_{11}^k}\sigma_z^k$$

The contributions of this lamina to the stress resultants are

$$N_z^k = \int\limits_{t_{k-1}}^{t_k} \sigma_z^k dy, \ N_x^k = \int\limits_{t_{k-1}}^{t_k} \sigma_x^k dy, \ N_x^k = \int\limits_{t_{k-1}}^{t_k} \tau_{xy}^k dy$$

$$M_z^k = \int\limits_{t_{k-1}}^{t_k} \sigma_z^k y \, dy, \ M_x^k = \int\limits_{t_{k-1}}^{t_k} \sigma_x^k y \, dy, \ M_{xy}^k = \int\limits_{t_{k-1}}^{t_k} \tau_{xy}^k y \, dy \tag{8.16}$$

Integrating Eq. (8.16) over the entire laminate, we get

$$N_z = A_{11}\frac{\partial \mathbf{w}}{\partial z} - B_{11}\frac{\partial^2 \mathbf{v}}{\partial z^2}, \ N_x = A_{12}\frac{\partial \mathbf{w}}{\partial z} - B_{12}\frac{\partial^2 \mathbf{v}}{\partial z^2}, \ N_{xz} = A_{16}\frac{\partial \mathbf{w}}{\partial z} - B_{16}\frac{\partial^2 \mathbf{v}}{\partial z^2} \tag{8.17a}$$

$$M_z = B_{11}\frac{\partial \mathbf{w}}{\partial z} - D_{11}\frac{\partial^2 \mathbf{v}}{\partial z^2}, \ M_x = B_{12}\frac{\partial \mathbf{w}}{\partial z} - D_{12}\frac{\partial^2 \mathbf{v}}{\partial z^2}, \ M_{xz} = B_{16}\frac{\partial \mathbf{w}}{\partial z} - D_{16}\frac{\partial^2 \mathbf{v}}{\partial z^2} \tag{8.17b}$$

For a stability problem, $N_z = -N$ and $M_z = 0$.
The first term of Eq. (8.17b) yields

$$\frac{\partial \mathbf{w}}{\partial z} = \frac{D_{11}}{B_{11}}\frac{\partial^2 \mathbf{v}}{\partial z^2} \tag{8.18}$$

Substituting this equation in the first term of Eq. (8.17a), we find that

$$\frac{\partial^2 \mathbf{v}}{\partial z^2} = -\frac{B_{11}}{A_{11}D_{11} - B_{11}^2} N \tag{8.19}$$

For the surface to remain flat, $\partial^2 v/\partial z^2$ must be zero. This implies that the quantity $B_{11}/(A_{11}D_{11} - B_{11}^2)$ must be zero for a given value of N. Assume that at buckling, the column undergoes deformation and the moment $M_z(z) = -N\mathbf{v}(z)$.

Further, since the curvature is negative, the governing equation can be obtained by putting $N_z = 0$ in the first term of Eq. (8.17a) and then substituting the result so obtained in the expression for $\partial \mathbf{w}/\partial z$ (given by the first term of Eq. (8.17b)) we get

$$\frac{d^2 \mathbf{v}}{dz^2} + \frac{A_{11}}{A_{11}D_{11} - B_{11}^2} N\mathbf{v} = 0 \tag{8.20}$$

Equation (8.20) is valid for a laminated anisotropic Euler column. In terms of non-dimensional parameter ξ, defined as $\xi = z/L$, Eq. (8.20) can be written as

$$\frac{d^2 \mathbf{v}}{d\xi^2} + \frac{A_{11}}{A_{11}D_{11} - B_{11}^2} L^2 N\mathbf{v} = 0 \tag{8.21}$$

Equation (8.21) is similar to the isotropic Euler column, and hence its solution can be written as

$$\mathbf{v}(z) = A \sin \lambda \xi + B \cos \lambda \xi \qquad (8.22)$$

where

$$\lambda^2 = \frac{A_{11}L^2}{A_{11}D_{11} - B_{11}^2}$$

For a simply supported anisotropic column, the critical load N for the first mode of buckling is [3]

$$N_{cr} = \frac{A_{11}D_{11} - B_{11}^2}{A_{11}} \frac{\pi^2}{L^2} \qquad (8.23)$$

For any other boundary conditions, a similar procedure can be followed, however, with a higher order differential equation.

8.3.2 Vibrations of Anisotropic Composite Columns

The analysis described earlier for composite beams and columns can easily be extended to the study of vibration problems. Composites which are largely employed in structural design are subjected to dynamic loads. Hence, knowledge of natural frequencies of such columns is very much needed for satisfactory design.

The equilibrium equation for an anisotropic composite column undergoing free oscillations is written as

$$b \frac{A_{11}D_{11} - B_{11}^2}{A_{11}} \frac{\partial^2 \mathbf{v}}{\partial z^4} = -\rho A \frac{\partial^2 \mathbf{v}}{\partial t^2} \qquad (8.24)$$

where ρ is the mass density of the beam material.

Equation (8.24) is an even order partial differential equation in space and time with a constant coefficient. One can look for a variable separable solution. Assuming a simply supported beam, the solution for Eq. (8.24) is written as

$$\mathbf{v}(z, t) = V_m \sin \frac{m \pi z}{L} \cos \omega_m t \qquad (8.25)$$

where m is the number of half sine waves in which the beam bends and ω_m is the mth natural frequency. Substituting the relation (8.25) into Eq. (8.24) and simplifying, we get

$$\omega_m = \frac{m^2 \pi^2}{L^2} \sqrt{\left(\frac{A_{11}D_{11} - B_{11}^2}{A_{11}[\rho A]} \right) b} \qquad (8.26)$$

where b is the width of the beam.

If the composite beam is symmetric, B_{11} is identically zero, and Eq. (8.26) reduces to

$$\omega_m = \frac{m^2 \pi^2}{L^2} \sqrt{\left(\frac{bD_{11}}{[\rho A]}\right)} \tag{8.27}$$

Equation (8.26) indicates that the presence of coupling is to bring down the natural frequency.

By applying a similar procedure, one can derive the natural frequency relations for different ideal boundary conditions.

8.3.3 Buckling of Sandwich Columns

Typically, a sandwich column is constructed using three layers. The top and bottom sheets are of high strength material, while the central thick layer is a material of low strength and density. The top and bottom sheets are termed as facings, while the middle portion is called the *core*. The facings of the sandwich column serve many purposes depending on the application. The stiffness, stability, configuration and, to a large extent, the strength are determined by the characteristics of the facings. The face sheet is either an elastic material or a fiber reinforced composite material. Details of the derivation presented here are also given by Pagano [2].

Consider a column shown in Fig. 8.4 subjected to an axial in-plane load along the z-direction resulting in compressive stresses. As the width of the column is much smaller than the length of the column, the strains along the y-direction are neglected. Further, there is no y-direction dependence. The constitutive equation for the one-dimensional structure reduces to

$$\begin{Bmatrix} N_z \\ M_z \end{Bmatrix} = \begin{bmatrix} A_{11} & B_{11} \\ B_{11} & D_{11} \end{bmatrix} \begin{Bmatrix} \varepsilon_z^0 \\ \kappa_z \end{Bmatrix} \tag{8.28}$$

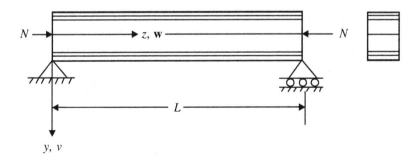

Fig. 8.4 Buckling of a sandwich column.

Assume that the sandwich section possesses mid-plane symmetry; B_{11} is identically zero. Hence

$$N_z = A_{11}\varepsilon_z^0 \text{ and } M_z = D_{11}\kappa_z = -D_{11}\frac{d^2\mathbf{v}}{dz^2} \qquad (8.29)$$

Therefore, at the onset of buckling, equating the transverse forces, we get

$$D_{11}\frac{d^4\mathbf{v}}{dz^4} + N_z\frac{d^2\mathbf{v}}{dz^2} = 0 \qquad (8.30)$$

Assuming the column to be simply supported at the two ends and the mode shape of the sandwich column to be the same as a solid column, we can write

$$\mathbf{v}(z) = \sum_{m=1}^{\infty} A_m \sin\frac{m\pi z}{L} \qquad (8.31)$$

Substitution of Eq. (8.31) into Eq. (8.30) results in

$$\sum_{m=1}^{\infty} A_m \left[D_{11}\frac{m^4\pi^4}{L^4} - N_x\frac{m^2\pi^2}{L^2} \right] \sin\frac{m\pi z}{L} = 0 \qquad (8.32)$$

Equation (8.32) is to be valid for all integer values of m. Hence, the minimum value of N_x, which corresponds to the first critical load of the sandwich column, is obtained when $m = 1$. Hence

$$N_{x\,cr} = D_{11}\frac{\pi^2}{L^2} \qquad (8.33)$$

8.4 Buckling of Sandwich Panels

If the facing of the sandwich column is sufficiently wide, the analysis has to be carried out as a panel. The following assumptions are made in deriving the governing equation for the sandwich panels with anisotropic facings [4]:

(a) The thickness is constrained to be sufficiently small, so that the local bending of the faces about its own centroidal axis can be neglected.
(b) The transverse shear effects due to the faces are neglected.
(c) The effect of transverse normal strain on the transverse displacement is neglected.
(d) The load carrying capacity of the core in the plane of the plate as compared to the facings is small and as a consequence, the normal stresses compared to the core on the plane of the plate and the shear stresses in the core on the planes perpendicular to the facings and in directions parallel to the facings may be neglected.

In view of the assumptions (c) and (d), the transverse displacement $\mathbf{w}(x, y)$ is constant through the thickness and the normal strain ε_z is zero.

8.4.1 Symmetric Sandwich Panel

A symmetric sandwich panel is one in which the faces on ether side of the mid-surface have identical material at identical locations along the z-direction. Generally, symmetric lay up will comprise one of the following types:

(i) Specially orthotropic laminate, for example, symmetric cross-ply laminate.
(ii) Symmetric angle-ply laminate.

The displacement field at any point (x, y, z) in the facings is given by

$$\mathbf{u}(x, y, z) = \mathbf{u}^0(x, y) + z\beta_x(x, y)$$
$$\mathbf{v}(x, y, z) = \mathbf{v}^0(x, y) + z\beta_y(x, y) \qquad (8.34)$$
$$\mathbf{w}(x, y, z) = \mathbf{w}(x, y, 0)$$

The constitutive relations in terms of the material axes (L, T, T') is given by

$$
\begin{Bmatrix} \sigma_L \\ \sigma_T \\ \sigma_{T'} \\ \sigma_{TT'} \\ \sigma_{LT'} \\ \sigma_{LT} \end{Bmatrix}
=
\begin{bmatrix}
Q_{11} & Q_{12} & Q_{13} & 0 & 0 & 0 \\
Q_{12} & Q_{22} & Q_{23} & 0 & 0 & 0 \\
Q_{13} & Q_{23} & Q_{33} & 0 & 0 & 0 \\
0 & 0 & 0 & Q_{44} & 0 & 0 \\
0 & 0 & 0 & 0 & Q_{55} & 0 \\
0 & 0 & 0 & 0 & 0 & Q_{66}
\end{bmatrix}
\begin{Bmatrix} \varepsilon_L \\ \varepsilon_T \\ \varepsilon_{T'} \\ \gamma_{TT'} \\ \gamma_{LT'} \\ \gamma_{LT} \end{Bmatrix}
\qquad (8.35)
$$

The strain-displacement relations for a small deformation are written as

$$\varepsilon_x = \frac{\partial \mathbf{u}}{\partial x} = \frac{\partial \mathbf{u}^0}{\partial x} + z\frac{\partial \beta}{\partial x}$$

$$\varepsilon_y = \frac{\partial \mathbf{v}}{\partial y} = \frac{\partial \mathbf{v}^0}{\partial y} + z\frac{\partial \beta}{\partial y}$$

$$\gamma_{xy} = \frac{\partial \mathbf{u}}{\partial y} + \frac{\partial \mathbf{v}}{\partial x} = \frac{\partial \mathbf{u}^0}{\partial y} + \frac{\partial \mathbf{v}^0}{\partial x} + 2z\left(\frac{\partial \beta}{\partial x} + \frac{\partial \beta}{\partial y}\right) \qquad (8.36)$$

$$\gamma_{xz} = \beta_x + \frac{\partial \mathbf{w}}{\partial x}, \gamma_{yz} = \beta_y + \frac{\partial \mathbf{w}}{\partial y}$$

The constitutive relations for the kth lamina of the facings and the core can be written as

$$
\begin{Bmatrix} \sigma_x \\ \sigma_y \\ \tau_{xy} \end{Bmatrix}
=
\begin{bmatrix}
\bar{Q}_{11} & \bar{Q}_{12} & \bar{Q}_{16} \\
\bar{Q}_{12} & \bar{Q}_{22} & \bar{Q}_{26} \\
\bar{Q}_{16} & \bar{Q}_{26} & \bar{Q}_{66}
\end{bmatrix}
\begin{Bmatrix} \varepsilon_x^0 + z\kappa_x \\ \varepsilon_y^0 + z\kappa_y \\ \varepsilon_{xy}^0 + z\kappa_{xy} \end{Bmatrix}
\qquad (8.37a)
$$

$$\begin{Bmatrix} \tau_{xz} \\ \tau_{yz} \end{Bmatrix} = \begin{bmatrix} Q_{44} & 0 \\ 0 & Q_{55} \end{bmatrix} \begin{Bmatrix} \beta_x + \dfrac{\partial w}{\partial x} \\ \beta_y + \dfrac{\partial w}{\partial y} \end{Bmatrix} \tag{8.37b}$$

The stress resultants are written as

$$\begin{Bmatrix} N_x \\ N_y \\ N_{xy} \end{Bmatrix} = \int\limits_{-d/2}^{d/2} \begin{Bmatrix} \sigma_x \\ \sigma_y \\ \tau_{xy} \end{Bmatrix} dz \tag{8.38a}$$

$$\begin{Bmatrix} M_x \\ M_y \\ M_{xy} \end{Bmatrix} = \int\limits_{-d/2}^{d/2} \begin{Bmatrix} \sigma_x \\ \sigma_y \\ \tau_{xy} \end{Bmatrix} z\, dz \tag{8.38b}$$

$$\begin{Bmatrix} Q_x \\ Q_y \end{Bmatrix} = \int\limits_{-d/2}^{d/2} \begin{Bmatrix} \tau_{xz} \\ \tau_{yz} \end{Bmatrix} dz \tag{8.38c}$$

The shear strain in the core is given by the following relations

$$\gamma_{xzc} = \left[\frac{d}{c} \right] \gamma_{xz} \tag{8.39a}$$

$$\gamma_{yzc} = \left[\frac{d}{c} \right] \gamma_{yz} \tag{8.39b}$$

$d = c + t$ (for equal face thickness).

In general, the face thickness t is very small compared to thickness of the core

$$\gamma_{xzc} = \gamma_{xz} \text{ and } \gamma_{yzc} = \gamma_{yz} \tag{8.40}$$

Hence, the shear resultants are given by

$$\begin{Bmatrix} Q_x \\ Q_y \end{Bmatrix} = \begin{bmatrix} A_{44} & 0 \\ 0 & A_{55} \end{bmatrix} \begin{Bmatrix} \beta_x + \dfrac{\partial w}{\partial x} \\ \beta_y + \dfrac{\partial w}{\partial y} \end{Bmatrix} \tag{8.41}$$

In which $A_{44} = dG_{xz}$ and $A_{55} = dG_{yz}$.

The total potential Π_p is written as

$$\Pi_p = \int\limits_v W\, dV - \int\limits_s T_i U_i\, ds - \int\limits_v F_i U_i\, dV \tag{8.42}$$

where W is the strain energy density function of the facings and core, V is the volume of the elastic body, T_i is the ith component of the surface traction, U_i is the ith component of the displacement and F_i is the ith component of the body force.

For the system to be in equilibrium

$$\delta\Pi_p = 0 \tag{8.43}$$

For the sandwich panel, W is

$$W = \frac{1}{2}\sum_{k=1}^{NL}\int\int_A\int_{h_{k-1}}^{h_k}[\sigma_x\varepsilon_x + \sigma_y\varepsilon_y + \tau_{xy}\gamma_{xy} + \tau_{xz}\gamma_{xz} + \tau_{yz}\gamma_{yz}]\,dA\,dz \tag{8.44}$$

For the facings, τ_{xz} and τ_{yz} are zero.

For the orthotropic sandwich panel, after substituting the expressions for stresses and strains, we get

$$
\begin{aligned}
W_f = \int_A\Bigg[&\frac{A_{11}}{2}\left(\frac{\partial u^0}{\partial x}\right)^2 + \frac{D_{11}}{2}\left(\frac{\partial\beta_x}{\partial x}\right)^2 + A_{12}\left(\frac{\partial v^0}{\partial y}\right)\left(\frac{\partial u^0}{\partial x}\right) + B_{11}\left(\frac{\partial u^0}{\partial x}\right)\left(\frac{\partial\beta_x}{\partial x}\right) \\
&+ B_{12}\left\{\left(\frac{\partial u^0}{\partial x}\right)\left(\frac{\partial\beta_y}{\partial y}\right) + \left(\frac{\partial v^0}{\partial y}\right)\left(\frac{\partial\beta_x}{\partial x}\right)\right\} + D_{12}\left(\frac{\partial\beta_x}{\partial x}\right)\left(\frac{\partial\beta_y}{\partial y}\right) \\
&+ A_{16}\left\{\left(\frac{\partial u^0}{\partial x}\right)\left(\frac{\partial u^0}{\partial y}\right) + \left(\frac{\partial u^0}{\partial x}\right)\left(\frac{\partial v^0}{\partial y}\right)\right\} + D_{16}\left\{\left(\frac{\partial\beta_x}{\partial y}\right)\left(\frac{\partial\beta_y}{\partial y}\right) + \left(\frac{\partial\beta_x}{\partial x}\right)\left(\frac{\partial\beta_y}{\partial x}\right)\right\} \\
&+ B_{16}\left\{\left(\frac{\partial u^0}{\partial x}\right)\left(\frac{\partial\beta_y}{\partial y}\right) + \left(\frac{\partial u^0}{\partial x}\right)\left(\frac{\partial\beta_y}{\partial x}\right) + \left(\frac{\partial v^0}{\partial y}\right)\left(\frac{\partial\beta_x}{\partial x}\right) + \left(\frac{\partial v^0}{\partial x}\right)\left(\frac{\partial\beta_x}{\partial x}\right)\right\} \\
&+ \frac{A_{22}}{2}\left(\frac{\partial v^0}{\partial y}\right) + B_{22}\left(\frac{\partial v^0}{\partial y}\right)\left(\frac{\partial\beta_y}{\partial y}\right) + \frac{D_{22}}{2}\left(\frac{\partial\beta_y}{\partial y}\right)^2 + A_{26}\left\{\left(\frac{\partial u^0}{\partial y}\right)\left(\frac{\partial v^0}{\partial y}\right) + \left(\frac{\partial v^0}{\partial x}\right)\left(\frac{\partial v^0}{\partial y}\right)\right\} \\
&+ B_{26}\left\{\left(\frac{\partial v^0}{\partial x}\right)\left(\frac{\partial\beta_x}{\partial y}\right) + \left(\frac{\partial v^0}{\partial y}\right)\left(\frac{\partial\beta_y}{\partial x}\right) + \left(\frac{\partial u^0}{\partial y}\right)\left(\frac{\partial\beta_y}{\partial y}\right) + \left(\frac{\partial v^0}{\partial x}\right)\left(\frac{\partial\beta_y}{\partial y}\right)\right\} \\
&+ D_{26}\left\{\left(\frac{\partial\beta_x}{\partial y}\right)\left(\frac{\partial\beta_y}{\partial y}\right) + \left(\frac{\partial\beta_x}{\partial x}\right)\left(\frac{\partial\beta_y}{\partial x}\right)\right\} + A_{66}\left\{\frac{1}{2}\left(\frac{\partial u^0}{\partial y}\right)^2 + \left(\frac{\partial u^0}{\partial y}\right)\left(\frac{\partial v^0}{\partial x}\right) + \frac{1}{2}\left(\frac{\partial v^0}{\partial x}\right)^2\right\} \\
&+ B_{66}\left\{\left(\frac{\partial u^0}{\partial y}\right)\left(\frac{\partial\beta_y}{\partial y}\right) + \left(\frac{\partial u^0}{\partial y}\right)\left(\frac{\partial\beta_y}{\partial x}\right) + \left(\frac{\partial v^0}{\partial x}\right)\left(\frac{\partial\beta_x}{\partial y}\right) + \left(\frac{\partial v^0}{\partial x}\right)\left(\frac{\partial\beta_y}{\partial x}\right)\right\} \\
&+ D_{66}\left\{\frac{1}{2}\left(\frac{\partial\beta_x}{\partial y}\right)^2 + \left(\frac{\partial\beta_x}{\partial y}\right)\left(\frac{\partial\beta_y}{\partial y}\right) + \frac{1}{2}\left(\frac{\partial\beta_x}{\partial x}\right)^2\right\}\Bigg]\,dA
\end{aligned} \tag{8.45}
$$

The strain energy of the core is given by

$$U_c = \frac{1}{2}\int_A \left[A_{44}\left(\beta_x + \frac{\partial w}{\partial x}\right)^2 + A_{55}\left(\beta_y + \frac{\partial w}{\partial y}\right)^2 \right] dA \qquad (8.46)$$

The total strain energy of the panel is

$$U = U_f + U_c \qquad (8.47)$$

The potential energy of the external in-plane load is

$$V = -\frac{N_x}{2}\int_A \left(\frac{\partial w}{\partial x}\right)^2 dA \qquad (8.48)$$

The total potential Π_p, is given by

$$\Pi_p = U + V \qquad (8.49)$$

For a symmetric sandwich panel, the coupling matrix $[B]$ is identically a null matrix and as a result, the in-plane displacements are uncoupled from the plate curvatures. As a consequence, Eq. (8.45) reduces to

$$U_f = \frac{1}{2}\int_A \left[D_{11}\left(\frac{\partial\beta_x}{\partial x}\right)^2 + D_{22}\left(\frac{\partial\beta_y}{\partial y}\right)^2 + D_{66}\left(\frac{\partial\beta_x}{\partial y}\right)\left(\frac{\partial\beta_y}{\partial x}\right) + 2D_{12}\left(\frac{\partial\beta_x}{\partial y}\right)\left(\frac{\partial\beta_y}{\partial y}\right) \right.$$
$$\left. + 2D_{16}\left(\frac{\partial\beta_x}{\partial x}\right)\left(\frac{\partial\beta_x}{\partial y} + \frac{\partial\beta_y}{\partial x}\right) + 2D_{26}\left(\frac{\partial\beta_y}{\partial y}\right)\left(\frac{\partial\beta_x}{\partial y} + \frac{\partial\beta_y}{\partial x}\right) \right] dx\,dy \qquad (8.50)$$

The expression for the strain energy of the core and the potential energy remains the same.

Computations have been carried out for a simply supported symmetric sandwich panel with the following material properties

Face	Core
$E_L = 10.335 \times 10^4$ N/mm^2	$G_{xz}/G_{yz} = 0.4$
$E_T = 6.89 \times 10^4$ N/mm^2	$a/b = 1.03$
$G_{LT}/E_T = 0.384$	$t_f = 1.016$ mm

Figure 8.5 shows the buckling response of an orthotropic sandwich panel. It is observed that the minimum buckling load obtained is less than for an isotropic sandwich panel. For an orthotropic square panel with an orthotropy ratio of 1.5 (D_{11}/D_{22}), the buckling parameter k is found to be 3.71. As the shear parameter increases, the buckling response also changes. For an extreme parameter value of 0.2, the minimum buckling load is independent of the aspect ratio.

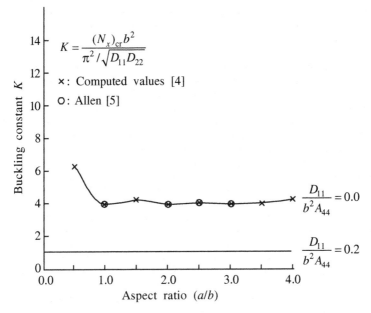

Fig. 8.5 Variation of buckling constant with aspect ratio for isotropic sandwich panel.

Summary

- The governing equation for a composite beam is derived.
- This equation is applied to a number of beam problems.
- The equation is modified to include wide beams and sandwich panels.
- Specific cases of vibration and buckling are also studied.

Problems

8.1 A graphite/epoxy laminated beam of the configuration $[0°/90°/45°/0°]_s$ is 1 mm thick, 20 mm wide. All the plies are of equal thickness. Material properties and strength properties of the lamina are

E_L = 180 GPa, E_T = 10 GPa, G_{LT} = 7.0 GPa, ν_{LT} = 0.25, σ_{LU} = 1700 MPa, σ'_{LU} = 1400 MPa, σ_{TU} = 40 MPa, σ'_{TU} = 230 MPa, τ_{LTU} = 100 MPa

(i) Determine the effective flexural modulus of the beam.
(ii) How can the modulus be improved without changing the ply orientations, ply material and number of plies?

8.2 Obtain the normal and shear stress distribution through the thickness of a laminated beam of ply thickness of 0.2 mm. The stacking sequence is given as $[0°/90°/0°]_s$. The material properties are E_L = 35 GPa, E_T = 10 GPa.

8.3 A simply supported laminated beam of length L with the stacking $[0°/45°]_s$ is subjected to a uniform loading. Obtain the transverse displacement $w(x)$ and the maximum stress which occurs at the section of maximum moment.

8.4   A laminated cantilever beam is subjected to a bending moment at the tip. Obtain the deflection of the beam at the tip. The stacking sequence of the plies is given in Problem 8.3.

8.5   A filament wound laminated pressure vessel is to be made from unidirectional graphite/epoxy. The winding angle is $0° \leq \theta \leq 90°$. The pressure vessel has closed ends. An internal pressure P is applied. In addition, a torque T is applied. This torque will result in a shear stress. The diameter d of the pressure vessel is 75 cm. Plot the required lamina thickness as a function of angle θ to ensure a safe design.

References

1.   Hoff, N.J., 1949, "Strength of Laminates and Sandwich Structural Element", *Engineering Laminate*, Dietz, A.G.H. (Ed.), John Wiley and Sons, New York.

2.   Pagano, N.J., 1967, "Analysis of Flexural Test of Bidirectional Composites", *Journal of Composite Materials*, 1, pp. 336-342.

3.   Iyengar, N.G.R., 2007, *Elastic Stability of Structural Elements*, MacMillan India Ltd.

4.   Ramprasad, S., 1988, "Bending and Buckling Analysis of Anisotropic Sandwich Panels", *M.Tech. Thesis*, Indian Institute of Technology Kanpur.

5.   Allen, H.G., 1969, *Analysis and Design of Structural Sandwich Panels*, Pergamon Press, London.

9. Analysis of Composite Laminates

In this chapter, we will:

• Discuss the response of the laminated fiber reinforced composites subjected to in-plane as well as out-of-plane loads (transverse).
• Consider the constitutive equations developed in earlier chapters for the laminate behavior.
• Derive the governing equations for the laminate assuming small transverse deformations for the laminate.
• Derive equations for higher order theories.

9.1 Introduction

The constitutive equations for the laminates developed in Chapter 6 relate the in-plane forces and moments to the mid-plane strains and curvatures developed in the laminate. Knowing the strains and curvatures, it is possible to find the laminate stresses and strains. In view of the advantages that composites offer over metallic materials, the use of composites as primary load carrying members is on the increase especially in aerospace structures. In many situations, the applied load is generally transverse loading (load normal to the x-y plane). As a consequence, in-plane forces and bending moments develop in the laminates which are functions of the spatial coordinate. The equations developed in Chapter 6 should be modified to take into account the dependence of moments and forces on the spatial coordinates. A large number of theories have been developed by assuming the laminates to be two-dimensional. We shall discuss some of these theories in this chapter.

9.2 Classical Laminate Theory

Analyses of plates fabricated from isotropic materials are well understood and are well documented [1, 2]. The derivations here are based on the assumption that the plate is thin, i.e., the thickness of the laminate is very small compared to the other two dimensions of the laminate. Hence, the displacement of any point in the laminate through the thickness is given by the displacement of the mid-surface of the laminate. The constitutive equations for the laminated composite plates assumed to be two-dimensional are quite involved as the number of material constants are more than for isotropic plates. This makes the analysis for a composite laminate more difficult [3-7].

The *classical laminate theory* is an extension of the plate theory developed for isotropic plates.

9.2.1 Assumptions

In addition to the assumptions regarding the thickness of the laminate, the following assumptions are made for deriving the governing equations based on the classical laminate theory:

- The laminate is made up of a number of plies with arbitrary fiber orientations. The material axes of the ply or lamina do not necessarily coincide with the reference axes of the laminate.
- The lamina behavior is linear and elastic.
- Each lamina is of constant thickness, in-plane as well as out-of-plane.
- The lamina interface bonding is perfect and there is no slip between the lamina. This implies that the strains are continuous across the interface of the lamina.
- Transverse normal stress σ_z is neglected.
- Transverse shear strains γ_{xz} and γ_{yz} are negligible.
- Transverse shear stresses τ_{xz} and τ_{yz} exist to balance the applied transverse load.

Figure 9.1 shows the deformations of an element due to the application of the transverse load [8].

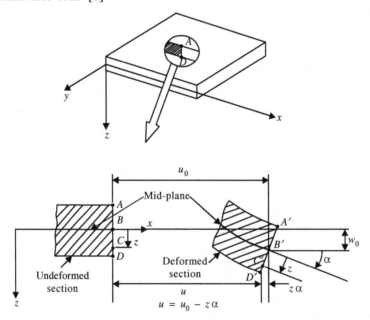

Fig. 9.1 Deformation of a line element during bending of the laminate in the *xz*-plane.

Equilibrium Equations

Figure 9.2 shows the differential element with in-plane forces and moments.

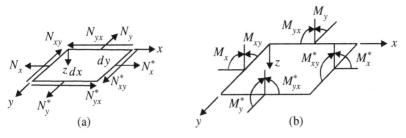

Fig. 9.2 Differential element with in-plane forces and moments.

$$N_x^* = N_x + \frac{\partial}{\partial x} N_x dx$$

For equilibrium, the resultant forces and moments on a differential element must be zero. Therefore, in the absence of body forces, the in-plane force equilibrium equations along the x and y directions are written as

$$\frac{\partial N_x}{\partial x} + \frac{\partial N_{xy}}{\partial y} = 0$$

$$\frac{\partial N_{xy}}{\partial x} + \frac{\partial N_y}{\partial y} = 0 \qquad (9.1)$$

Defining transverse shear forces as

$$(Q_x, Q_y) = \int_{-t/2}^{t/2} (\tau_{xz}, \tau_{yz}) \, dz \qquad (9.2)$$

the equilibrium of forces along the z-direction results in

$$\frac{\partial Q_x}{\partial x} + \frac{\partial Q_y}{\partial y} + q(x, y) = 0 \qquad (9.3)$$

In addition to the force equilibrium, we have two moment equilibrium equations about the x- and y-axes. By summing up the moments about the x-axis, we obtain

$$M_y dx - \left(M_y + \frac{\partial M_y}{\partial y} dy \right) dx + M_{xy} dy - \left(M_{xy} + \frac{\partial M_{xy}}{\partial x} dx \right) dy$$

$$+ \left(Q_{xz} + \frac{\partial Q_{xz}}{\partial x} dx \right) dy \frac{dy}{2} + \left(Q_{yz} + \frac{\partial Q_{yz}}{\partial y} dy \right) dx dy$$

$$- Q_{xz} dy \frac{dy}{2} + q(x, y) dx dy \frac{dy}{2} = 0 \qquad (9.4)$$

Neglecting terms of higher order, we obtain

$$\frac{\partial M_{xy}}{\partial x} + \frac{\partial M_y}{\partial y} - Q_y = 0 \qquad (9.5)$$

Similarly, by summing up the moments about the y-axis and simplifying, we get

$$\frac{\partial M_x}{\partial x} + \frac{\partial M_{xy}}{\partial y} - Q_x = 0 \qquad (9.6)$$

Thus, we have 5 equilibrium equations for the laminate. However, these can be reduced to 3 by combining Eqs. (9.3), (9.5) and (9.6). Equation (9.1) represents the in-plane equilibrium equations, while the third equation can be written as

$$\frac{\partial^2 M_x}{\partial x^2} + 2\frac{\partial^2 M_{xy}}{\partial x \partial y} + \frac{\partial^2 M_y}{\partial y^2} = -q(x, y) \qquad (9.7)$$

It can be observed that Eqs. (9.1) and (9.7) do not involve material property and hence can be applied to isotropic, orthotropic and anisotropic plates. However, when these equations are written in terms of displacements u, v and w, they become functions of material properties. Further, by properly replacing $q(x, y)$, Eq. (9.7) can be applied for analysis of free vibration and buckling problems as well.

9.2.2 Equilibrium Equations in Terms of Displacements

Equations (9.1) and (9.7) can also be written in terms of mid-plane displacements, u_0, v_0 and w, By making use of Eqs. (6.26) and (6.27) and replacing the mid-plane strains and curvatures by mid-plane displacements, the equations are written as

$$A_{11}\frac{\partial^2 u^0}{\partial x^2} + 2A_{16}\frac{\partial^2 u^0}{\partial x \partial y} + A_{66}\frac{\partial^2 u^0}{\partial y^2} + A_{16}\frac{\partial^2 v^0}{\partial x^2} + (A_{12} + A_{66})\frac{\partial^2 v^0}{\partial x \partial y}$$

$$+ A_{26}\frac{\partial^2 v^0}{\partial y^2} - B_{11}\frac{\partial^3 w}{\partial x^3} - 3B_{16}\frac{\partial^3 w}{\partial x^2 \partial y} - (B_{12} + 2B_{66})\frac{\partial^3 w}{\partial x \partial y^2} - B_{26}\frac{\partial^3 w}{\partial y^3} = 0 \quad (9.8)$$

$$A_{16}\frac{\partial^2 u^0}{\partial x^2} + (A_{12} + A_{66})\frac{\partial^2 u^0}{\partial x \partial y} + A_{26}\frac{\partial^2 u^0}{\partial y^2} + A_{66}\frac{\partial^2 v^0}{\partial x^2} + 2A_{26}\frac{\partial^2 v^0}{\partial x \partial y}$$

$$+ A_{22}\frac{\partial^2 v^0}{\partial y^2} - B_{16}\frac{\partial^3 w^0}{\partial x^3} - (B_{12} + 2B_{66})\frac{\partial^3 w^0}{\partial x^2 \partial y} - 3B_{26}\frac{\partial^3 w^0}{\partial x \partial y^2} - B_{22}\frac{\partial^3 w^0}{\partial y^3} = 0 \quad (9.9)$$

$$D_{11}\frac{\partial^4 w^0}{\partial x^4} + 4D_{16}\frac{\partial^4 w^0}{\partial x^3 \partial y} + 2(D_{12}+2D_{66})\frac{\partial^4 w^0}{\partial x^2 \partial y^2} + 4D_{26}\frac{\partial^4 w^0}{\partial x \partial y^3}$$

$$+ D_{22}\frac{\partial^4 w^0}{\partial y^4} - B_{11}\frac{\partial^3 u^0}{\partial x^3} - 3B_{16}\frac{\partial^3 u^0}{\partial x^2 \partial y} - (B_{12}+2B_{66})\frac{\partial^3 u^0}{\partial x \partial y^2}$$

$$- B_{26}\frac{\partial^3 u^0}{\partial y^3} - B_{16}\frac{\partial^3 v^0}{\partial x^3} - (B_{12}+2B_{66})\frac{\partial^3 v^0}{\partial x^2 \partial y} - 3B_{26}\frac{\partial^3 v^0}{\partial x \partial y^2} - B_{22}\frac{\partial^3 v^0}{\partial y^3} = q(x,y)$$

$$(9.10)$$

Equations (9.8) to (9.10) are the three governing equations for a general laminated plate subjected to transverse loading. These are three coupled partial differential equations with constant coefficients. It can be noted that in-plane displacements u^0, v^0 are coupled with transverse displacement w, when the elements of coupling stiffness elements, B_{ij} are present. For symmetric laminates, $B_{ij} = 0$ and hence the governing equation for transverse displacements Eq. (9.10), becomes independent of Eqs. (9.8) and (9.9). It is not possible to obtain a closed form solution for a general laminate. However, closed form solution of the series type is possible for a few special classes of laminates subjected to a few simple boundary conditions. We shall illustrate the technique through some examples.

Example 9.1

Obtain the expression for transverse displacements of an especially orthotropic rectangular laminate, simply supported along all edges and subjected to a transverse loading $q(x, y)$. The laminate dimensions are a, b along the x and y directions.

Solution

For a specially orthotropic laminate, the following stiffness coefficients are zero

$$B_{ij} = 0, \ i, j = 1, \ 2, \ 6; \ A_{16} = A_{26} = D_{16} = D_{26} = 0 \qquad (9.11)$$

As a consequence, Eq. (9.10) reduces to

$$D_{11}\frac{\partial^4 w}{\partial x^4} + 2(D_{12}+2D_{66})\frac{\partial^4 w}{\partial x^2 \partial y^2} + D_{22}\frac{\partial^4 w}{\partial y^4} = q(x,y) \qquad (9.12)$$

Defining

$$x = a\xi, \ y = b\zeta \text{ and } p = a/b \text{ (aspect ratio)}$$

Equation (9.12) can be rewritten in a non-dimensional form as

$$D_{11}\frac{\partial^4 w}{\partial \xi^4} + 2(D_{12}+2D_{66})p^2\frac{\partial^4 w}{\partial \xi^2 \partial \eta^2} + D_{22}p^4\frac{\partial^4 w}{\partial \eta^4} = q(\xi,\eta) \qquad (9.13)$$

For a simply supported laminate, the displacements and moments along all the edges must vanish. They are also written in non-dimensional form as

$$w(0, \eta) = 0, \ w(1, \eta) = 0, \ w(\xi, 0) = 0, \ w(\xi, 1) = 0, \ \xi = 0, 1, \eta = 0, 1 \quad (9.14)$$

$$M_\xi = -\left(D_{11} \frac{\partial^2 w}{\partial \xi^2} + p^2 D_{12} \frac{\partial^2 w}{\partial \eta^2} \right) = 0$$

$$M_\eta = -\left(D_{12} \frac{\partial^2 w}{\partial \xi^2} + p^2 D_{22} \frac{\partial^2 w}{\partial \eta^2} \right) = 0 \quad (9.15)$$

Since the transverse displacement is zero everywhere along the edges, Eqs. (9.15) reduce to

$$\frac{\partial^2 w}{\partial \xi^2} = 0 \text{ at } \xi = 0 \text{ and } 1 \text{ for all } \eta$$

$$\frac{\partial^2 w}{\partial \eta^2} = 0 \text{ at } \eta = 0 \text{ and } 1 \text{ for all } \xi \quad (9.16)$$

It is possible to choose a function in the form of a series for $w(\xi, \eta)$ which satisfies all the boundary conditions and the governing differential equation, provided the external transverse load is also expressed in terms of the same functional form. If the transverse load is represented as

$$q(\xi, \eta) = \sum_{m=1}^{\infty} \sum_{n=1}^{\infty} q_{mn} \sin m\pi\xi \sin n\pi\eta \quad (9.17)$$

where

$$q_{mn} = 4 \int_0^1 \int_0^1 q(\xi, \eta) \sin m\pi\xi \ n\pi\eta \ d\xi \ d\eta$$

the transverse displacement is represented as (Navier's solution)

$$w(\xi, \eta) = \sum_{m=1}^{\infty} \sum_{n=1}^{\infty} W_{mn} \sin m\pi\xi \sin n\pi\eta \quad (9.18)$$

where W_{mn} are the arbitrary displacement coefficients to be determined, and m and n are positive integers.

The boundary conditions and the governing differential equation is identically satisfied. Further, as these functions are orthogonal functions, the integration becomes simpler and the convergence of the series is well established. These are also fast converging functions.

Substituting Eq. (9.17) in Eq. (9.13) and simplifying, we get

$$W_{mn} = \frac{a^4 q_{mn}}{\pi^4 \left[D_{11} m^4 + 2(D_{12} + 2D_{66})(mnp)^2 + D_{22}(np)^4 \right]} \quad (9.19)$$

Hence, the displacement

$$w(\xi, \eta) = \sum_{m=1}^{\infty} \sum_{n=1}^{\infty} \frac{a^4 q_{mn} \sin(m\pi\xi)\sin(n\pi\eta)}{\pi^4 \left[D_{11}m^4 + 2(D_{12} + 2D_{66})(mnp)^2 + D_{22}(np)^4 \right]} \quad (9.20)$$

The transverse load coefficient $q_{mn}(\xi, \eta)$ can be obtained from the given load distribution. For a uniformly distributed load across the laminate, $q(\xi, \eta) = q_0$, the load coefficient q_{mn} can be determined from the relation

$$q_{mn} = 4 \int_0^1 \int_0^1 q(\xi, \eta) \sin m\pi\xi \sin n\pi\eta \, d\xi \, d\eta \quad (9.21)$$

Substituting for $q(\xi, \eta)$, and integrating, we get

$$q_{mn} = \frac{16q_0}{\pi^2 mn} \quad \text{for } m \text{ and } n \text{ odd}$$

$$= 0 \quad \text{for } m \text{ and } n \text{ even}$$

The transverse displacement, for the uniformly distributed load is given as

$$w(\xi, \eta) = \frac{16q_0}{\pi^6} \sum_{m=1,3..}^{\infty} \sum_{n=1,3..}^{\infty} \frac{\sin(m\pi\xi)\sin(n\pi\eta)}{mn \left[D_{11}m^4 + 2(D_{12} + 2D_{66})(mnp)^2 + D_{22}(np)^4 \right]} \quad (9.22)$$

With the knowledge of $w(\xi, \eta)$, the stresses σ_x, σ_y and τ_{xy} for any lamina are obtained from the relation

$$\begin{Bmatrix} \sigma_x \\ \sigma_y \\ \tau_{xy} \end{Bmatrix}_k = -z \begin{bmatrix} \bar{Q}_{11} & \bar{Q}_{12} & 0 \\ \bar{Q}_{12} & \bar{Q}_{22} & 0 \\ 0 & 0 & \bar{Q}_{66} \end{bmatrix}_k \begin{Bmatrix} \dfrac{\partial^2 w}{\partial x^2} \\ \dfrac{\partial^2 w}{\partial y^2} \\ \dfrac{\partial^2 w}{\partial x \partial y} \end{Bmatrix} \quad (9.23)$$

With the knowledge of stresses in each ply, one can find whether any of the lamina has failed for the given loading by applying the failure theories.

Example 9.2

A simply supported symmetric rectangular laminate with dimensions a and b is subjected to a uniform transverse load $q(x, y)$. Derive an expression for $w(x, y)$.

Solution

As the laminate is symmetric, the governing equation for $w(x, y)$ is independent of $u(x, y)$ and $v(x, y)$. Hence, it can be solved independently. The governing equation for $w(x, y)$ is obtained by setting all elements of $[B]$ to zero. This is written as

$$D_{11}\frac{\partial^4 w}{\partial x^4} + 4D_{16}\frac{\partial^4 w}{\partial x^3 \partial y} + 2(D_{12}+2D_{66})\frac{\partial^4 w}{\partial x^2 \partial y^2}$$

$$+ 4D_{26}\frac{\partial^4 w}{\partial x \partial y^3} + D_{22}\frac{\partial^4 w}{\partial y^4} = q(x, y) \qquad (9.24)$$

For simply supported edges, the boundary conditions are written as

$$x = 0, a: \quad w = 0, \quad M_x = -\left(D_{11}\frac{\partial^2 w}{\partial x^2} + 2D_{16}\frac{\partial^2 w}{\partial x \partial y} + D_{12}\frac{\partial^2 w}{\partial y^2}\right) = 0$$

$$y = 0, b: \quad w = 0, \quad M_y = -\left(D_{12}\frac{\partial^2 w}{\partial x^2} + 2D_{26}\frac{\partial^2 w}{\partial x \partial y} + D_{22}\frac{\partial^2 w}{\partial y^2}\right) = 0 \quad (9.25)$$

Navier's technique applied in the previous problem cannot be applied here, as both the differential equations and the boundary conditions contain even and odd derivatives. This is due to the presence of bending-twisting coupling terms such as D_{16} and D_{26}. Hence, the differential equation approach cannot be employed to solve such problems. One can obtain an approximate technique, like the energy technique, for seeking a solution.

The strain energy of the laminate is given by [8]

$$U_{bp} = \frac{1}{2}\int_0^a\int_0^b\left[D_{11}\left(\frac{\partial^2 w}{\partial x^2}\right)^2 + 2D_{12}\frac{\partial^2 w}{\partial x^2}\frac{\partial^2 w}{\partial y^2} + D_{22}\left(\frac{\partial^2 w}{\partial y^2}\right)^2 \right.$$

$$\left. + 4\left(D_{16}\frac{\partial^2 w}{\partial x^2} + D_{16}\frac{\partial^2 w}{\partial y^2}\right)\frac{\partial^2 w}{\partial x \partial y} + 4D_{66}\left(\frac{\partial^2 w}{\partial x \partial y}\right)^2 \right]\partial x \partial y \quad (9.26)$$

The potential energy due to the work done by the load is given as

$$V = -\int_0^a\int_0^b q(x, y)\, w(x, y)\, \partial x \partial y \qquad (9.27)$$

The total potential Π is given by

$$\Pi = U_{bp} + V \qquad (9.28)$$

The solution of Eq. (9.28) is obtained by employing the Ritz technique. This requires that the displacement function chosen should satisfy only the geometric boundary conditions.

The solution is assumed to be of the following form

$$w(x, y) = \sum_{m=1}^{\infty}\sum_{n=1}^{\infty} A_{mn} X_m(x) Y_n(y) \qquad (9.29)$$

Substituting Eq. (9.29) into Eqs. (9.26) and (9.27) and minimizing the total potential with respect to the coefficients A_{mn} yields the following set of equations

$$\sum_{i=1}^{\infty}\sum_{j=1}^{\infty}\left\{D_{11}\int_{0}^{a}\frac{d^2X_i}{dx^2}\frac{d^2X_m}{dx^2}dx\int_{0}^{b}Y_jY_ndy+D_{12}\left[\int_{0}^{a}X_m\frac{d^2X_i}{dx^2}dx\int_{0}^{b}Y_j\frac{d^2Y_n}{dy^2}dy\right.\right.$$

$$+\int_{0}^{a}X_i\frac{d^2X_m}{dx^2}dx\int_{0}^{b}Y_n\frac{d^2Y_j}{dy^2}dy\right]+D_{22}\int_{0}^{a}X_iX_mdx\int_{0}^{b}\frac{d^2Y_j}{dy^2}\frac{d^2Y_n}{dy^2}dy$$

$$+4D_{66}\int_{0}^{a}\frac{dX_i}{dx}\frac{dX_m}{dx}dx\int_{0}^{b}\frac{dY_j}{dy}\frac{dY_n}{dy}dy+2D_{16}\left[\int_{0}^{a}\frac{d^2X_i}{dx^2}\frac{dX_m}{dx}dx\int_{0}^{b}Y_n\frac{dY_n}{dy}dy\right.$$

$$+\int_{0}^{a}\frac{dX_i}{dx}\frac{d^2X_m}{dx^2}dx\int_{0}^{b}Y_n\frac{dY_j}{dy}dy\right]+2D_{26}\left[\int_{0}^{a}X_m\frac{dX_i}{dx}dx\int_{0}^{b}\frac{dY_j}{dy}\frac{d^2Y_n}{dy^2}dy\right.$$

$$\left.\left.+\int_{0}^{a}X_i\frac{dX_m}{dx}dx\int_{0}^{b}\frac{d^2Y_j}{dy^2}\frac{dY_n}{dy}dy\right]\right\}A_{ij}=q(x,y)\int_{0}^{a}X_mdx\int_{0}^{b}Y_ndy$$

$$(9.30)$$

$$m=1,\ 2,........\infty$$
$$n=1,\ 2,........\infty$$

For the purpose of engineering accuracy, a finite number of terms in the series representation are considered. Further, we choose orthogonal functions such as trigonometric functions which assure fast convergence of the solution and also result in simplification of integration.

For the present case, the following solutions are assumed

$$X_m(x)=\sin\frac{m\pi x}{a},\quad Y_n(y)=\sin\frac{n\pi y}{b} \qquad (9.31)$$

For a uniformly loaded square plate with $D_{22}/D_{11}=1$, $(D_{12}+2D_{66})/D_{11}=1.5$ and $(D_{16}/D_{11})=(D_{26}/D_{11})=-0.5a$, the maximum deflection at the centre [9]

$$W_{\text{max}}=\frac{0.00425a^4q_0}{D_{11}} \qquad (9.32)$$

Example 9.3
Obtain the expression for transverse deflection of a simply supported, rectangular, anti-symmetric angle-ply laminate subjected to a sinusoidal loading.

Solution
As the laminate is anti-symmetric, some elements of [B] matrix exist, non-zero elements being B_{16} and B_{26}. All other elements of [B] are zero. In view of this, all the three governing differential equations are coupled. They have to be solved simultaneously. Closed form solution of the series type is possible only for a few boundary conditions. For other boundary conditions, one has to take recourse either to the energy approach or some numerical

techniques such as finite element and finite difference. We shall consider a set of specific boundary conditions for which the closed form solution exists.

For this lay up, the following stiffness coefficients vanish

$$A_{16} = A_{26} = B_{11} = B_{12} = B_{22} = B_{66} = D_{16} = D_{26} = 0 \quad (9.33)$$

After incorporation of these conditions, the governing differential equations reduce to

$$A_{11}\frac{\partial^2 u^0}{\partial x^2} + A_{66}\frac{\partial^2 u^0}{\partial y^2} + (A_{12} + A_{66})\frac{\partial^2 v^0}{\partial x \partial y} - 3B_{16}\frac{\partial^3 w}{\partial x^2 \partial y} - B_{26}\frac{\partial^3 w}{\partial y^3} = 0 \quad (9.34a)$$

$$A_{22}\frac{\partial^2 v^0}{\partial y^2} + A_{66}\frac{\partial^2 v^0}{\partial x^2} + (A_{12} + A_{66})\frac{\partial^2 u^0}{\partial x \partial y} - B_{16}\frac{\partial^3 w}{\partial x^3} - 3B_{26}\frac{\partial^3 w}{\partial x \partial y^2} = 0 \quad (9.34b)$$

$$D_{11}\frac{\partial^4 w}{\partial x^4} + 2(D_{12} + 2D_{66})\frac{\partial^4 w}{\partial x^2 \partial y^2} + D_{22}\frac{\partial^4 w}{\partial y^4}$$

$$- B_{16}\left(3\frac{\partial^3 u^0}{\partial x^2 \partial y} + \frac{\partial^3 v^0}{\partial x^3}\right) - B_{26}\left(\frac{\partial^3 u^0}{\partial y^3} + 3\frac{\partial^3 v^0}{\partial x \partial y^2}\right) = q(x, y) \quad (9.34c)$$

The boundary conditions are given by

$x = 0, a$ for all y

$$w = 0, \quad M_x = B_{16}\left(\frac{\partial u^0}{\partial y} + \frac{\partial v^0}{\partial x}\right) - D_{11}\frac{\partial^2 w}{\partial x^2} - D_{12}\frac{\partial^2 w}{\partial y^2} = 0$$

$$u = 0, \quad N_{xy} = A_{66}\left(\frac{\partial u^0}{\partial y} + \frac{\partial v^0}{\partial x}\right) - B_{16}\frac{\partial^2 w}{\partial x^2} - B_{26}\frac{\partial^2 w}{\partial y^2} = 0$$

$y = 0, b$ for all x $\qquad\qquad\qquad\qquad\qquad\qquad\qquad (9.35)$

$$w = 0, \quad M_y = B_{26}\left(\frac{\partial u^0}{\partial y} + \frac{\partial v^0}{\partial x}\right) - D_{12}\frac{\partial^2 w}{\partial x^2} - D_{22}\frac{\partial^2 w}{\partial y^2} = 0$$

$$v = 0, \quad N_{xy} = A_{66}\left(\frac{\partial u^0}{\partial y} + \frac{\partial v^0}{\partial x}\right) - B_{16}\frac{\partial^2 w}{\partial x^2} - B_{26}\frac{\partial^2 w}{\partial y^2} = 0$$

The expressions for u^0, v^0 and w are chosen such that they satisfy both static and natural boundary conditions at all edges of the laminate. These are given as

$$u^0(x, y) = \sum_{m=1}^{\infty}\sum_{n=1}^{\infty} U_{mn} \sin\frac{m\pi x}{a}\cos\frac{n\pi y}{b}$$

$$v^0(x, y) = \sum_{m=1}^{\infty}\sum_{n=1}^{\infty} V_{mn} \cos\frac{m\pi x}{a}\sin\frac{n\pi y}{b} \qquad (9.36)$$

$$w(x, y) = \sum_{m=1}^{\infty}\sum_{n=1}^{\infty} W_{mn} \sin\frac{m\pi x}{a}\sin\frac{n\pi y}{b}$$

Substituting relations (9.36) in Eqs. (9.34) and equating terms of the trigonometric functions, we get

$$-A_{11}U_{mn}\left(\frac{m\pi}{a}\right)^2 - A_{66}U_{mn}\left(\frac{n\pi}{b}\right)^2 - (A_{12}+A_{66})V_{mn}\left(\frac{m\pi}{a}\right)\left(\frac{n\pi}{b}\right)$$

$$+3B_{16}W_{mn}\left(\frac{m\pi}{a}\right)^2\left(\frac{n\pi y}{b}\right) + B_{26}W_{mn}\left(\frac{n\pi}{b}\right)^3 = 0$$

$$-(A_{12}+A_{66})U_{mn}\left(\frac{m\pi}{a}\right)\left(\frac{n\pi}{b}\right) - A_{66}\left(\frac{m\pi}{a}\right)^2 V_{mn} - A_{22}\left(\frac{n\pi}{b}\right)^2 V_{mn}$$

$$+B_{16}\left(\frac{m\pi}{a}\right)^3 W_{mn} + 3B_{26}\left(\frac{m\pi}{a}\right)\left(\frac{n\pi}{b}\right)^2 = 0$$

$$D_{11}\left(\frac{m\pi}{a}\right)^4 W_{mn} + 2(D_{12}+2D_{66})\left(\frac{m\pi}{a}\right)^2\left(\frac{n\pi}{b}\right)^2 W_{mn} + D_{22}\left(\frac{n\pi}{b}\right)^4 W_{mn}$$

$$-B_{16}\left[3\left(\frac{m\pi}{a}\right)^2\left(\frac{n\pi}{b}\right)U_{mn} + \left(\frac{m\pi}{a}\right)^3 V_{mn}\right]$$

$$-B_{26}\left[\left(\frac{n\pi}{b}\right)^3 U_{mn} + 3\left(\frac{m\pi}{a}\right)\left(\frac{n\pi}{b}\right)^2 V_{mn}\right] = q_{mn} \quad (9.37)$$

These result in three simultaneous, non-homogenous algebraic equations in terms of U_{mn}, V_{mn} and W_{mn}. Solving, we get

$$U_{mn} = q_{mn}\frac{p^4 b^3 n}{\pi^3 D_{mn}}\Big[(A_{66}m^2 + A_{22}n^2 p^2)(3B_{16}m^2 + B_{26}n^2 p^2)$$

$$- m^2(A_{12}+A_{66})(B_{16}m^2 + 3B_{26}n^2 p^2)\Big]$$

$$V_{mn} = q_{mn}\frac{p^3 b^3 m}{\pi^3 D_{mn}}\Big[(A_{11}m^2 + A_{66}n^2 p^2)(B_{16}m^2 + 3B_{26}n^2 p^2)$$

$$- n^2 p^2(A_{12}+A_{66})(3B_{16}m^2 + B_{26}n^2 p^2)\Big]$$

$$W_{mn} = q_{mn}\frac{p^4 b^4}{\pi^4 D_{mn}}\Big[(A_{11}m^2 + A_{66}n^2 p^2)(A_{16}m^2 + A_{22}n^2 p^2)$$

$$- (A_{12}+A_{66})^2 m^2 n^2 p^2\Big] \quad (9.38)$$

where

$$D_{mn} = \Big\{\big[(A_{11}m^2 + A_{66}n^2 p^2)(A_{66}m^2 + A_{22}n^2 p^2) - (A_{12}+A_{66})^2 m^2 n^2 p^2\big]$$

$$\times\big[D_{11}m^4 + 2(D_{12}+2D_{66})m^2 n^2 p^2 + D_{22}n^4 R^4\big] + 2m^2 n^2 p^2(A_{12}+A_{66})$$

$$\times(3B_{16}m^2 + B_{26}n^2 p^2)(B_{16}m^2 + 3B_{26}n^2 p^2) - n^2 p^2(A_{66}m^2 + A_{22}n^2 p^2)$$

$$\times(B_{16}m^2 + B_{26}n^2 p^2)^2 - m^2(A_{11}m^2 + A_{66}n^2 p^2)(B_{16}m^2 + 3B_{26}n^2 p^2)^2\Big\}$$

$$(9.39)$$

and

$$p = a/b$$

9.3 Governing Equations for Laminated Plates Subjected to Compressive In-plane Loads

When thin laminated plates are subjected to in-plane loads, in-plane displacements (u^0 and v^0) are developed. This is true up to some in-plane load levels. When the in-plane load reaches a critical value, then in addition to in-plane displacements, out-of-plane displacement is developed. The in-plane load at which the out-of-plane displacement is developed is called *buckling load*. In buckling, out-of-plane displacements are caused by in-plane loads. The governing Eqs. (9.8) to (9.10) cannot be used to predict the buckling load as these do not take into account the interaction between in-plane loads and out-of-plane displacements. While deriving Eqs. (9.8) to (9.10), it was assumed that the out-of-plane displacements are small such that the resultant forces N_x, N_y and N_{xy} do not have any component in the z-direction and hence did not contribute to the forces along the z-direction. We need to take into account their contribution along the z-direction.

Transverse Equilibrium

To obtain the equilibrium of forces in the z-direction, we have to consider the deflected shape of the plate. Because of the transverse deflection, the forces N_x, N_y, N_{yx} and N_{xy} will have components along the z-direction. Figure 9.3 shows the deflected shape of the element along with the component forces $\partial w^*/\partial x$ and $\partial w^*/\partial y$ defined as

$$\frac{\partial w^*}{\partial x} = \frac{\partial w}{\partial x} + \frac{\partial}{\partial x}\left(\frac{\partial w}{\partial x}\right)dx$$

$$\frac{\partial w^*}{\partial y} = \frac{\partial w}{\partial y} + \frac{\partial}{\partial y}\left(\frac{\partial w}{\partial y}\right)dy$$

(9.40)

In view of the assumption on small deformation

$$\sin\left(\frac{\partial w}{\partial x}\right) \approx \frac{\partial w}{\partial x}$$

$$\cos\left(\frac{\partial w}{\partial x}\right) \approx 1$$

(9.41)

The component of the forces N_x along the positive z-axis is

$$\left(N_x + \frac{\partial N_x}{\partial x}dx\right)\left(\frac{\partial w}{\partial x} + \frac{\partial^2 w}{\partial x^2}dx\right)dy - N_x\frac{\partial w}{\partial x}dy$$

By retaining terms involving N_x, we can simplify these terms to

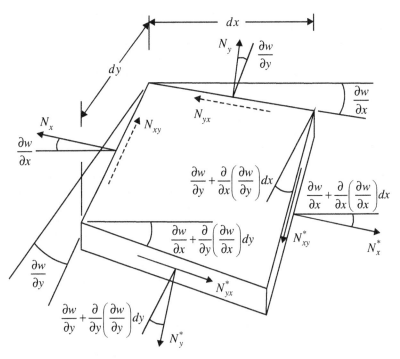

Fig. 9.3 In-plane forces acting on a deflected plate element.

$$N_x \frac{\partial^2 w}{\partial x^2} dx\, dy$$

The component of the forces N_y and along the positive z-axis is

$$\left(N_y + \frac{\partial N_y}{\partial y} dy\right)\left(\frac{\partial w}{\partial y} + \frac{\partial^2 w}{\partial y^2} dy\right) dx - N_y \frac{\partial w}{\partial y} dx$$

This, on simplification, reduces to

$$N_y \frac{\partial^2 w}{\partial y^2} dx\, dy$$

The z-component of the forces N_{xy} and N_{xy}^* is

$$\left(N_{xy} + \frac{\partial N_{xy}}{\partial x} dx\right)\left(\frac{\partial w}{\partial y} + \frac{\partial^2 w}{\partial x \partial y} dx\right) dy - N_{xy} \frac{\partial w}{\partial y} dy$$

This, on simplification, reduces to

$$N_{xy} \frac{\partial^2 w}{\partial x \partial y} dx\, dy$$

Similarly, the component of the force N_{yx} along the z-direction is

$$N_{xy} \frac{\partial^2 w}{\partial x \partial y} dx\, dy$$

The equilibrium of moments about the z-axis yields $N_{yx} = N_{xy}$. Hence, by adding all the components of forces, we get the resultant force along the z-axis due to the in-plane forces as

$$\left[N_x \frac{\partial^2 w}{\partial x^2} + N_y \frac{\partial^2 w}{\partial y^2} + 2N_{xy} \frac{\partial^2 w}{\partial x \partial y} \right] dx\, dy \tag{9.42}$$

To establish the contribution of the shear forces along the z-direction, consider Fig. 9.4 which shows the moments and shear forces acting on the plate element. The shear forces and moments on the x and y faces at a distance dx and dy from the origin are defined as

$$Q_x^* = Q_x + \frac{\partial}{\partial x} Q_x \, dx$$

$$Q_y^* = Q_y + \frac{\partial}{\partial y} Q_y \, dy$$

$$M_x^* = M_x + \frac{\partial}{\partial x} M_x \, dx \tag{9.43}$$

$$M_y^* = M_y + \frac{\partial}{\partial y} M_y \, dy$$

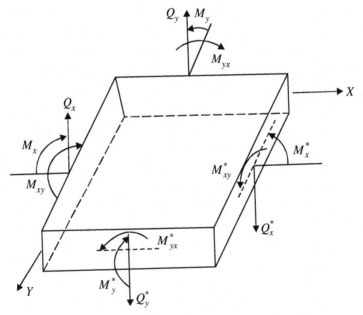

Fig. 9.4 Shear forces and moments acting on plate element.

$$M^*_{xy} = M_{xy} + \frac{\partial}{\partial x} M_{xy} \, dx$$

$$M^*_{yx} = M_{yx} + \frac{\partial}{\partial y} M_{yx} \, dy$$

The component of the shear forces along the z-direction is

$$Q^*_x \, dy - Q_x \, dy + Q^*_y \, dx - Q_y \, dx$$

Substituting for Q^*_x and Q^*_y and simplifying, we obtain

$$\left(\frac{\partial Q_x}{\partial x} + \frac{\partial Q_y}{\partial y} \right) dx \, dy \tag{9.44}$$

Combining Eqs. (9.41) and (9.43), we get the equilibrium of forces along the z-direction as

$$\frac{\partial Q_x}{\partial x} + \frac{\partial Q_y}{\partial y} + N_x \frac{\partial^2 w}{\partial x^2} + N_y \frac{\partial^2 w}{\partial y^2} + 2N_{xy} \frac{\partial^2 w}{\partial x \partial y} = 0 \tag{9.45}$$

Combining Eq. (9.44) with Eqs. (9.5) and (9.6), Eq. (9.7) in the absence of a transverse load can be rewritten as

$$\frac{\partial^2 M_x}{\partial x^2} + 2 \frac{\partial^2 M_{xy}}{\partial x \partial y} + \frac{\partial^2 M_y}{\partial y^2} - N_x \frac{\partial^2 w}{\partial x^2} - N_y \frac{\partial^2 w}{\partial y^2} - 2N_{xy} \frac{\partial^2 w}{\partial x \partial y} = 0 \tag{9.46}$$

In Eq. (9.45), N_x, N_y and N_{xy} are the in-plane forces developed at any point inside the plate and are treated as positive. If the stresses developed are negative, one needs to change the sign.

Thus, Eqs. (9.1) and (9.46) govern the initial buckling of the composite laminate.

Example 9.4
Obtain an expression for the initial buckling load of a specially orthotropic rectangular laminate subjected to in-plane compressive load along the x-direction only. The laminate is simply supported along edges.

Solution
The laminate is assumed to be flat till the critical load is reached. At the onset of buckling

$$N_x = -N_{x0} \tag{9.47}$$

Equation (9.45) reduces to

$$\frac{\partial^2 M_x}{\partial x^2} + 2 \frac{\partial^2 M_{xy}}{\partial x \partial y} + \frac{\partial^2 M_y}{\partial y^2} + N_{x0} \frac{\partial^2 w}{\partial x^2} = 0 \tag{9.48}$$

Writing M_x, M_y and M_{xy} in terms of displacement w, we get

$$D_{11}\frac{\partial^4 w}{\partial x^4} + 2(D_{12} + 2D_{66})\frac{\partial^4 w}{\partial x^2 \partial y^2} + D_{22}\frac{\partial^4 w}{\partial y^4} + N_{x0}\frac{\partial^2 w}{\partial x^2} = 0 \qquad (9.49)$$

In terms of the non-dimensional parameters ξ and η, defined as

$$x = a\xi, \ y = b\eta \qquad (9.50)$$

Equation (9.49) can be written as

$$D_{11}\frac{\partial^4 w}{\partial \xi^4} + 2p^2(D_{12} + 2D_{66})\frac{\partial^4 w}{\partial \xi^2 \partial \eta^2} + D_{22}p^4\frac{\partial^4 w}{\partial \eta^4} + N_{x0}a^2\frac{\partial^2 w}{\partial \xi^2} = 0 \quad (9.51)$$

The simply supported boundary conditions along the edges give

$$w = 0, \quad M_\xi = -\left[D_{11}\frac{\partial^2 w}{\partial \xi^2} + D_{12}p^2\frac{\partial^2 w}{\partial \eta^2}\right] = 0 \ (\xi = 0, 1) \qquad (9.52)$$

$$w = 0, \quad M_\eta = -\left[D_{12}\frac{\partial^2 w}{\partial \xi^2} + D_{22}p^2\frac{\partial^2 w}{\partial \eta^2}\right] = 0 \ (\eta = 0, 1) \qquad (9.53)$$

Equation (9.51) is a fourth order partial differential equation with constant coefficients. As stated in the problem of the bending of a laminate subjected to transverse loading, the boundary conditions (9.52) and (9.53) are homogenous with second order derivatives. Therefore, we can look for a variable separable solution. We assume that the displacement function $w(x, y)$ in terms of the following double Fourier sine series

$$w(x, y) = \sum_{m=1}^{\infty}\sum_{n=1}^{\infty} W_{mn} \sin m\pi\xi \sin n\pi\eta \qquad (9.54)$$

Equation (9.54) satisfies the boundary conditions (9.52) and (9.53) exactly.

Substitution of Eq. (9.54) into Eq. (9.51) results in a non-trivial solution given as

$$N_x = \frac{\pi^2}{a^2}\left[D_{11}m^2 + 2p^2(D_{11} + 2D_{66})n^2 + D_{22}p^4\frac{n^4}{m^2}\right] \qquad (9.55)$$

It is observed from Eq. (9.52), that the buckling load is a function of material property and a combination of m and n. The smallest value of N_x for any value of m occurs when $n = 1$. The lowest critical load depends on material properties and the aspect ratio of the laminate. If the material property is known, it is possible to find the value of m as a function of the aspect ratio at which minimum critical load is obtained. For a specific case, when

$$\frac{D_{11}}{D_{22}} = 10, \quad \frac{D_{12} + 2D_{66}}{D_{22}} = 1$$

Equation (9.55) reduces to

$$N_x = \pi^2 \frac{D_{22}}{a^2}\left[10m^2 + 2p^2 + \frac{p^4}{m^2} \right] \qquad (9.56)$$

The transition from m to $(m + 1)$ half waves occurs at a discrete value of p with the value of N_x remaining the same, i.e.

$$10(m+1)^2 + \frac{p^4}{(m+1)^2} = 10m^2 + \frac{p^4}{m^2}$$
$$p = (10)^{1/4}\sqrt{m(m+1)} \qquad (9.57)$$

Substituting $m = 1$, we get $p = 2.5$. At this aspect ratio, we have the transition from one half wave to two half waves. By setting $m = 2$, we can obtain the transition from two to three half waves and this occurs when $p = 4.35$.

The buckling load for a square plate ($p = 1$) is obtained by putting $m = 1$

$$(N_x)_{cr} = \frac{13\pi^2 D_{22}}{b^2} = k\frac{\pi^2 D_{22}}{b^2} \qquad (9.58)$$

Table 9.1 gives the value of k for different values of aspect ratio.

Table 9.1 Variation of k with aspect ratio p

m	1	1	1	2	2	3
p	0.5	1	2	3	4	5
k	42.25	13.00	8.50	8.96	8.50	8.37

9.4 Higher Order Single Governing Equation

The method of analysis we have given is useful if we can find out a suitable function separately for each of the three displacement components u^0, v^0, and w. However, to obtain a more general solution of equilibrium, the equations are to be combined into a single, higher order partial differential equation. In Sharma et al. [10-12], the technique of doing this has been discussed in detail and applied to different sets of boundary conditions. In what follows, we shall reduce the three governing equations to a single, eighth-order partial differential equation in terms of a single displacement function. Equations (9.34a), (9.34b) and (9.46) can be rewritten as

$$L_{11}u^0 + L_{12}v^0 + L_{13}w = 0 \qquad (9.59a)$$
$$L_{21}u^0 + L_{22}v^0 + L_{23}w = 0 \qquad (9.59b)$$

$$L_{31}u^0 + L_{32}v^0 + L_{33}w = -N_x \frac{\partial^2 w}{\partial x^2} - N_y \frac{\partial^2 w}{\partial x^2} - 2N_{xy} \frac{\partial^2 w}{\partial x \partial y} \qquad (9.59c)$$

where

$$L_{11} = A_{11} \frac{\partial^2}{\partial x^2} + 2A_{16} \frac{\partial^2}{\partial x \partial y} + A_{66} \frac{\partial^2}{\partial y^2} \qquad (9.60a)$$

$$L_{21} = L_{12} = A_{16} \frac{\partial^2}{\partial x^2} + (A_{12} + A_{66}) \frac{\partial^2}{\partial x \partial y} + A_{26} \frac{\partial^2}{\partial y} \qquad (9.60b)$$

$$L_{31} = L_{13} = -B_{11} \frac{\partial^3}{\partial x^3} - B_{16} \frac{\partial^3}{\partial x^2 \partial y} - (B_{12} + 2B_{66}) \frac{\partial^3}{\partial x \partial y^2} - B_{26} \frac{\partial^3}{\partial y^3} \qquad (9.60c)$$

$$L_{22} = A_{66} \frac{\partial^2}{\partial x^2} + 2A_{26} \frac{\partial^2}{\partial x \partial y} + A_{22} \frac{\partial^2}{\partial y^2} \qquad (9.60d)$$

$$L_{32} = L_{23} = -B_{16} \frac{\partial^3}{\partial x^3} - (B_{12} + 2B_{66}) \frac{\partial^3}{\partial x^2 \partial y} - 3B_{26} \frac{\partial^3}{\partial x \partial y^2} - B_{22} \frac{\partial^3}{\partial y^3} \qquad (9.60e)$$

$$L_{33} = D_{11} \frac{\partial^4}{\partial x^4} + 4D_{16} \frac{\partial^4}{\partial x^3 \partial y} + 2(D_{12} + 2D_{66}) \frac{\partial^4}{\partial x^2 \partial y^2}$$

$$+ 4D_{26} \frac{\partial^4}{\partial x \partial y^3} + D_{22} \frac{\partial^4}{\partial y^4} \qquad (9.60f)$$

Let us assume

$$u^0 = -L_{12} \, \eta \, (x, y) - L_{13} \, \psi \, (x, y) \qquad (9.61a)$$

$$v^0 = -L_{11} \, \eta \, (x, y) \qquad (9.61b)$$

$$w = -L_{11} \, \psi \, (x, y) \qquad (9.61c)$$

The functions $\eta \, (x, y)$ are continuous and differentiable in the domain of the laminate. These displacements satisfy Eq. (9.59a) identically, whereas Eq. (9.59b) becomes

$$(L_{22}L_{11} - L_{12}L_{21})\eta + (L_{23}L_{11} - L_{13}L_{21})\psi = 0 \qquad (9.62)$$

Equation (9.62) can be rewritten as

$$P \, \eta + Q \, \psi = 0 \qquad (9.63)$$

If we now choose

$$\eta = -Q \, \Phi(x, y) \qquad (9.64a)$$

$$\psi = P \, \Phi(x, y) \qquad (9.64b)$$

Equation (9.59b) is identically satisfied. This enables us to express Eqs. (9.61a), (9.61b) and (9.61c), in terms of ϕ alone, as

$$u^0 = (L_{12}L_{23} - L_{13}L_{22})\ \phi(x,\ y) \tag{9.65a}$$

$$v^0 = (L_{13}L_{21} - L_{23}L_{11})\ \phi(x,\ y) \tag{9.65b}$$

$$w = (L_{11}L_{22} - L_{12}L_{21})\ \phi(x,\ y) \tag{9.65c}$$

where

$$\phi(x,\ y) = L_{11}\Phi(x,\ y)$$

Substituting Eqs. (9.65) in Eq. (9.59c), we have

$$(L_{11}L_{22}L_{33} - L_{21}L_{12}L_{33} - L_{32}L_{23}L_{11} + L_{31}L_{12}L_{23} + L_{32}L_{13}L_{21} - L_{31}L_{13}L_{22})$$

$$\times \phi(x,\ y) = \left(N_x \frac{\partial^2}{\partial x^2} + 2N_{xy}\frac{\partial^2}{\partial x \partial y} + N_y \frac{\partial^2}{\partial y^2} \right)(L_{11}L_{22} - L_{12}L_{21})\phi \tag{9.66}$$

Equation (9.55), when expanded, becomes

$$A_1 \frac{\partial^8 \phi}{\partial x^8} + A_2 \frac{\partial^8 \phi}{\partial x^7 \partial y} + A_3 \frac{\partial^8 \phi}{\partial x^6 \partial y^2} + A_4 \frac{\partial^8 \phi}{\partial x^5 \partial y^3} + A_5 \frac{\partial^8 \phi}{\partial x^4 \partial y^4}$$

$$+ A_6 \frac{\partial^8 \phi}{\partial x^3 \partial y^5} + A_7 \frac{\partial^8 \phi}{\partial x^2 \partial y^6} + A_8 \frac{\partial^8 \phi}{\partial x \partial y^7} + A_9 \frac{\partial^8 \phi}{\partial y^8}$$

$$= \left(N_x \frac{\partial^2}{\partial x^2} + 2N_{xy}\frac{\partial^2}{\partial x \partial y} + N_y \frac{\partial^2}{\partial y^2} \right) \left\{ (A_{11}A_{66} - A_{16}^2)\frac{\partial^4 \phi}{\partial x^4} \right\}$$

$$+ 2(A_{11}A_{26} - A_{12}A_{16})\frac{\partial^4 \phi}{\partial x^3 \partial y}$$

$$+ (A_{11}A_{22} + 2A_{16}A_{26} - 2A_{12}A_{66} - A_{16}^2)\frac{\partial^4 \phi}{\partial x^2 \partial^2}$$

$$+ 2(A_{16}A_{22} - A_{12}A_{26})\frac{\partial^4 \phi}{\partial x \partial y^3}(A_{66}A_{22} - A_{16}^2)\frac{\partial^4 \phi}{\partial y^4} \tag{9.67}$$

where

$$A_1 = B_{11}(2A_{16}B_{16} - A_{66}B_{11}) - A_{11}B_{16}^2 + D_{11}(A_{11}A_{66} + A_{16}^2) \tag{9.68a}$$

$$A_2 = 2B_{11}(A_{16}G + B_{16}E_1 - 3A_{66}B_{16} - A_{26}B_{11}) + 2B_{16}(2A_{16}B_{16}$$
$$- A_{11}G) + 2D_{11}(A_{16}A_{66} + A_{11}A_{26} - A_{16}E_1) + 4D_{16}(A_{11}A_{66} - A_{16}^2) \tag{9.68b}$$

$$A_3 = 2B_{11}(3A_{16}B_{26} + E_1G - 5A_{26}B_{16} - A_{66}G) + 2B_{16}(2A_{16}g + 3E_1B_{16}$$
$$- 5A_{66}B_{16} - 3A_{11}B_{26}) + D_{11}(A_{16}A_{26} + A_{66}^2 + A_{11}A_{22} - E_1^2) + 8D_{16}$$
$$(A_{16}A_{66} + A_{11}A_{26} - A_{16}E_1) - A_{22}B_{11}^2 - A_{11}G^2 + 2F(A_{11}A_{66} - A_{16}^2) \tag{9.68c}$$

$$A_4 = 2B_{11}(A_{16}B_{22} + 3E_1B_{26} - A_{66}B_{66} - 3A_{22}B_{16} - A_{26}G) - 2B_{22}B_{16}A_{11}$$
$$- 4B_{16}(2A_{16}B_{26} - 2E_1G - 3A_{26}B_{16} - 2A_{66}G) - 6B_{26}A_{11}G$$

$$+ 2D_{11}(A_{22}A_{16} + A_{22}A_{66} - A_{26}E_1) + 4D_{16}(A_{66}^2 + A_{11}A_{22}$$
$$+ 2A_{16}A_{26} - E_1^2) + 4D_{26}(A_{11}A_{66} - A_{16}^2) + 4F(A_{16}A_{66} + A_{11}A_{26} - A_{16}E_1)$$
$$\hspace{9cm}(9.68d)$$

$$\begin{aligned}
A_5 =\ & 2B_{11}(E_1B_{22} + A_{26}B_{26} - A_{22}G) + 2B_{22}(A_{16}B_{16} - A_{11}G) - 4B_{16}A_{26}G \\
& - 4B_{26}A_{16}G - 9A_{22}B_{16}^2 - 9A_{11}B_{26}^2 + 4B_{16}B_{26}(5E_1 - 3A_{66}) \\
& + D_{11}(A_{22}A_{66} - A_{26}^2) + D_{22}(A_{11}A_{66} - A_{16}^2) + 8D_{16}(A_{22}A_{16} \\
& + A_{26}A_{66} - A_{26}E_1) + 8D_{26}(A_{16}A_{66} - A_{11}A_{26} - A_{16}E_1) \\
& + 2G(E_1^2 - A_{66}G) + 2F(A_{66}^2 + A_{11}A_{22} + 2A_{16}A_{26} - E_1^2)
\end{aligned}$$
$$\hspace{12cm}(9.68e)$$

$$\begin{aligned}
A_6 =\ & 2B_{11}(A_{26}B_{22} - A_{22}B_{26}) + 2B_{22}(2E_1B_{16} - A_{16}G) - 3A_{11}B_{26} - A_{66}B_{16} \\
& + 2B_{16}(4A_{26}B_{26} - 3A_{22}G) + 4B_{26}(2E_1G - 2A_{66}G - 3A_{16}B_{26}) \\
& + 2D_{22}(A_{16}A_{66} + A_{11}A_{26} - A_{16}E_1) + 4D_{16}(A_{22}A_{66} - A_{26}^2) \\
& + 4D_{26}(A_{11}A_{22} + 2A_{16}A_{26} + 2A_{26}^2 - 2E_1^2) \\
& + 4F(A_{22}A_{16} + A_{22}A_{66} - A_{26}E_1)
\end{aligned}$$
$$\hspace{12cm}(9.68f)$$

$$\begin{aligned}
A_7 =\ & B_{22}(6A_{26}B_{16} + 2E_1G - 10A_{16}B_{26} - 2A_{66}G - A_{11}B_{22}) \\
& - 6B_{16}B_{26}A_{22} + 2B_{26}^2(3E_1 - 5A_{66}) + D_{22}(A_{11}A_{22} + A_{66}^2 - E_1^2 + 2A_{16}A_{26}) \\
& + 8D_{26}(A_{22}A_{16} + A_{26}A_{66} - A_{26}E_1) + 2F(A_{22}A_{66} - A_{26}^2) \\
& + G(4A_{26}B_{26} - A_{22}G)
\end{aligned}$$
$$\hspace{12cm}(9.68g)$$

$$\begin{aligned}
A_8 =\ & 2B_{22}(A_{26}G + B_{26}E_1 - 3A_{66}B_{26} - A_{16}B_{22}) + (2B_{26}A_{26} - GA_{22}) \\
& 2B_{26} + 2D_{22}(A_{22}A_{16} + A_{26}A_{66} - A_{26}E_1) + 4D_{26}(A_{66}A_{22} - A_{26}^2)
\end{aligned}$$
$$\hspace{11cm}(9.68h)$$

$$A_9 = B_{22}(2A_{26}B_{26} - B_{22}A_{66}) - A_{22}B_{26}^2 + D_{22}(A_{22}A_{66} - A_{26}^2) \hspace{2cm}(9.68i)$$

$$E_1 = A_{12} + A_{66},\ F = D_{12} + 2D_{66},\ G = B_{12} + 2B_{66} \hspace{2.5cm}(9.68j)$$

Equation (9.67) is the governing equation for the buckling of laminated plates. For an anti-symmetric angle-ply laminate, the following relation holds

$$A_{16} = A_{26} = B_{11} = B_{12} = B_{22} = B_{66} = D_{16} = D_{26} = 0 \hspace{1.5cm}(9.69)$$

In view of Eq. (9.69), the coefficients A_2, A_4, A_6 and A_8 vanish in Eq. (9.67). Thus, the governing equation for the buckling of an anti-symmetric angle-ply laminate is

$$A_1 \frac{\partial^8 \phi}{\partial x^8} + A_3 \frac{\partial^8 \phi}{\partial x^6 \partial y^2} + A_5 \frac{\partial^8 \phi}{\partial x^4 \partial y^4} + A_7 \frac{\partial^8 \phi}{\partial x^2 \partial y^6} + A_9 \frac{\partial^8 \phi}{\partial y^8}$$

$$= \left(N_x \frac{\partial^2}{\partial x^2} + 2N_{xy} \frac{\partial^2}{\partial x \partial y} + N_y \frac{\partial^2}{\partial y^2} \right) \left\{ (A_{11}A_{22} - 2A_{12}A_{66} - A_{12}^2) \frac{\partial^4 \phi}{\partial x^2 \partial y^2} \right.$$

$$\left. + A_{66}A_{22} \frac{\partial^4 \phi}{\partial y^4} + A_{11}A_{66} \frac{\partial^4 \phi}{\partial x^4} \right\} \hspace{3cm}(9.70)$$

where A_1, A_3, A_5, A_7, A_9 are obtained after substituting Eq. (9.69) in Eq. (9.67). For an anti-symmetric cross-ply laminate, the following relations hold

$$A_{16} = A_{26} = D_{16} = D_{26} = B_{12} = B_{16} = B_{26} = B_{66} = 0$$
$$B_{22} = - B_{11} \hspace{6cm}(9.71)$$

In view of the relations (9.71), the coefficients A_2, A_4, A_6, A_8 vanish in Eq. (9.67). The governing equation for buckling of an anti-symmetric laminate is the same as Eq. (9.70). However, the coefficients are not the same.

9.5 Governing Equations for Free Vibration

Knowledge of the natural frequency of laminates is of considerable interest for designers. This is necessary so that the excitation frequencies are kept far away from the natural frequencies, otherwise it may lead to resonance resulting in large amplitude oscillations.

When the laminate undergoes oscillations, it develops inertial forces. The inertial forces in general will be in-plane and out-of-plane. Laminates are generally stiff for in-plane motions and hence for small amplitudes of oscillations, the in-plane inertia forces can be neglected when compared to out-of-plane inertia forces.

The governing differential equation for a general laminate undergoing transverse oscillations is written as

$$A_{11}\frac{\partial^2 u^0}{\partial x^2} + 2A_{16}\frac{\partial^2 u^0}{\partial x \partial y} + A_{66}\frac{\partial^2 u^0}{\partial y^2} + A_{16}\frac{\partial^2 v}{\partial x^2} + (A_{12}+A_{66})\frac{\partial^2 v}{\partial x \partial y} + A_{26}\frac{\partial^2 v}{\partial y^2}$$
$$- B_{11}\frac{\partial^3 w}{\partial x^3} - 3B_{16}\frac{\partial^3 w}{\partial^2 x \partial y} - (B_{12}+2B_{66})\frac{\partial^3 w}{\partial x \partial y^2} - B_{26}\frac{\partial^3 w}{\partial y^3} = 0 \qquad (9.72)$$

$$A_{16}\frac{\partial^2 u^0}{\partial x^2} + 2A_{26}\frac{\partial^2 v^0}{\partial x \partial y} + A_{66}\frac{\partial^2 v^0}{\partial y^2} + A_{16}\frac{\partial^2 v^0}{\partial x^2} + (A_{12}+A_{66})\frac{\partial^2 u^0}{\partial x \partial y} + A_{26}\frac{\partial^2 u^0}{\partial y^2}$$
$$- B_{16}\frac{\partial^3 w}{\partial x^3} - 3B_{26}\frac{\partial^3 w}{\partial x \partial y^2} - (B_{12}+2B_{66})\frac{\partial^3 w}{\partial x^2 \partial y} - B_{22}\frac{\partial^3 w}{\partial y^3} = 0 \qquad (9.73)$$

$$D_{11}\frac{\partial^4 w}{\partial x^4} + 4D_{16}\frac{\partial^4 w}{\partial x^3 \partial y} + 2(D_{12}+2D_{66})\frac{\partial^4 w}{\partial x^2 \partial y^2} + 4D_{26}\frac{\partial^4 w}{\partial x \partial y^3} + D_{22}\frac{\partial^4 w}{\partial y^4}$$
$$- B_{11}\frac{\partial^3 u^0}{\partial x^3} - 3B_{16}\frac{\partial^3 u^0}{\partial x^2 \partial y} - (B_{12}+2B_{66})\frac{\partial^3 u^0}{\partial x \partial y^2} - B_{26}\frac{\partial^3 u^0}{\partial y^3} - B_{16}\frac{\partial^3 v}{\partial x^3}$$
$$- (B_{12}+2B_{66})\frac{\partial^3 v}{\partial x^2 \partial y} - 3B_{26}\frac{\partial^3 v}{\partial x \partial y^2} - B_{22}\frac{\partial^3 v}{\partial y^3} + \rho\frac{\partial^2 w}{\partial t^2} = 0 \qquad (9.74)$$

ρ is the mass per unit area of the laminate given by

$$\rho = \sum_{k=1}^{N} \int_{h_k}^{h_{k-1}} \rho^k dz \qquad (9.75)$$

where ρ^k is the mass per unit volume of the kth ply of the laminate.

It is impossible to obtain a closed form solution for a general laminate. The closed form solution exists for some cases with simplified boundary conditions.

Example 9.5

Obtain the natural frequencies of a specially orthotropic laminate subjected to simply supported boundary conditions along all the edges.

Solution

Since we are looking for the bending frequencies, Eq. (9.74) reduces to

$$D_{11}\frac{\partial^4 w}{\partial x^4} + 2(D_{12} + 2D_{66})\frac{\partial^4 w}{\partial x^2 \partial y^2} + D_{22}\frac{\partial^4 w}{\partial y^4} + \rho\frac{\partial^2 w}{\partial t^2} = 0 \qquad (9.76)$$

Equation (9.76) is a homogeneous even order partial differential equation in space and time. We can choose a variable separable solution in space and time. Further, as the free vibration displacements are harmonic in time, $w(x, y, t)$ is assumed to be of the form

$$w(x, y, t) = W(x, y)\, e^{i\omega t} \qquad (9.77)$$

where $W(x, y)$ is the mode shape and ω is the natural frequency of oscillation.

Substitution of Eq. (9.77) in Eq. (9.76) results in

$$D_{11}\frac{\partial^4 W}{\partial x^4} + 2(D_{12} + 2D_{66})\frac{\partial^4 W}{\partial x^2 \partial y^2} + D_{22}\frac{\partial^4 W}{\partial y^4} - \rho\,\omega^2\, W = 0 \qquad (9.78)$$

A solution of the type

$$W(x, y) = \sum_{m=1}^{\infty}\sum_{n=1}^{\infty} W_{mn}\sin\left(\frac{m\pi x}{a}\right)\sin\left(\frac{n\pi y}{b}\right) \qquad (9.79)$$

satisfies the differential equation and the boundary conditions exactly for all time.

Substitution of Eq. (9.79) in Eq. (9.78) gives

$$\omega^2 = \frac{\pi^4}{\rho}\left[D_{11}\left(\frac{m}{a}\right)^4 + 2(D_{12} + 2D_{66})\left(\frac{m}{a}\right)^2\left(\frac{n}{b}\right)^2 + D_{22}\left(\frac{n}{b}\right)^4\right] \qquad (9.80)$$

It can be observed that the frequency increases as m and n increase. The lowest value is obtained when $m = n = 1$. Thus, the lowest natural frequency is given by

$$\omega^2 = \frac{\pi^4}{\rho b^4}\left[D_{11}\left(\frac{b}{a}\right)^4 + 2(D_{12} + 2D_{66})\left(\frac{b}{a}\right)^2 + D_{22}\right] \qquad (9.81)$$

The associated mode shape is given by

$$W(x, y) = W_{11}\sin\frac{\pi x}{a}\sin\frac{\pi y}{b} \qquad (9.82)$$

Example 9.6

Obtain the expression for natural frequency of an anti-symmetric angle-ply laminate.

Solution

Neglecting in-plane inertia, the governing differential equations of motion of the angle-ply laminate are given by

$$A_{11}\frac{\partial^2 U}{\partial x^2} + A_{66}\frac{\partial^2 U}{\partial y^2} + (A_{12} + A_{66})\frac{\partial^2 V}{\partial x \partial y} - 3B_{16}\frac{\partial^3 W}{\partial x^2 \partial y} - B_{26}\frac{\partial^3 W}{\partial y^3} = 0$$

$$(A_{12} + A_{66})\frac{\partial^2 U}{\partial x \partial y} + A_{66}\frac{\partial^2 V}{\partial x^2} + A_{22}\frac{\partial^2 V}{\partial y^2} - B_{16}\frac{\partial^3 W}{\partial x^3} - 3B_{26}\frac{\partial^3 W}{\partial x \partial y^2} = 0 \qquad (9.83)$$

$$D_{11}\frac{\partial^4 W}{\partial x^4} + 2(D_{12} + 2D_{66})\frac{\partial^4 W}{\partial x^2 \partial y^2} + D_{22}\frac{\partial^4 W}{\partial y^4}$$

$$-B_{16}\left(3\frac{\partial^3 U}{\partial x^2 \partial y} + \frac{\partial^3 V}{\partial x^3}\right) - B_{26}\left(\frac{\partial^3 U}{\partial y^3} + 3\frac{\partial^3 V}{\partial x \partial y^2}\right) - \rho\omega^2 W = 0 \qquad (9.84)$$

For simply supported edges, free in the tangential direction, boundary conditions are

At $x = 0, a$

$$W = 0, \; B_{16}\left(\frac{\partial U}{\partial y} + \frac{\partial V}{\partial x}\right) - D_{11}\frac{\partial^2 W}{\partial x^2} - D_{12}\frac{\partial^2 W}{\partial y^2} = 0$$

$$U = 0, \; A_{66}\left(\frac{\partial U}{\partial y} + \frac{\partial V}{\partial x}\right) - B_{16}\frac{\partial^2 W}{\partial x^2} - B_{26}\frac{\partial^2 W}{\partial y^2} = 0 \qquad (9.85)$$

For $y = 0, b$

$$W = 0, \; B_{26}\left(\frac{\partial U}{\partial y} + \frac{\partial V}{\partial x}\right) - D_{12}\frac{\partial^2 W}{\partial x^2} - D_{22}\frac{\partial^2 W}{\partial y^2} = 0$$

$$V = 0, \; A_{66}\left(\frac{\partial U}{\partial y} + \frac{\partial V}{\partial x}\right) - B_{16}\frac{\partial^2 W}{\partial x^2} - B_{26}\frac{\partial^2 W}{\partial y^2} = 0 \qquad (9.86)$$

where

$$U = U_{mn} \sin\frac{m\pi x}{a}\cos\frac{n\pi y}{b}$$

$$V = V_{mn} \cos\frac{m\pi x}{a}\sin\frac{n\pi y}{b}$$

$$W = W_{mn} \sin\frac{m\pi x}{a}\sin\frac{n\pi y}{b} \qquad (9.87)$$

Substitution of relations (9.87) in Eq. (9.84) leads to the following homogenous algebraic equations

$$\begin{bmatrix} \bar{A}_{mn} & \bar{B}_{mn} & \bar{C}_{mn} \\ \bar{B}_{mn} & \bar{D}_{mn} & \bar{E}_{mn} \\ \bar{C}_{mn} & \bar{D}_{mn} & \bar{F}_{mn} - \lambda_{mn} \end{bmatrix} \begin{Bmatrix} U_{mn} \\ V_{mn} \\ W_{mn} \end{Bmatrix} = \begin{Bmatrix} 0 \\ 0 \\ 0 \end{Bmatrix} \tag{9.88}$$

In which

$$\bar{A}_{mn} = A_{11}m^2 + A_{66}n^2 p^2$$

$$\bar{B}_{mn} = (A_{12} + A_{66})mnp$$

$$\bar{C}_{mn} = -\frac{n\pi}{b}(3B_{16}m^2 + B_{26}n^2 p^2)$$

$$\bar{D}_{mn} = A_{66}m^2 + A_{11}n^2 p^2$$

$$\bar{E}_{mn} = -\frac{m\pi}{pb}(B_{16}m^2 + 3B_{26}n^2 p^2)$$

$$\bar{F}_{mn} = \frac{\pi^2}{p^2 b^2}\left[D_{11}(m^4 + n^4 p^4) + 2(D_{12} + 2D_{66}m^2 n^2 p^2)\right]$$

$$\lambda_{mn} = \frac{\omega_{mn}^2 p^2 b^2}{\pi^2}\rho, \; p = a/b$$

For a non-trivial solution, the determinant of the coefficient of U_{mn}, V_{mn} and W_{mn} is zero.

This leads to

$$\omega_{mn}^2 = \frac{\pi^4}{\rho p^4 b^4}\left\{ D_{11}m^4 + 2(D_{12} + 2D_{66})m^2 n^2 p^2 + D_{22}n^4 p^4 - \frac{1}{J_0} \right.$$

$$\left. [m(B_{16}m^2 + 3B_{26}p^2 n^2)J_1 + np(3B_{16}m^2 + B_{26}n^2 p^2)J_2] \right\} \tag{9.89}$$

where

$$J_0 = (A_{11}m^2 + A_{66}n^2 p^2)(A_{66}m^2 + A_{22}n^2 p^2) - (A_{11} + A_{66})^2 m^2 n^2 p^2$$

$$J_1 = (A_{11}m^2 + A_{66}n^2 p^2)(B_{16}m^2 + 3B_{26}n^2 p^2) - n^2 p^2 (A_{12} + A_{66})$$
$$(3B_{16}m^2 + B_{26}n^2 p^2)$$

$$J_2 = (A_{66}m^2 + A_{22}n^2 p^2)(3B_{16}m^2 + B_{26}n^2 b^2) - n^2 p^2 (A_{12} + A_{66})$$
$$(B_{16}m^2 + 3B_{26}n^2 p^2)$$

9.6 Governing Equations Employing Higher Order Theories

Experimental observations reveal that the results obtained for laminates employing the classical laminate plate theory (CLPT) explained in the earlier sections is quite good so long as the laminates are thin (i.e. $a/t \approx 100$). Further, elasticity solutions also agree with the CLPT. For thick and moderately thick laminates (a/t less than 10), the results very much differ from CLPT. This is due to the fact that in the case of composites, shear modulus is an independent property and is, in general, very much different from the elastic modulus. It has been shown that the deflection of laminates is much higher than what is predicted by the CLPT [13, 14]. In other words, CLPT assumption makes the laminates stiffer. Similarly, the natural frequency and buckling load is much higher than the actual value. Various theories have been put forward to account for this discrepancy. These discrepancies are attributed to the presence of transverse shear strains which cannot be neglected for moderately thick laminates.

9.6.1 First Order Shear Deformation Theory (FSDT)

While deriving the governing equation for laminates under CLPT it was assumed that the transverse shear strains γ_{xz} and γ_{yz} are zero. This is based on the assumption that plane sections before deformation remain plane after deformation and perpendicular to the laminate mid-plane. In the *first order shear deformation theory*, plane sections remain plane after deformation but do not remain perpendicular to the laminate mid-plane. This is taken into account by modifying the displacement function by including the rotations of the laminate section as an independent variable. The displacement function is represented as

$$u(x, y, z) = u^0(x, y) + z\,\psi_x(x, y)$$
$$v(x, y, z) = v^0(x, y) + z\,\psi_y(x, y) \qquad (9.90)$$
$$w(x, y, z) = w(x, y, 0)$$

where ψ_x and ψ_y are the rotations about the y- and x-axes, respectively. The shear strains γ_{xz} and γ_{yz} are given as

$$\gamma_{xz} = \frac{\partial u}{\partial z} + \frac{\partial w}{\partial x} = \psi_x + \frac{\partial w}{\partial x}$$
$$\gamma_{yz} = \frac{\partial v}{\partial z} + \frac{\partial w}{\partial y} = \psi_y + \frac{\partial w}{\partial y} \qquad (9.91)$$

It is observed that the transverse shear strains are constant over the laminate thickness. Thus, the FSDT predicts uniform transverse shear stresses through the thickness of the laminate. This violates the condition that the shear stresses τ_{xz} and τ_{yz} must vanish at the top and bottom surfaces of the laminate as they are free surfaces. Furthermore, within the lamina they have to be parabolic. Thus, the first order theory though simple to apply, suffers from this inaccuracy. However, for most cases, it is observed that this theory predicts better results than the classical laminate theory. Employing Eq. (9.90), we can write the strains as

$$\begin{Bmatrix} \varepsilon_x \\ \varepsilon_y \\ \gamma_{xy} \end{Bmatrix} = \begin{Bmatrix} \varepsilon_x^0 \\ \varepsilon_y^0 \\ \gamma_{xy}^0 \end{Bmatrix} + z \begin{Bmatrix} \kappa_x \\ \kappa_y \\ \kappa_{xy} \end{Bmatrix} \tag{9.92}$$

where $\varepsilon_x^0, \varepsilon_y^0, \gamma_{xy}^0$ are the mid-plane strains. Further

$$\kappa_x = \frac{\partial \psi_x}{\partial x}, \kappa_y = \frac{\partial \psi_y}{\partial y}, \kappa_{xy} = \frac{\partial \psi_x}{\partial y} + \frac{\partial \psi_y}{\partial x} \tag{9.93}$$

The transverse shear force resultants are written as

$$\begin{Bmatrix} R_{yz} \\ R_{xz} \end{Bmatrix} = \begin{bmatrix} A_{44} & A_{45} \\ A_{45} & A_{55} \end{bmatrix} \begin{Bmatrix} \gamma_{yz} \\ \gamma_{xz} \end{Bmatrix} \tag{9.94}$$

$$A_{ij} = K \int\limits_{-t/2}^{t/2} \bar{Q}_{ij} dz$$

K is the shear correction factor to account for the variation of transverse shear stresses through the thickness. It also depends on the cross-sectional shape, the ply properties and stacking sequence. Generally its value is taken to be 5/6 for most cases.

Following the steps discussed for the case of CLPT, the governing equations can be written as

$$A_{11}\frac{\partial^2 u^0}{\partial x^2} + 2A_{16}\frac{\partial^2 u^0}{\partial x \partial y} + A_{66}\frac{\partial^2 u^0}{\partial y^2} + A_{16}\frac{\partial^2 v^0}{\partial x^2} + (A_{12}+A_{66})\frac{\partial^2 v^0}{\partial x \partial y} + A_{26}\frac{\partial^2 v^0}{\partial y^2}$$

$$+ B_{11}\frac{\partial^2 \psi_x}{\partial x^2} + 2B_{16}\frac{\partial^2 \psi_x}{\partial x \partial y} + B_{66}\frac{\partial^2 \psi_x}{\partial y^2} + B_{16}\frac{\partial^2 \psi_y}{\partial x^2} + (B_{12}+B_{66})\frac{\partial^2 \psi_y}{\partial x \partial y} + B_{26}\frac{\partial^2 \psi_y}{\partial y^2} = 0 \tag{9.95}$$

$$A_{16}\frac{\partial^2 u^0}{\partial x^2} + (A_{12}+A_{66})\frac{\partial^2 u^0}{\partial x \partial y} + A_{26}\frac{\partial^2 u^0}{\partial y^2} + A_{66}\frac{\partial^2 v^0}{\partial x^2} + 2A_{26}\frac{\partial^2 v^0}{\partial x \partial y} + A_{22}\frac{\partial^2 v^0}{\partial y^2}$$

$$+ B_{16}\frac{\partial^2 \psi_x}{\partial x^2} + (B_{12}+B_{66})\frac{\partial^2 \psi_x}{\partial x \partial y} + B_{26}\frac{\partial^2 \psi_x}{\partial y^2} + B_{66}\frac{\partial^2 \psi_y}{\partial x^2} + 2B_{26}\frac{\partial^2 \psi_y}{\partial x \partial y} + B_{22}\frac{\partial^2 \psi_y}{\partial y^2} = 0 \tag{9.96}$$

$$B_{11}\frac{\partial^2 u^0}{\partial x^2} + 2B_{16}\frac{\partial^2 u^0}{\partial x \partial y} + B_{66}\frac{\partial^2 u^0}{\partial y^2} + B_{16}\frac{\partial^2 v^0}{\partial x^2} + (B_{12}+B_{66})\frac{\partial^2 v^0}{\partial x \partial y} + B_{26}\frac{\partial^2 v^0}{\partial x^2}$$

$$+ D_{11}\frac{\partial^2 \psi_x}{\partial x^2} + 2D_{16}\frac{\partial^2 \psi_x}{\partial x \partial y} + D_{66}\frac{\partial^2 \psi_x}{\partial y^2} + D_{16}\frac{\partial^2 \psi_y}{\partial x^2} + (D_{12}+D_{66})\frac{\partial^2 \psi_y}{\partial x \partial y} + D_{26}\frac{\partial^2 \psi_y}{\partial y^2}$$

$$- A_{44}\left(\psi_x + \frac{\partial w}{\partial x}\right) - A_{45}\left(\psi_y + \frac{\partial w}{\partial y}\right) = 0 \tag{9.97}$$

$$B_{16}\frac{\partial^2 u^0}{\partial x^2}+(B_{12}+B_{66})\frac{\partial^2 u^0}{\partial x\partial y}+B_{26}\frac{\partial^2 u^0}{\partial y^2}+B_{66}\frac{\partial^2 v^0}{\partial x^2}+2B_{26}\frac{\partial^2 v^0}{\partial x\partial y}+B_{22}\frac{\partial^2 v^0}{\partial x^2}$$

$$+D_{16}\frac{\partial^2 \psi_x}{\partial x^2}+(D_{12}+D_{66})\frac{\partial^2 \psi_x}{\partial x\partial y}+D_{26}\frac{\partial^2 \psi_x}{\partial y^2}+D_{26}\frac{\partial^2 \psi_y}{\partial x^2}+2D_{26}\frac{\partial^2 \psi_y}{\partial x\partial y}D_{22}\frac{\partial^2 \psi_y}{\partial y^2}$$

$$-A_{45}\left(\psi_x+\frac{\partial w}{\partial x}\right)-A_{55}\left(\psi_y+\frac{\partial w}{\partial y}\right)=0$$

$$(9.98)$$

$$A_{55}\left(\frac{\partial \psi_y}{\partial y}+\frac{\partial^2 w}{\partial y^2}\right)+A_{45}\left(\frac{\partial \psi_x}{\partial y}+\frac{\partial \psi_y}{\partial x}+2\frac{\partial^2 w}{\partial x\partial y}\right)+A_{44}\left(\frac{\partial \psi_x}{\partial x}+\frac{\partial^2 w}{\partial x^2}\right)+q(x,y)=0$$

$$(9.99)$$

For a specially orthotropic laminate subjected to a transverse load $q(x, y)$, where bending is predominant, the governing equations can be reduced to

$$D_{11}\frac{\partial^2 \psi_x}{\partial x^2}+(D_{12}+D_{66})\frac{\partial^2 \psi_y}{\partial x\partial y}+D_{66}\frac{\partial^2 \psi_x}{\partial y^2}-A_{44}\left(\psi_x+\frac{\partial w}{\partial x}\right)=0$$

$$D_{22}\frac{\partial^2 \psi_y}{\partial y^2}+(D_{12}+D_{66})\frac{\partial^2 \psi_x}{\partial x\partial y}+D_{66}\frac{\partial^2 \psi_y}{\partial x^2}-A_{55}\left(\psi_y+\frac{\partial w}{\partial y}\right)=0 \quad (9.100)$$

$$A_{55}\left(\frac{\partial \psi_y}{\partial y}+\frac{\partial^2 w}{\partial y^2}\right)+A_{44}\left(\frac{\partial \psi_x}{\partial x}+\frac{\partial^2 w}{\partial x^2}\right)+q(x,y)=0$$

For a simply supported laminate, the following representations for ψ_x, ψ_y, w satisfy the boundary conditions exactly.

$$\psi_x(x,y)=\sum_{m=1}^{\infty}\sum_{n=1}^{\infty}\Psi_{xmn}\cos\frac{m\pi x}{a}\sin\frac{n\pi y}{b}$$

$$\psi_y(x,y)=\sum_{m=1}^{\infty}\sum_{n=1}^{\infty}\Psi_{ymn}\sin\frac{m\pi x}{a}\cos\frac{n\pi y}{b} \qquad (9.101)$$

$$w(x,y)=\sum_{m=1}^{\infty}\sum_{n=1}^{\infty}W_{mn}\sin\frac{m\pi x}{a}\cos\frac{n\pi y}{b}$$

The transverse load $q(x, y)$ can also be written as

$$q(x,y)=\sum_{m=1}^{\infty}\sum_{n=1}^{\infty}Q_{mn}\sin\frac{m\pi x}{a}\sin\frac{n\pi y}{b} \qquad (9.102)$$

Substitution of Eqs. (9.101) and (9.102) in Eq. (9.100) leads to the following set of equations

$$D_{11}\left(\frac{m\pi}{a}\right)^2 \Psi_{xmn} + (D_{12} + D_{66})\left(\frac{m\pi}{a}\right)\left(\frac{n\pi}{b}\right)\Psi_{ymn}$$

$$+ D_{66}\left(\frac{m\pi}{b}\right)\Psi_{xmn} + A_{55}\Psi_{xmn} + A_{55}\left(\frac{m\pi}{a}\right)W_{mn} = 0$$

$$(D_{12} + D_{66})\left(\frac{m\pi}{a}\right)\left(\frac{n\pi}{b}\right)\Psi_{xmn} + D_{22}\left(\frac{n\pi}{b}\right)^2 \Psi_{ymn}$$

$$+ D_{66}\left(\frac{m\pi}{a}\right)^2 \Psi_{ymn} + A_{44}\Psi_{ymn} + A_{44}\left(\frac{n\pi}{b}\right)W_{mn} = 0$$

$$A_{55}\left(\frac{m\pi}{a}\right)\Psi_{xmn} + A_{44}\left(\frac{n\pi}{b}\right)\Psi_{ymn} + A_{44}\left(\frac{n\pi}{b}\right)^2 W_{mn} + A_{55}\left(\frac{m\pi}{a}\right)^2 W_{mn} = -Q_{mn}$$

These are a set of non-homogenous algebraic equations. We can obtain Ψ_{xmn}, Ψ_{ymn}, W_{mn} in terms of the external load. Knowing these coefficients, it is possible to obtain transverse deflection along the x, y directions.

Whitney [15] obtained the deflection of a simply supported square cross-ply symmetric laminate subjected to a sinusoidal distributed transverse load, $q(x, y) = q_0 \sin\dfrac{\pi x}{a} \sin\dfrac{\pi x}{b}$. The following lamina properties are used

$$E_L = 175\,\text{GPa}, E_T = E_{T'} = 7\,\text{GPa}, G_{LT} = G_{LT'} = 3.5\,\text{GPa},$$
$$G_{TT'} = 1.4\,\text{GPa}, \nu_{LT} = \nu_{LT'} = 0.25$$

Figure 9.5 shows the variation of non-dimensional central deflection of the laminate as a function of side-to-thickness ratio along with the results

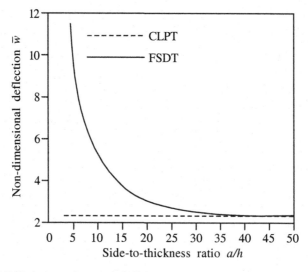

Fig. 9.5 Variation of central deflection with side-to-thickness ratio for a symmetric cross-ply laminate [15].

obtained employing CLPT. It is observed that for low thickness ratios, CLPT differs quite a bit from FSDT. However, as the ratio increases CLPT is closer to HSDT.

9.6.2 Higher Order Shear Deformation Theory (HSDT)

In the previous section, it was shown that FSDT is better than the CLPT. However, it suffers from the disadvantage that one should know the shear correction factor K. Reddy [16] put forward a simple higher order theory for laminate composites which accounts not only for transverse shear strains, but also the parabolic variation of transverse shear strains through the thickness of the laminate. It does not require any shear correction factor. The displacement field is expressed as

$$u(x, y, z) = u^0(x, y) + z\psi_x(x, y) + z^2\xi_x(x, y) + z^3\zeta_x(x, y)$$

$$v(x, y, z) = v^0(x, y) + z\psi_y(x, y) + z^2\xi_y(x, y) + z^3\zeta_y(x, y) \qquad (9.103)$$

$$w(x, y, z) = w(x, y, 0)$$

It is assumed that the laminate thickness is small is comparison to the other two dimensions. Hence, the transverse displacement of any point in the laminate is given by the mid-surface displacement. The functions $\xi_x, \xi_y, \zeta_x, \zeta_y$ are determined by the condition that the transverse shear stresses τ_{xz}, τ_{yz} vanish at the top and bottom surfaces of the laminate. The presence of the cubic term in the displacement ensures that the transverse shear strains are parabolic in nature.

$$\gamma_{xz} = \frac{\partial u}{\partial z} + \frac{\partial w}{\partial x} = \psi_x + 2z\xi_x + 3z^2\zeta_x + \frac{\partial w}{\partial x}$$

$$\gamma_{yz} = \frac{\partial v}{\partial z} + \frac{\partial w}{\partial y} = \psi_y + 2z\xi_y + 3z^2\zeta_y + \frac{\partial w}{\partial y} \qquad (9.104)$$

From the condition that the transverse shear stresses have to be zero at the top and bottom surface of the laminate, we get

$$\xi_x = 0, \ \xi_y = 0$$

and $\qquad \zeta_x = -\frac{4}{3t^2}\left(\psi_x + \frac{\partial w}{\partial x}\right), \zeta_y = -\frac{4}{3t^2}\left(\psi_y + \frac{\partial w}{\partial y}\right) \qquad (9.105)$

Substituting these values in Eq. (9.103), the displacement relations can be written as

$$u(x, y, z) = u^0(x, y) + z\left[\psi_x - \frac{4}{3}\left(\frac{z}{t}\right)^2\left(\psi_x + \frac{\partial w}{\partial x}\right)\right] \qquad (9.106a)$$

$$v(x, y, z) = v^0(x, y) + z\left[\psi_y - \frac{4}{3}\left(\frac{z}{t}\right)^2\left(\psi_y + \frac{\partial w}{\partial y}\right)\right] \qquad (9.106b)$$

$$w(x, y, z) = w(x, y, 0) \tag{9.106c}$$

The strains associated with these displacements are

$$\varepsilon_x = \varepsilon_x^0 + z(\kappa_x^0 + z^2\kappa_x^2)$$

$$\varepsilon_y = \varepsilon_y^0 + z(\kappa_y^0 + z^2\kappa_y^2)$$

$$\varepsilon_z = 0$$

$$\gamma_{yz} = \gamma_{yz}^0 + z^2\kappa_{yz}^2 \tag{9.107}$$

$$\gamma_{xz} = \gamma_{xz}^0 + z^2\kappa_{xz}^2$$

$$\gamma_{xy} = \gamma_{xy}^0 + z(\kappa_{xy}^0 + z^2\kappa_{xy}^2)$$

where

$$\kappa_x^0 = \frac{\partial \psi}{\partial x}, \kappa_x^2 = -\frac{4}{3t^2}\left(\frac{\partial \psi_x}{\partial x} + \frac{\partial^2 w}{\partial x^2}\right); \kappa_y^0 = \frac{\partial \psi}{\partial y}, \kappa_y^2 = -\frac{4}{3t^2}\left(\frac{\partial \psi_y}{\partial y} + \frac{\partial^2 w}{\partial y^2}\right)$$

$$\gamma_{yz}^0 = \psi_y + \frac{\partial w}{\partial y}, \kappa_{yz}^2 = -\frac{4}{t^2}\left(\psi_y + \frac{\partial w}{\partial y}\right); \gamma_{xz}^0 = \psi_x + \frac{\partial w}{\partial x}, \kappa_{xz}^2 = -\frac{4}{t^2}\left(\psi_x + \frac{\partial w}{\partial x}\right)$$

$$\kappa_{xy}^0 = \frac{\partial \psi_x}{\partial y} + \frac{\partial \psi_y}{\partial x}, \kappa_{xy}^2 = -\frac{4}{3t^2}\left(\frac{\partial \psi_x}{\partial y} + \frac{\partial \psi_y}{\partial x} + 2\frac{\partial^2 w}{\partial x \partial y}\right) \tag{9.108}$$

One can then obtain the governing equations as

$$\frac{\partial N_x}{\partial x} + \frac{\partial N_{xy}}{\partial y} = 0, \frac{\partial N_{xy}}{\partial x} + \frac{\partial N_y}{\partial y} = 0$$

$$\frac{\partial Q_x}{\partial x} + \frac{\partial Q_y}{\partial y} + q(x, y) - \frac{4}{t^2}\left(\frac{\partial R_x}{\partial x} + \frac{\partial R_y}{\partial y}\right) + \frac{4}{3t^2}\left(\frac{\partial^2 P_x}{\partial x^2} + 2\frac{\partial^2 P_{xy}}{\partial x \partial y} + \frac{\partial^2 P_y}{\partial x^2}\right) = 0$$

$$\frac{\partial M_x}{\partial x} + \frac{\partial M_{xy}}{\partial y} - Q_x + \frac{4}{t^2}R_x - \frac{4}{3t^2}\left(\frac{\partial P_x}{\partial x} + \frac{\partial P_{xy}}{\partial y}\right) = 0 \tag{9.109}$$

$$\frac{\partial M_{xy}}{\partial x} + \frac{\partial M_y}{\partial y} - Q_y + \frac{4}{t^2}R_y - \frac{4}{3t^2}\left(\frac{\partial P_{xy}}{\partial x} + \frac{\partial P_y}{\partial y}\right) = 0$$

The resultant forces and moments are defined as

$$(N_i, M_i, P_i) = \int_{-t/2}^{t/2} \sigma_i(1, z, z^3)dz, \quad i = x, y, xy$$

$$(Q_x, R_x) = \int_{-t/2}^{t/2} \tau_{xz}(1, z^2)dz \tag{9.110}$$

$$(Q_y, R_y) = \int_{-t/2}^{t/2} \tau_{yz}(1, z^2)dz$$

On integration, we get

$$
\begin{Bmatrix} N_x \\ N_y \\ N_{xy} \\ M_x \\ M_y \\ M_{xy} \\ P_x \\ P_y \\ P_{xy} \end{Bmatrix} =
\begin{bmatrix}
A_{11} & A_{12} & A_{16} & B_{11} & B_{12} & B_{16} & E_{11} & E_{12} & E_{16} \\
A_{12} & A_{22} & A_{26} & B_{12} & B_{22} & B_{26} & E_{12} & E_{22} & E_{26} \\
A_{16} & A_{26} & A_{66} & B_{16} & B_{26} & B_{66} & E_{16} & E_{26} & E_{66} \\
B_{11} & B_{12} & B_{16} & D_{11} & D_{12} & D_{16} & F_{11} & F_{12} & F_{16} \\
B_{12} & B_{22} & B_{26} & D_{12} & D_{22} & D_{26} & F_{12} & F_{22} & F_{26} \\
B_{16} & B_{26} & B_{66} & D_{16} & D_{26} & D_{66} & F_{16} & F_{26} & F_{66} \\
E_{11} & E_{12} & E_{16} & F_{11} & F_{12} & F_{16} & H_{11} & H_{12} & H_{16} \\
E_{12} & E_{22} & E_{26} & F_{12} & F_{22} & F_{26} & H_{12} & H_{22} & H_{26} \\
E_{16} & E_{26} & E_{66} & F_{16} & F_{26} & F_{66} & H_{16} & H_{26} & H_{66}
\end{bmatrix}
\begin{Bmatrix} \varepsilon_x^0 \\ \varepsilon_y^0 \\ \gamma_{xy}^0 \\ \kappa_x^0 \\ \kappa_y^0 \\ \kappa_{xy}^0 \\ \kappa_x^2 \\ \kappa_y^2 \\ \kappa_{xy}^2 \end{Bmatrix}
\qquad (9.111a)
$$

$$
\begin{Bmatrix} Q_y \\ Q_x \\ R_y \\ R_x \end{Bmatrix} =
\begin{bmatrix}
A_{44} & A_{45} & D_{44} & D_{45} \\
A_{45} & A_{55} & D_{45} & D_{55} \\
D_{44} & D_{45} & F_{44} & F_{45} \\
D_{45} & D_{55} & F_{45} & F_{55}
\end{bmatrix}
\begin{Bmatrix} \gamma_{yz}^0 \\ \gamma_{xz}^0 \\ \kappa_{yz}^2 \\ \kappa_{xz}^2 \end{Bmatrix}
\qquad (9.111b)
$$

where

$$
(A_{ij}, B_{ij}, D_{ij}, E_{ij}, F_{ij}, H_{ij}) = \int_{-t/2}^{t/2} \overline{Q}_{ij}(1, z, z^2, z^3, z^4, z^6)\,dz \quad (i, j = 1, 2, 6)
$$

and
$$
(A_{ij}, D_{ij}, F_{ij}) = \int_{-t/2}^{t/2} \overline{Q}_{ij}(1, z^2, z^4)\,dz \quad (i, j = 4, 5) \qquad (9.112)
$$

There are very few problems for which a closed form solution is possible. Reddy [17] employed the Navier approach to obtain the solution for a simply supported, symmetric cross-ply rectangular laminate. Details of the solution are given by Reddy [16]. Figure 9.6 shows the variation of non-dimensional central deflection of a square laminate with modular ratio.

It is observed that CLPT under-predicts the deflection even at lower values of modular ratio. The difference between the higher order theory and CLPT increases as the modular ratio increases. Thus, for thick laminates or laminates with higher modular ratio, one needs to resort to higher order transverse deformation theory for better prediction of responses.

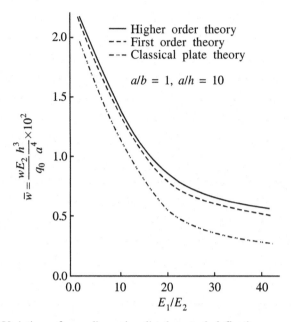

Fig. 9.6 Variation of non-dimensionalized central deflection as a function of modular ratio from a four layer cross-ply square laminate.

Summary

- Governing equations for the laminate have been derived assuming small deformation.
- In general, the results obtained on the basis of this theory are conservative.
- The governing equations are modified by the inclusion of higher order terms in the displacement relations.
- Several examples are solved to illustrate the technique.

Problems

9.1 Obtain the expression for critical buckling load of a laminated orthotropic rectangular plate simply supported along the loading edges and clamped along the opposite edges.

9.2 A 4 mm thick laminated composite plate is made up of carbon/PEEK. The material properties are as follows: Volume fraction, $V_f = 0.66$, $E_L = 134$ GPa, $E_T = 8.96$ GPa, $G_{LT} = 5.10$ GPa, $\nu_{LT} = 0.28$, $\sigma_{LU} = 2130$ MPa, $\sigma_{TU} = 80$ MPa, $\sigma'_{LU} = 1100$ MPa, $\sigma'_{TU} = 200$ MPa, $\tau_{LTU} = 160$ MPa.

(a) Calculate the [A], [B] and [D] matrices for the lay-up sequence $[0°/45°]_s$.
(b) Obtain the central deflection of a square laminate simply supported along all edges and subjected to a uniformly distributed transverse load P.

9.3 For the data given in Problem 9.2, obtain the central deflection of a clamped-clamped laminate.

9.4 For the data given in Problem 9.3, obtain the first natural frequency of the square laminate.

9.5 Repeat Problems 9.4 and 9.5, if the lay up sequence is $[0°/90°]_s$.

9.6 For a simply supported square anti-symmetric laminate relatively thick, obtain the expression for the initial buckling load using higher order shear deformation theory.

9.7 A simply supported, specially orthotropic laminate is subjected to an in-plane compressive load N_x along the x-direction and along the y-direction, it is subjected to a tensile load, $N_y = 0.5\ N_x$. Derive the expression for the first buckling load.

9.8 For the laminate defined in Problem 9.7, obtain the fundamental natural frequency of the laminate.

9.9 What will happen to the natural frequency, if the axial load N_x is tensile?

9.10 A simply supported specially orthotropic laminate is subjected to an in-plane compressive load N_x. Obtain the expression for natural frequency, assuming thin plate theory. What will happen to natural frequency if axial load is gradually increased?

References

1. Timoshenko, S. and Woinowsky-Kriger, S., 1959, *Theory of Plates and Shells*, 2nd Edition, McGraw Hill, New York.

2. Szilord, R., 1973, *Theory and Analysis of Plates: Classical and Numerical Methods*, Prentice Hall, Englewood Cliffs, New Jersey.

3. Ashton, J.E. and Whitney, J.M., 1970, *Theory of Laminated Plates*, Technomic, Westport, Connecticut.

4. Vinson, J.R. and Sierakowski, R.L., 2002, *The Behavior of Structures Composed of Composite Materials*, 2nd Edition, Kluwer Academic, The Netherlands.

5. Reddy, J.N., 2004, *Mechanics of Laminated Composite Plates and Shells*, 2nd Edition, CRC Press, Boca Raton, Florida.

6. Christensen, R.M., 1979, *Mechanics of Composite Materials*, Wiley, New York.

7. Kollar, L.P. and Springer, G.S., 2003, *Mechanics of Composite Structures*, Cambridge University Press, New York.

8. Iyengar, N.G.R., 2007, *Elastic Stability of Structural Elements*, Macmillan India Ltd., New Delhi.

9. Ashton, J.E., 1967, "Approximate Plate Analysis", *General Dynamics Research and Engineering*, FZM 4099.

10. Sharma, S., Iyengar, N.G.R. and Murthy, P.N., 1980, "The Buckling of Anti-symmetrically Laminated Angle-ply and Cross-ply Plates", *J. Fibre Sc. Tech.*, 13, pp. 29-48.

11. Sharma, S., Iyengar, N.G.R. and Murthy, P.N., 1980, "The Buckling of Anti-symmetrically Laminated Cross and Angle-ply Rectangular Plates on Winkler-Pasternak Foundation", *Proc. ASCE J. Engg. Mech. Div.*, 106, pp. 161-176.

12. Sharma, S., Iyengar, N.G.R. and Murthy, P.N., 1980, "The Buckling of Anti-symmetrically Cross and Angle-ply Laminated Plates", *Int. J. Mech. Sc.*, 22, pp. 607-620.

13. Reissner, E., 1945, "The Effect of Transverse Shear Deformation in Bending of Elastic Plates", *J. Appl. Mech.*, 18, pp. 69-77.
14. Mindlin, R.D., 1951, "Influence of Rotary Inertia and Shear on Flexural Motions of Isotropic Elastic Plates", *J. Appl. Mech.*, 18, pp. 326-343.
15. Whitney, J.M., 1969, "The Effect of Transverse Shear Deformation on the Bending of Laminated Plates", *J. Compos. Mater.*, 3, pp. 534-547.
16. Reddy, J.N., 1984, "A Simple Higher-Order Theory for Laminated Composite Plates", *J. Appl. Mech.*, 51, pp. 745-752.
17. Reddy, J.N., 1984, *Energy and Variational Methods in Applied Mechanics*, Wiley, New York.

10. Short Fiber Composites

In this chapter, we will:

- Learn about short fibers generally used in secondary structures in which the load carrying capacity is less.
- Analyze short fiber composites, where the fibers are aligned.
- Discuss the various models proposed.

10.1 Introduction

In earlier chapters, we discussed the behavior of continuous fiber composites. It was assumed that the fibers extend from one end of the lamina to the other end and remain intact throughout. When the load exceeds the failure load, the fibers break. In actual situations, fiber breaks take place even at small loads. In many applications, the direction of load generated is not clearly known, and hence the stress induced in the laminate is also not known. There can be situations when the state of stress generated is approximately the same in all directions. In such situations, unidirectional continuous fiber reinforcement may not be the best solution. One can design a laminate with a number of layers having different fiber orientations, which results in in-plane isotropy, but these will not exhibit isotropic behavior for bending loads. An alternate way to produce an isotropic layer is to use *random short fiber composites*. These are also termed as *discontinuous fiber composites*.

Short fiber reinforced composites are not as strong and stiff as continuous fiber composites. This is the reason they are not employed to withstand primary structural loads. They can be very efficiently used in components having complex geometrical shapes and contours. Depending on the method of manufacture, short or discontinuous composites are of different types. They are available as (i) aligned short fiber composites, (ii) off-axis aligned short fiber composites and (iii) random short fiber composites. Figure 10.1 shows the three forms of short or discontinuous composites. Short fiber composites are generally used in automobiles, storage tanks in chemical industries and in situations where isotropic behavior is required.

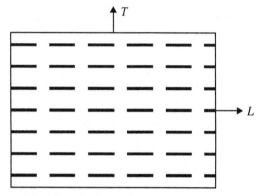

(a) Aligned short fiber composite

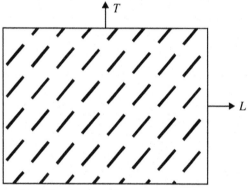

(b) Off-axis short fiber composite

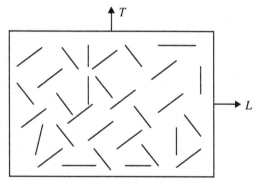

(c) Random short fiber composite

Fig. 10.1 Three forms of short fiber composites.

The analysis of short fiber composites is a little more involved when compared to continuous fiber composites. In this case, the load is not directly applied on the fibers. It is taken by the matrix and then transferred from the matrix to the fiber end and the cylindrical surface.

10.2 Analysis of Aligned Short Fiber Composites

As in the case of continuous fiber composite lamina, here again we consider a representative volume element (RVE) of aligned short fiber composite as shown in Fig. 10.2. The element is subjected to an axial tensile load. Depending on the type of element considered, the deformation pattern after the application of the load will be different. Figure 10.3 shows the behavior pattern for a case when the fiber is well within the RVE. It is observed that after the application of the axial load, deformation in the matrix is uniform up to the point where the matrix comes in contact with the fiber. There is a mismatch between the modulus of the fiber and the matrix. This leads to a large shear deformation near the ends of the fiber. At the middle of the fiber, there is no shear deformation. The axial load applied is transferred as interfacial shear between fiber and matrix. The interfacial shear has its peak value at the ends of the fiber. The analysis presented here is based on the work by Gibson [1]. Early investigations regarding the study of stress

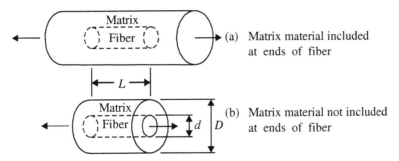

(a) Matrix material included at ends of fiber

(b) Matrix material not included at ends of fiber

Fig. 10.2 Representative volume element for aligned short fiber [1].

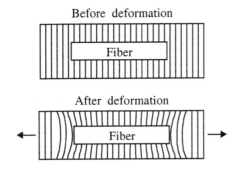

Fig. 10.3 Deformation behavior when fiber is well within RVE.

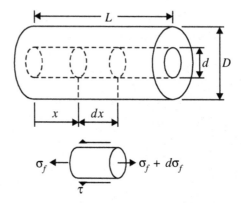

Fig. 10.4 Free body diagram of an element.

transfer are due to Cox [2] and Outwater [3]. Rosen [4] modified the analysis of Dow [5] based on shear lag analysis. Figure 10.4 shows the free body diagram of a differential element of the fiber after the load is applied.

Considering the equilibrium of forces along the fiber direction, we have

$$\left(\sigma_f + \frac{d\sigma_f}{dz}dz\right)\frac{\pi d^2}{4} - \sigma_f \frac{\pi d^2}{4} - \tau(z)\pi d\, dz = 0 \qquad (10.1)$$

where σ_f is the fiber stress along the axial direction, $\dfrac{d\sigma_f}{dz}$ is the variation of axial stress, d is the diameter of the fiber treated as a constant along the length of the fiber, τ is the interfacial shear stress at a distance z from one end of the fiber.

Simplifying Eq. (10.1), we get

$$\frac{d\sigma_f}{dz} = \frac{4\tau(z)}{d} \qquad (10.2)$$

Equation (10.2) indicates that for a uniform fiber diameter, the rate of variation of fiber stress is proportional to the shear stress. The fiber stress at any z can be obtained by integrating Eq. (10.2). Separating variables and integrating, we obtain

$$\int_{\sigma_0}^{\sigma_f} d\sigma_f = \frac{4}{d}\int_0^z \tau(z)\, dz \qquad (10.3)$$

This can be further simplified as

$$\sigma_f - \sigma_0 = \frac{4}{d}\int_0^z \tau(z)\, dz \qquad (10.4)$$

σ_0 is the stress at the fiber end. It is generally negligible since large stress concentration exists at the fiber end and as a consequence, the fiber end may separate from the matrix. The fiber stress at any z can be written as

$$\sigma_f = \frac{4}{d}\int_0^z \tau(z)\,dz \tag{10.5}$$

The fiber stress at any z can only be obtained if we know the variation of shear stress τ as a function of z. Kelly and Tyson [6] assumed that the matrix behaves as a rigid plastic material as shown in Fig. 10.5(a). Cox [2] assumed a linear elastic behavior as shown in Fig. 10.5(b). The Kelly and Tyson model is simple to apply when compared to the Cox model. Applying the Kelly and Tyson model, we obtain

$$\sigma_f = \frac{4}{d}\tau_{ym}z \tag{10.6}$$

where τ_{ym} is the matrix yield stress. Equation (10.6) indicates that the fiber stress varies linearly with the distance from the fiber end. Further, fiber stress distribution should be symmetric at the middle of the fiber, i.e. $z = L/2$.

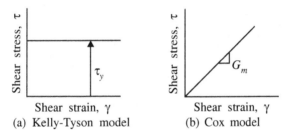

Fig. 10.5 Behavior of matrix material.

Therefore

$$(\sigma_f)_{\max} = \frac{2\tau_{ym}L}{d} \tag{10.7}$$

The ultimate fiber stress cannot exceed the value given by a continuous fiber of diameter d. It has been shown earlier that for a fiber subjected to an axial load, the maximum fiber stress is given by

$$(\sigma_f)_{\max} = \frac{E_f}{E_c}\sigma_c \tag{10.8}$$

where σ_c is the applied composite stress and E_c is the composite modulus. The minimum fiber length in which the maximum stress given by Eq. (10.8) is achieved has been referred to as the *ineffective length* [4] or the *load-transfer length* [7].

Equating Eqs. (10.7) and (10.8), we get

$$\frac{2\tau_{ym}L_i}{d} = \frac{E_f}{E_c}\sigma_c \qquad (10.9)$$

Simplifying

$$L_i = \frac{d\,E_f\,\sigma_c}{2\tau_{ym}\,E_c} \qquad (10.10)$$

In view of Eq. (10.8), the *load-transfer length* L_i is a function of applied stress.

Figure 10.6 shows the variation of shear and fiber stress along the fiber length. It is observed that the load transfer takes place over the length L_i. Further, up to this length, the fiber stress is less than the maximum value. Beyond this length, fiber stress remains constant and there is no load transfer from the matrix to the fiber.

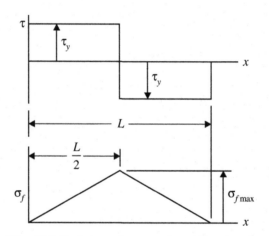

Fig. 10.6 Variation of stress along fiber length.

Another limiting stress is the fiber strength. This can be defined as

$$\sigma_{fu} \geq \frac{E_f\sigma_c}{E_c} \qquad (10.11)$$

The fiber length corresponding to this ultimate fiber stress is defined as

$$L_c = \frac{\sigma_{fu}\,d}{2\,\tau_{ym}} \qquad (10.12)$$

This length is referred to as the *critical length*. L_c is independent of the applied stress. Equation (10.12) can also be rewritten to give the interfacial shear strength corresponding to the critical length as

$$\tau_{ym} = \frac{\sigma_{fu}\, d}{2\, L_c} \tag{10.13}$$

Drzal et al. [8, 9] employed Eq. (10.13) to determine the interfacial shear strength from measurements of *critical length*.

We can also obtain the longitudinal modulus of the aligned discontinuous fiber composite using the model proposed by Kelly [10]. A similar model has been developed by Rosen [4], which is referred to as the *shear lag* model. Equation (10.12) indicates that the rate of change of axial stress in the fiber along the length is a linear function of interfacial shear stress. We present here some pertinent results. Readers may refer to details of the Kelly [10] model which is available in Gibson [1].

Following the steps given by Gibson [1], the governing differential equation for the fiber axial load F is given as

$$\frac{d^2 F}{dz^2} - \lambda^2 F = -\kappa\varepsilon \tag{10.14}$$

where κ is a constant and a function of geometrical and material properties, ε is the strain in the matrix in the absence of the fiber and

$$\lambda^2 = \frac{\kappa}{A_f E_f} \tag{10.15}$$

Equation (10.14) is a non-homogenous second order differential equation, the solution of which consists of a particular solution and a complimentary solution. The complimentary solution involves two arbitrary constants to be determined from the boundary conditions. The boundary conditions are obtained from the fact that the axial load F is zero at $z = 0$ and $z = L$. After going through the various steps indicated by Gibson [1], the resulting fiber stress at any z is given as

$$\sigma_f = E_f \varepsilon \left[1 - \frac{\cosh\lambda(0.5L - z)}{\cosh(0.5\lambda L)} \right] \tag{10.16}$$

Following the rule of mixtures and assuming that the stress-strain relation for fiber and the matrix is linearly elastic, the longitudinal modulus is written as

$$E_c = E_f \left[1 - \frac{\tanh(\lambda L/2)}{(\lambda L/2)} \right] \mathbf{v_f} + E_m (1 - \mathbf{v_f}) \tag{10.17}$$

Kelly [10] has further shown that

$$\lambda^2 = \frac{2\pi G_m}{A_f E_f \ln(D/d)} \tag{10.18}$$

and

$$\tau = \frac{E_f \, d \, \varepsilon \lambda}{4} \left[\frac{\sinh[\lambda(0.5L - z)]}{\cosh(0.5\lambda L)} \right] \qquad (10.19)$$

where G_m is the matrix shear modulus and D is the external diameter of the element. Figure 10.7 shows the variation of interfacial shear stress τ and the fiber stress σ_f as a function of fiber length $L < L_i$. Gibson et al. [11] used the relations for σ_f and τ in a strain energy method. The longitudinal modulus E_c calculated from the energy approach was found to agree closely with the Cox model.

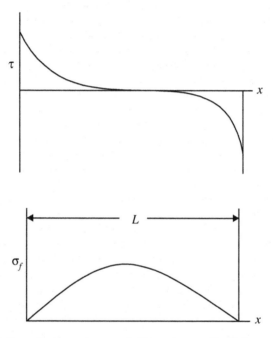

Fig. 10.7 Variation of shear stress and fiber stress with distance (Cox model).

10.3 Improved Analysis for Aligned Short Fiber Composites

Figure 10.8 shows the stress distributions obtained by assuming a perfectly plastic model for the matrix. Matrix material generally exhibits elastic-plastic behavior. The stress analysis on the basis of elastic-plastic behavior of the matrix is quite involved. One can get a solution close to the exact solution by employing a numerical technique such as the finite element analysis. Finite element analysis has been carried out by a number of investigators [12-21]. Some of them have treated the matrix material to be elastic while others have treated it as elastic-plastic. Some results of these analyses are presented here. For a more detailed discussion, the readers may refer to the cited references. Figure 10.7 shows the variation of fiber stress and interfacial shear stress along the length obtained by treating the matrix

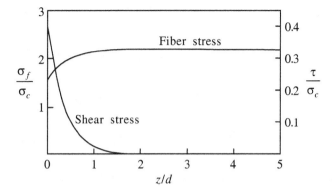

Fig. 10.8 FEA analysis result. Fiber stress and shear stress along fiber length [7].

to be elastic. Figure 10.8 shows the variation of matrix stresses as a function of fiber length. It is observed that there is a stress concentration at the fiber ends. This may lead to the separation of fibers from the matrix at the fiber ends. Bansal [19] obtained the stress distribution by treating the matrix as elastic-plastic. His analysis shows that the stress transfer through the fiber end is negligible. Interfacial shear stress near the fiber end is not a constant.

Suarez et al. [22], on the basis of experiments conducted on aligned discontinuous graphite/epoxy composites, observed that measured moduli are very much lower than the predicted curve from the Cox model. To improve the experimental results, they introduced the concept of an *effective fiber aspect ratio*, which would account for the fact that the reinforcement was not a single fiber, but a bundle of fibers having an aspect ratio lower than that of a single fiber.

The effective fiber aspect ratio is defined as

$$(L/d)_{\text{eff}} = \chi(L/d) \tag{10.20}$$

where χ is a curve fitting parameter.

10.4 Semi-empirical Relations for Prediction of Modulus and Strength of Aligned Short Fiber Composites

The analytical and finite element methods discussed in the earlier sections for prediction of longitudinal and transverse moduli is quite involved. The Halpin-Tsai equations for continuous fibers are modified to predict the longitudinal and transverse moduli for discontinuous fibers and can be written as

$$\frac{E_L}{E_m} = \frac{1 + (2L/d)\eta_L \mathbf{v}_f}{1 - \eta_L \mathbf{v}_f} \tag{10.21}$$

$$\frac{E_T}{E_m} = \frac{1 + 2\eta_T \mathbf{v}_f}{1 - \eta_T \mathbf{v}_f} \qquad (10.22)$$

where

$$\eta_L = \frac{(E_f / E_m) - 1}{(E_f / E_m) + 2(L/d)}$$

$$\eta_T = \frac{(E_f / E_m) - 1}{(E_f / E_m) + 2} \qquad (10.23)$$

It is observed that the longitudinal elastic modulus depends on the ratio of fiber length to fiber diameter. However, the transverse modulus is not affected by the (L/d) ratio.

The average fiber stress $\bar{\sigma}_f$ is given as

$$\bar{\sigma}_f = \frac{1}{L} \int_0^L \sigma_f \, dz \qquad (10.24)$$

The average fiber stress depends on whether the fiber length is less than or greater than the *load transfer length* L_i.

The average stress $\bar{\sigma}_f$ can then be found from the figure as

$$\bar{\sigma}_f = \frac{1}{2}(\sigma_f)_{max} = \frac{\tau_{my} L}{d}, \quad L < L_i \qquad (10.25)$$

$$\bar{\sigma}_f = (\sigma_f)_{max}\left(1 - \frac{L_i}{2L}\right), \quad L > L_i \qquad (10.26)$$

Thus, for aligned short fiber composites, the composite stress can be written following the rule of mixtures as

$$\sigma_c = \frac{1}{2}(\sigma_f)_{max} \mathbf{v}_f + \sigma_m \mathbf{v}_m, \quad L < L_i \qquad (10.27)$$

$$\sigma_c = (\sigma_f)_{max}\left(1 - \frac{L_i}{2L}\right)\mathbf{v}_f + \sigma_m \mathbf{v}_m, \quad L > L_i \qquad (10.28)$$

If the length of the fiber in the given composite is less than the *critical length,* L_c, the fibers will not break since the maximum fiber stress is less than the average fiber strength. Composite strength is governed by the failure of the matrix or interface. The ultimate strength is given by

$$\sigma_{cul} = \frac{\tau_{my} L}{d}\mathbf{v}_f + \sigma_{mu} \mathbf{v}_m, \quad L < L_c \qquad (10.29)$$

If the length of the fiber is greater than the *critical length* L_c, the fiber failure initiates when the maximum fiber stress is equal to the ultimate strength of the fibers. The ultimate composite stress is then given by

$$\sigma_{cul} = \sigma_{ful}\left(1 - \frac{L_c}{2L}\right)\mathbf{v}_f + (\sigma_m):\varepsilon_f^* \mathbf{v}_m, \quad L > L_c \tag{10.30}$$

When the length of the fiber is very much greater than the *critical length*, one can employ the continuous fiber equations to determine the ultimate strength of the composite.

While writing up Eqs. (10.29) and (10.30), it is assumed that the fiber volume fraction is greater than $\mathbf{v}_{min}$.

Similar to continuous fibers, one can also define fiber volume fraction $\mathbf{v}_{min}$ and $\mathbf{v}_{cric}$ for discontinuous fibers. When the fiber volume fraction in a composite is less than $\mathbf{v}_{min}$, the fibers will be stretched beyond their failure strain. The entire load is supported by the matrix. The ultimate strength of the composite is governed by the matrix alone and is given by

$$\sigma_{cul} = \sigma_{mul}(1 - \mathbf{v}_f) \tag{10.31}$$

At fiber volume fraction of $\mathbf{v}_{min}$, the composite failure will be a combination of fiber failure and matrix. A limiting value can be found when the failure mode changes and is given by equating

$$\overline{\sigma}_f \mathbf{v}_f + (\sigma_m):\varepsilon_f^*(1 - \mathbf{v}_f) = \sigma_{mul}(1 - \mathbf{v}_f) \tag{10.32}$$

Simplifying Eq. (10.32)

$$\mathbf{v}_{min} = \frac{\sigma_{mul} - (\sigma_m):\varepsilon_f^*}{\overline{\sigma}_f - \sigma_{mul} - (\sigma_m):\varepsilon_f^*} \tag{10.33}$$

In this case, there is no strengthening of the composite as the ultimate stress of the composite can at the most be σ_{mul}. One can also define $\mathbf{v}_{cric}$, the fiber volume fraction beyond which there will be strengthening of the composite as

$$\sigma_{cul} = \overline{\sigma}_f \mathbf{v}_f + (\sigma_m):\varepsilon_m^*(1 - \mathbf{v}_f) \geq \sigma_{mul}$$

or $$\mathbf{v}_{cric} = \frac{\sigma_{mul} - (\sigma_m):\varepsilon_f^*}{\overline{\sigma}_f - (\sigma_m):\varepsilon_f^*} \tag{10.34}$$

10.5 Off-axis Aligned Short Fiber Composites

The analysis of off-axis aligned short fiber composites is more involved as the failure mode changes when the fiber orientation is varied from 0° to 90°. Chon and Sun [23] employed a shear-lag analysis. Readers may refer to Gibson [1] for details. Some of the important observations of the analysis are:

(i)   The maximum interfacial stress τ_{max} occurs at some orientation of the fibers.

(ii) τ_{max} decreases with increasing E_f/G_m, and the fiber orientation angle increases with increasing E_f/G_m.

(iii) Fiber volume fractions play an important role in strength of composites.

(iv) The off-axis elastic constants are influenced more by fiber orientation than fiber length.

10.6 Randomly Oriented Short Fiber Composites

If the requirements of the design are to obtain isotropic properties in all directions by using short fiber composites, then it is necessary that the short fibers are randomly oriented in the plane. This technique is adopted when the direction of external loads acting on the structure is not clearly known and also in situations when the external loads acting do not have any preferred direction. Details of the investigations carried out by others to predict the modulus of the random short fiber composites are given by Gibson [1]. Tsai and Pagano [24] developed the following approximate relations to predict the modulus:

$$\tilde{E} = \frac{3}{8}E_L + \frac{5}{8}E_T, \quad \tilde{G} = \frac{1}{8}E_L + \frac{1}{4}E_T \tag{10.35}$$

$\tilde{E}$, $\tilde{G}$ are the elastic modulus and shear modulus of random short fiber composites. E_L and E_T are the longitudinal and transverse moduli of aligned short fiber composites having the same L/d and fiber volume fraction as the random short fiber composites. E_L and E_T are defined by Eq. (10.21) and Eq. (10.22), respectively. Knowing $\tilde{E}$ and $\tilde{G}$, one can obtain Poisson's ratio by using the constraint relation for isotropic materials.

10.7 Aligned Ribbon Reinforced Composites

Another form of short fiber composites is a ribbon reinforced or tape reinforced composite. As compared to conventional short fiber composites, this has better stiffness and strength in the longitudinal and transverse directions. The ribbon has rectangular cross-section with length greater than the width. Since there are reinforcements along both directions, they are generally more resistant to impact. Further, moisture absorption is less compared to fiber reinforced composites. Following Fig. 10.9 and from Agarwal et al. [25], the volume fraction of reinforcement is given by

$$v_{fr} = \frac{L_{fr}w_{fr}}{(L_{fr} + L_m)(w_{fr} + w_m)} = \frac{1}{[1 + (L_m/L_{fr})][1 + (w_m/w_{fr})]} \tag{10.36}$$

where L_{fr} and w_{fr}, are the length and width of the ribbon, respectively, L_m is the length of the matrix between the ribbon and w_m is the spacing

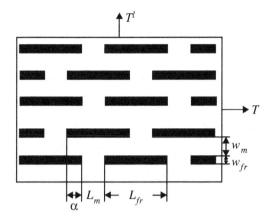

Fig. 10.9 Cross-section of ribbon reinforced composite.

between the ribbons in the width direction. Equation (10.36) can be rewritten considering the overlap between the ribbon in the thickness directions as

$$\mathbf{v_{fr}} = \frac{1}{2[1-(\alpha/L_{fr})][1+(w_m/w_{fr})]} \tag{10.37}$$

where α is the overlap length.

The Halpin-Tsai relations for the estimation of longitudinal and transverse moduli for a ribbon reinforced composite are

$$E_{Lr} = E_{fr}\mathbf{v_{fr}} + E_m\mathbf{v}_m \tag{10.38a}$$

$$E_{Tr} = \frac{1+(2L_{fr}/w_{fr})\eta_r \mathbf{v_{fr}}}{1-\eta_r \mathbf{v_{fr}}} \tag{10.38b}$$

$$\eta_r = \frac{(E_{fr}/E_m)-1}{(E_{fr}/E_m)+2(L_{fr}/w_{fr})} \tag{10.39}$$

Ribbon reinforced composites will give high ultimate strength provided the adhesion between the matrix and ribbon is good.

Summary
- Various models proposed for aligned and random short fiber composites have been discussed.
- Empirical relations for random short fiber composites are discussed.

References

1. Gibson, R.F., 1994, *Principles of Composite Material Mechanics*, McGraw-Hill, International Edition.
2. Cox, H.L., 1952, "The Elasticity and Strength of Paper and Other Fibrous Materials", *Br. J. Appl. Phy.*, 3, pp. 72-79.

3. Outwater, J.O., 1956, "New Predictions and Interpretations", *Mod. Plast.*, 56.
4. Rosen, B.W., 1964, *Mechanics of Composite Strengthening in Fiber Composite Materials*, Ch. 3, American Society for Metals, Metals Park, Ohio, p. 37.
5. Dow, N.F., 1963, "Study of Stresses Near a Discontinuity in a Filament Reinforced Composite Material", General Electric Company Report No. TISR63SD612.
6. Kelly, A. and Tyson, W.R., 1965, "Tensile Properties of Fiber Reinforced Metals: Copper/Tungsten and Copper/Molybdenum", *J. of Mech. and Phys. of Solids*, 13, pp. 329-350.
7. Agarwal, B.D. and Broutman, L.J., 1990, *Analysis and Performance of Fiber Composites*, 2nd Edition, John Wiley and Sons, New York.
8. Drzal, L.T., Rich, M.J. and Llyod, P.F., 1982, "Adhesion of Graphite Fibers to Epoxy Matrices II: The Role of Fiber Surface Treatment", *Journal of Adhesion*, 16, pp. 1-30.
9. Drzal, L.T., Rich, M.J., Koenig, M.F. and Llyod, M.F., 1983, "Adhesion of Graphite Fibers to Epoxy Matrices II: The Effect of Fiber Finish", *Journal of Adhesion*, 16, pp. 133-152.
10. Kelly, A., 1973, *Strong Solids*, 2nd Edition, Clarendon Press, Oxford, England.
11. Gibson, R.F., Chaturvedi, S.K. and Sun, C.T., 1982, "Complex Moduli of Aligned Discontinuous Fiber Reinforced Polymer Composites", *Journal of Material Science*, 17, pp. 3499-3509.
12. Carrara, A.S. and McGarry, F.J., 1968, "Matrix and Interface Stresses in a Discontinuous Fiber Composite Model", *J. Composite Materials*, 2, pp. 222-243.
13. Maclaughlin, T.F. and Barker, R.M., 1972, "Effect of Modulus Ratio on Stress Near a Discontinuous Fiber", *Experimental Mechanics*, 12, pp. 178-183.
14. Owen, D.R. and Lyness, J.F., 1972, "Investigations of Bond Failure in Fiber Reinforced Material by Finite Element Methods", *Fiber Science Technology*, 5, pp. 129-141.
15. Broutman, L.J. and Agarwal, B.D., 1974, "A Theoretical Study of the Effect of an Interfacial Layer on the Properties of Composites", *Polymer Engg. Science*, 14, pp. 581-588.
16. Lin, T.H., Salinas, D. and Ito, Y.M., 1972, "Elastic-plastic Analysis of Unidirectional Composites", *J. Composite Materials*, 6, pp. 48-60.
17. Agarwal, B.D., 1972, "Micro Mechanics Analysis of Composite Materials using Finite Element Methods", *Ph.D. Thesis*, IIT, Chicago, Illinois.
18. Agarwal, B.D., Lifshitz, J.M. and Broutman, L.J., 1974, "Elastic-plastic Finite Element Analysis of Short Fiber Composites", *Fiber Science Technology*, 7, pp. 45-62.
19. Bansal, R.K., 1976, "Finite Element Analysis of Fiber Interaction in Discontinuous Fiber Reinforced Composites", *M.Tech. Thesis*, Indian Institute of Technology Kanpur.
20. Agarwal, B.D. and Bansal, R.K., 1977, "Plastic Analysis of Fiber Interactions in Discontinuous Fibre Composites", *Fibre Science Technology*, 10, pp. 281-297.
21. Agarwal, B.D. and Bansal, R.K., 1979, "Effect of an Interfacial Layer on the Properties of Fibrous Composites: A Theoretical Analysis", *Fibre Science Technology*, 12, pp. 149-158.

22. Suarez, S.A., Gibson, R.F., Sun, C.T. and Chaturvedi, S.K., 1986, "The Influence of Fiber Length and Fiber Orientation on Damping and Stiffness of Polymer Composite Materials", *Experimental Mechanics*, 26, pp. 175-184.

23. Chon, C.T. and Sun, C.T., 1980, "Stress Distribution Along a Short Fiber in Fiber Reinforced Platics", *Journal of Material Science*, 15, pp. 931-938.

24. Tsai, S.W. and Pagano, P.E., 1968, "Invariant Properties of Composites Materials", In Tsai, S.W., Halpin, J.C. and Paganao, N.J. (Eds.), *Composite Materials Workshop*, Technomic Publishing Co., Lancaster, Pennsylvania, pp. 233-252.

25. Agarwal, B.D., Broutman, L.J. and Chandrashekhara, K., 2006, *Analysis and Performance of Fiber Composites*, 3rd Edition, John Wiley and Sons, New York.

11. Optimization

In this chapter, we will discuss:

- Optimization of composite laminates subjected to various in-plane and out-of-plane loads.
- Different mathematical techniques employed for optimization.

11.1 Introduction

Decision making features predominantly in all fields of human activity, including scientific and technological, and affects every sphere of life. Engineering design which entails sizing, dimensioning and detailed planning is not exempt from influence. Composites are extensively used in aerospace, automobile and other industries. They are subjected to varieties of loads during their lifetime. In addition to offering advantages such as low weight, corrosion resistance, high fatigue strength and dimensional stability, they possess high specific modulus and specific strength.

The advantages of composites can be realized if they can be used optimally in order to meet weight and strength requirements.

The techniques of *linear programming*, *nonlinear programming*, *geometric programming*, *genetic algorithm* etc. can be used to find solutions for a certain class of optimization problems. Depending on the nature of the problem, we can directly or indirectly apply one or more of these techniques.

Very few structural engineering problems, irrespective of the materials employed, can be grouped under linear programming problems since either the objective function and/or the constraints are nonlinear functions of design variables. Around 1960, several numerical methods were developed to solve nonlinear optimization problems. The conjugate gradient methods of Fletcher and Reeves [1] and the variable metric methods of Davidon [2] and Fletcher-Powell [3] known as the DFP method have been employed for unconstrained minimization problems. For constrained minimization problems, Fiacco and McCormick's SUMT techniques [4] have been successfully used in a number of problems.

Around the same time, along with developments in gradient based methods, there were developments in non-gradient or direct methods. The pattern search method of Hooke and Jeeves [5] and Powell's method of conjugate directions [6] belong to this group. Most recent among the direct methods are genetic algorithms due to Holland [7] and Goldberg [8].

The use of nonlinear optimization techniques in structural design was pioneered by Schmidt. Since then, a number of techniques have been developed. Notable among them are due to Fox [9], Haug and Arora [10], Reklaitis et al. [11], Haftka [12] and Rao [13]. Most of these deal with isotropic materials. Iyengar and Gupta [14] have used nonlinear programming techniques for optimization of composite laminates subjected to a variety of loadings. These techniques have been extensively applied to minimize weight, increase stiffness, increase critical buckling load and natural frequencies by a proper choice of design parameters. Fiber reinforced composites have a large number of design parameters, for example, fiber orientation of plies, thickness of plies, lay up sequence, material properties of individual plies, number of plies and so on. In view of the large number of design variables, the optimization problem for composites is quite involved. Nonlinear programming techniques generally lead to a local optimum. One has to repeat the optimization procedure with different initial values to check whether the optimum obtained is local or global.

The use of genetic algorithm was pioneered by Holland [7]. It has become a powerful and robust tool for function optimization. These algorithms are computationally simple but powerful in their search for improvement in successive generations [8]. Genetic algorithm mimics some of the processes observed in natural evolution. The basic techniques of genetic algorithm are designed to simulate the mechanism of population genetics and natural rules for survival in pursuit of the ideas of adaptation. One of the big advantages of genetic algorithm is that it does not require differentiability of either the objective function or the constraints [8]. The constraint handling capacity of a genetic algorithm is better than other optimization techniques because of its population based approach [15]. One of the earliest applications of genetic algorithm for structural optimization was the design of a 10 bar truss [16]. Rao et al. [17] studied the optimal placement of actuators in actively controlled structures. Chaturvedi et al. [18] studied structural optimization using a real coded genetic algorithm.

In recent years, many research workers have employed genetic algorithm for the optimum design of composite structures. Nagendra et al. [19] considered the optimum laminate stacking sequence for buckling with strain constraints. Investigations on the design of composite laminates for maximizing laminate strength and stiffness with a fixed number of plies was presented by Callahan and Week [20]. Kogiso et al. [21] employed genetic algorithms for the design of composite laminates subject to strength, buckling and ply contiguity conditions. Mahesh et al. [22] have applied genetic algorithms for the optimal design of turbine blades. Muc and Gurba [23] used the genetic algorithm for layout optimization of composite laminates with finite element computation of the objective function. Lin and Hajela [24] described the implementation of genetic search methods for the optimal design of structural systems with a mix of continuous, integer and discrete design variables. Optimization of composite laminates with cutouts is a complex problem involving a non-differentiable objective function and constraints.

In this chapter, we will discuss a few problems of optimization of composite laminates employing nonlinear programming and genetic algorithm techniques.

Nonlinear Programming Techniques (Application)

Example 11.1
Find the minimum weight design of angle-ply laminated composite plates simply supported on all edges and subjected to transverse and/or dynamic loads.

Solution
In this study, we shall obtain the minimum weight design of a composite laminate for the following two cases:

(i) An anti-symmetric angle-ply laminate with fiber orientation sequence alternating between $+\theta$ and $-\theta$.
(ii) An orthotropic angle-ply laminate with an equal number of $+\theta$ and $-\theta$ fiber orientation in each lamina.

Based on the classical laminate theory, the equilibrium equations in terms of displacements u, υ and w for a general laminate are [25]

$$A_{11}\frac{\partial^2 u}{\partial x^2} + 2A_{16}\frac{\partial^2 u}{\partial x \partial y} + A_{66}\frac{\partial^2 u}{\partial y^2} + A_{16}\frac{\partial^2 \upsilon}{\partial x^2} + (A_{12} + A_{66})\frac{\partial^2 \upsilon}{\partial x \partial y} + A_{26}\frac{\partial^2 \upsilon}{\partial y^2}$$

$$- B_{11}\frac{\partial^3 w}{\partial x^3} - 3B_{16}\frac{\partial^3 w}{\partial x^2 \partial y} - (B_{12} + 2B_{66})\frac{\partial^3 w}{\partial x \partial y^2} - B_{26}\frac{\partial^3 w}{\partial y^3} = 0 \qquad (11.1.1)$$

$$A_{16}\frac{\partial^2 u}{\partial x^2} + 2A_{26}\frac{\partial^2 \upsilon}{\partial x \partial y} + A_{26}\frac{\partial^2 u}{\partial y^2} + A_{66}\frac{\partial^2 \upsilon}{\partial x^2} + (A_{12} + A_{66})\frac{\partial^2 u}{\partial x \partial y} + A_{22}\frac{\partial^2 \upsilon}{\partial y^2}$$

$$- B_{16}\frac{\partial^3 w}{\partial x^3} - 3B_{26}\frac{\partial^3 w}{\partial x \partial y^2} - (B_{12} + 2B_{66})\frac{\partial^3 w}{\partial x^2 \partial y} - B_{22}\frac{\partial^3 w}{\partial y^3} = 0 \qquad (11.1.2)$$

$$D_{11}\frac{\partial^4 w}{\partial x^4} + 4D_{16}\frac{\partial^4 w}{\partial x^3 \partial y} + 2(D_{12}+2D_{66})\frac{\partial^4 w}{\partial x^2 \partial y^2} + 4D_{26}\frac{\partial^4 w}{\partial x \partial y^3} + D_{22}\frac{\partial^4 w}{\partial y^4}$$

$$- B_{11}\frac{\partial^3 u}{\partial x^3} - 3B_{16}\frac{\partial^3 u}{\partial x^2 \partial y} - (B_{12}+2B_{66})\frac{\partial^3 u}{\partial x \partial y^2} - B_{22}\frac{\partial^3 \upsilon}{\partial y^3} - B_{16}\frac{\partial^3 \upsilon}{\partial x^3}$$

$$- (B_{12}+2B_{66})\frac{\partial^3 \upsilon}{\partial x^2 \partial y} - 3B_{26}\frac{\partial^3 \upsilon}{\partial x \partial y^2} = q(x,y) + \rho\frac{\partial^2 w}{\partial t^2} \tag{11.1.3}$$

The stiffness coefficients are defined in earlier chapters. On the right-hand side of Eq. (11.1.3), the first term corresponds to the transverse loading and the second term to the inertial forces that develop due to oscillations. In this study, the two terms on the right-hand side of Eq. (11.1.3) are considered one at a time.

For an anti-symmetric angle-ply laminate, the stiffness coefficients identically zero are

$$A_{11} = A_{26} = D_{16} = D_{26} = B_{11} = B_{22} = B_{12} = B_{66} = 0 \tag{11.1.4}$$

Further, for an orthotropic angle-ply laminate, the stiffness coefficients identically zero are

$$A_{16} = A_{26} = B_{16} = B_{26} = D_{16} = D_{26} = 0 \tag{11.1.5}$$

The boundary conditions for a laminate simply supported on all edges are

$$w(0, y) = 0, \ M_x(0, y) = 0, \ \upsilon(0, y) = 0 \ (x = 0)$$
$$w(a, y) = 0, \ M_x(0, y) = 0, \ \upsilon(a, y) = 0 \ (x = a)$$
$$w(x, 0) = 0, \ M_y(x, 0) = 0, \ u(x, 0) = 0 \ (y = 0)$$
$$w(x, b) = 0, \ M_y(x, b) = 0, \ u(x, b) = 0 \ (y = b) \tag{11.1.6}$$

The *objective function* $F(t_k)$ is defined by

$$F(t_k) = \text{minimize}\left(ab\rho\sum_{i=1}^{N} t_i \right) \tag{11.1.7}$$

where a, b are the laminate dimensions, ρ is the density of the material and is taken to be the same for all the plies in the laminate, and t_i is the thickness of the ith ply. Since the laminate size and lamina material are fixed, minimizing the weight implies minimizing the total thickness of the laminate.

The behavior constraints are:

(i)   The deflection of the laminate for a given load should be less than that prescribed, i.e.

$$0 \le W \le W_{max} \tag{11.1.8}$$

(ii) The first natural frequency of the laminate should be bounded, i.e.

$$f_{min} \le f \le f_{max} \tag{11.1.9}$$

(iii) The stress within the domain of the laminate should (to prevent lamina failure) be less than that predicted by the Tsai-Hill criterion, i.e.

$$\left(\frac{\sigma_L}{\sigma_{LU}}\right)^2 + \left(\frac{\sigma_T}{\sigma_{TU}}\right)^2 + \left(\frac{\tau_{LT}}{\tau_{LTU}}\right)^2 - \left(\frac{\sigma_L}{\sigma_{LU}}\right)\left(\frac{\sigma_T}{\sigma_{LU}}\right) < 1 \qquad (11.1.10)$$

The *side constraints* imposed on the thickness of each ply is

$$t_{\min} \le t_i \le t_{\max} \qquad (11.1.11)$$

The *optimization problem* as stated by Eqs. (11.1.7)-(11.1.10) is a constrained optimization problem. This constrained problem is transformed into a series of unconstrained problems by using the interior penalty function approach. The advantage of choosing the interior penalty function is that the solution is always in the feasible domain. The optimization problem then reduces to

$$P(r, t_i) = F(t_i) - r \sum_{j=1}^{m} g_j(\theta_i, t_i) \qquad (11.1.12)$$

where θ_i is the fiber orientation of the ith ply. Equation (11.1.12) is then solved by using the Davidon-Fletcher-Powell variable metric method. To evaluate the constraints (g_i) in Eq. (11.1.12), we have to solve the governing equations with the associated boundary conditions.

Static Case

The displacements u, υ and w which satisfy the boundary conditions (11.1.6) can be expressed as

$$u(x, y) = \sum_m \sum_n U_{mn} \sin\frac{m\pi x}{a}\cos\frac{n\pi y}{b} \qquad (11.1.13a)$$

$$\upsilon(x, y) = \sum_m \sum_n V_{mn} \cos\frac{m\pi x}{a}\sin\frac{n\pi y}{b} \qquad (11.1.13b)$$

$$w(x, y) = \sum_m \sum_n W_{mn} \sin\frac{m\pi x}{a}\sin\frac{n\pi y}{b} \qquad (11.1.13c)$$

and the transverse load $q(x, y)$ can be expressed as

$$q(x, y) = \sum_m \sum_n Q_{mn} \sin\frac{m\pi x}{a}\sin\frac{n\pi y}{b} \qquad (11.1.13d)$$

where m and n are the number of half sine waves along the x- and y-direction and the coefficient $Q_{mn} = 16q_0/(\pi^2 mn)$, q_0 being the intensity of loading. With this representation, the governing equations are identically satisfied. Substitution of Eq. (11.1.3) into the governing equation gives a set of simultaneous equations from which the displacement coefficients U_{mn}, V_{mn} and W_{mn} can be obtained as [14]

$$U_{mn} = q_0(T_{22}T_{13} - T_{12}T_{23})/R,\; V_{mn} = q_0(T_{11}T_{23} - T_{13}T_{12})/R,\; W_{mn} = q_0(T_{11}T_{22} - T_{12}^2)/R$$
$$(11.1.14)$$

where

$$R = T_{11}(T_{22}T_{33} - T_{23}^2) - T_{12}(T_{12}T_{33} - T_{13}T_{23}) - T_{13}(T_{12}T_{23} - T_{13}T_{23}) \quad (11.1.15)$$

In Eqs. (11.1.14) and (11.1.15), each of T_{11}, T_{12},... is a function of the stiffness coefficients and laminate dimensions.

$$T_{11} = A_{11}\left(\frac{m\pi}{a}\right)^2 + A_{66}\left(\frac{n\pi}{b}\right)^2,\; T_{12} = (A_{12} + A_{66})\left(\frac{m\pi}{a}\right)\left(\frac{n\pi}{b}\right) \quad (11.1.16a)$$

$$T_{13} = -\left[3B_{16}\left(\frac{m\pi}{a}\right)^2 + B_{26}\left(\frac{n\pi}{b}\right)^2\right]\left(\frac{n\pi}{b}\right),\; T_{22} = A_{66}\left(\frac{m\pi}{a}\right)^2 + A_{22}\left(\frac{n\pi}{b}\right)^2 \quad (11.1.16b)$$

$$T_{23} = -\left[B_{16}\left(\frac{m\pi}{a}\right)^2 + 3B_{26}\left(\frac{n\pi}{b}\right)^2\right]\left(\frac{m\pi}{a}\right) \quad (11.1.16c)$$

$$T_{33} = D_{11}\left(\frac{m\pi}{a}\right)^4 + 2(D_{12} + 2D_{66})\left(\frac{m\pi}{a}\right)^2\left(\frac{n\pi}{b}\right)^2 + D_{22}\left(\frac{n\pi}{b}\right)^4 \quad (11.1.16d)$$

For an orthotropic angle-ply laminate, the displacement coefficient equations remain unchanged along with coefficients T_{11} etc., except the coefficients T_{13} and T_{23}, which take the form

$$T_{13} = B_{11}\left(\frac{m\pi}{a}\right)^3 + (B_{12} + 2B_{66})\left(\frac{m\pi}{a}\right)\left(\frac{n\pi}{b}\right)^2$$

$$T_{23} = B_{22}\left(\frac{n\pi}{b}\right)^3 + (B_{12} + 2B_{66})\left(\frac{m\pi}{a}\right)^2\left(\frac{n\pi}{b}\right) \quad (11.1.17)$$

Dynamic Case

Since the equation governing the laminate vibration is an even order function of time, the time and spatial coordinates in it can be separated. We assume a functional form of the type

$$u(x, y, t) = U \sin\left(\frac{m\pi x}{a}\right)\cos\left(\frac{n\pi y}{b}\right)e^{i\omega t} \quad (11.1.18a)$$

$$\upsilon(x, y, t) = V \cos\left(\frac{m\pi x}{a}\right)\sin\left(\frac{n\pi y}{b}\right)e^{i\omega t} \quad (11.1.18b)$$

$$w(x, y, t) = W \sin\left(\frac{m\pi x}{a}\right)\sin\left(\frac{n\pi y}{b}\right)e^{i\omega t} \quad (11.1.18c)$$

This representation satisfies the boundary conditions exactly. Substitution of Eq. (11.1.18) in Eqs. (11.1.1)-(11.1.3) and subsequent simplification results in a set of homogenous algebraic equations. The eigen values corresponding to this set of equations are obtained as [14]

$$\omega^2 = \frac{1}{\rho}\left[T_{33} + \frac{2T_{12}T_{23}T_{13} - T_{22}T_{12}^2 - T_{12}T_{23}^2}{T_{11}T_{22} - T_{12}^2} \right] \qquad (11.1.19)$$

Numerical computations have been carried out for the angle-ply laminated rectangular plate of boron/epoxy composite, treating the number of plies (N) and aspect ratio (p) as parameters. The composite properties have been taken from Jones [25]. These are

E_L = 207.2 kN/mm^2, E_T = 20.7 kN/mm^2
G_{LT} = 68.6 kN/mm^2, ν_{LT} = 0.21, ρ = 0.208 × 10^{-5} kg/mm^3
σ_{LU} = 82.86 N/mm^2, σ_{TU} = 1.38 kN/mm^2, σ_{LCU} = 2.76 kN/mm^2
σ_{TCU} = 0.276 kN/mm^2, σ_{LTU} = 0.1243 kN/mm^2
q_0 = 0.1 kg/cm^2, b = 50.8 cm

The upper and lower bound values chosen for the behavior variables are

$$W_{max} = 0.3 \text{ cm}, \; W_{min} = 0.0 \text{ cm}, \; f_{max} = 100 \text{ Hz}$$

$$f_{min} = 16.5 \text{ Hz}, \; t_{max} = 0.5 \text{ cm}, \; t_{min} = 0.001 \text{ cm}$$

Table 11.1.1 gives the optimum ply thickness ratio obtained by the process of optimization for an anti-symmetric angle-ply laminate along with that obtained analytically for a specially orthotropic laminate. These results indicate that an anti-symmetric angle-ply laminate tends to be specially orthotropic at the optimum point.

Table 11.1.1 Variation of ply thickness with aspect ratio (anti-symmetric angle-ply laminate)

N	p	Optimum ply thickness ratio		Theoretical value (t_1/t_2)
		t_1/t_2	t_2/t_3	
4	1.0	0.41442		0.41421
	1.5	0.41470		0.41421
6	1.0	0.49076	0.71065	0.49074
	1.5	0.41909	0.70658	0.49030

Table 11.1.2 shows the constraints active at the optimum point for an anti-symmetric angle-ply laminate. As can be seen, the deflection constraint is active for all values of aspect ratio (p). However, at $p = 2$, the frequency is active in addition to the deflection constraint.

Table 11.1.2 Constraints active at optimum point (anti-symmetric angle-ply laminate)

N	p		
	1.0	1.5	2.0
2	Deflection	Deflection	Deflection and frequency
4	Deflection	Deflection	Deflection and frequency
6	Deflection	Deflection	Deflection and frequency
8	Deflection	Deflection	Deflection and frequency

Figure 11.1.1 depicts the variation of optimum weight with the number of plies and aspect ratio for an anti-symmetric angle-ply laminate. The increase in number of plies beyond four does not change the optimum weight of the laminate. Figure 11.1.2 shows the variation in the optimum weight of an orthotropic angle-ply laminate with the increase in aspect ratio and number of plies.

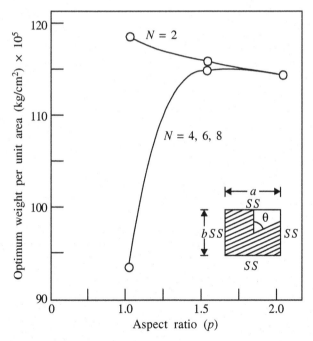

Fig. 11.1.1 Variation of optimum weight with aspect ratio (anti-symmetric angle-ply).

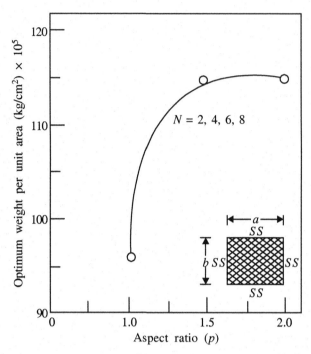

Fig. 11.1.2 Variation of optimum weight with aspect ratio (orthotropic angle-ply).

Table 11.1.3 lists the constraints active at the optimum point for an orthotropic angle-ply laminate.

Table 11.1.3 Constraints active at optimum point (orthotropic angle-ply laminate)

N	p		
	1.0	1.5	2.0
2	Deflection	Deflection and frequency	Frequency
4	Deflection	Deflection and frequency	Frequency
6	Deflection	Deflection and frequency	Frequency
8	Deflection	Deflection and frequency	Frequency

Table 11.1.4 gives the optimum fiber orientation and ply thickness for an orthotropic angle-ply laminate. Here, it can be noted that, at an optimum point, the laminate approaches a symmetric configuration. Further, since the contribution of the outermost plies to the bending stiffness is more than that of the plies closest to the mid-plane, the latter plies can have relaxed design constraints.

Table 11.1.4 Optimum fiber orientation and ply thickness (orthotropic angle-ply laminate)

N	p	Stacking sequence (± degree)	Ply thickness (cm)			Optimum weight (kg/cm)
			t_1/t_2	t_3/t_4	t_5/t_6	
2	1.0	45/45	0.2309/0.2308			0.00096
	1.5	90/90	0.2742/0.2959			0.00114
	2.0	90/90	0.2762/0.2767			0.00115
4	1.0	45/42.5/42.5/45	0.2132/0.0225	0.0241/0.2019		0.00096
	1.5	90/75/75/90	0.2526/0.0221	0.0221/0.2534		0.00114
	2.0	90/88/88/90	0.2226/0.0518	0.0521/0.2264		0.00115
6	1.0	44.7/45.1/32.7	0.1108/0.1142	0.0072/0.0077		0.00096
	1.5	90/97.5/56.5/ 62/95.6/90.2	0.2455/0.0233	0.0073/0.0075	0.0282/0.2384	0.00114
	2.0	90/91/43/43/ 91/90	0.1407/0.1305	0.0051/0.0051	0.1315/0.1401	0.00012

Example 11.2

Find the optimum design of laminated composite plates under multiple constraints.

Solution

Hirano [26], Khot et al. [27, 28] deal with optimum reinforcements treating strength or stiffness as a design criterion. Schmit and Farshi [29] arrived at the minimum weight design of symmetric composite plates by considering the thickness of each lamina as the only design variable. Singh [30] conducted theoretical and experimental studies on the optimal design of laminated beams and simply supported orthotropic plates.

In this example, we shall look at the feasibility of multiple behavior constraints in the minimum weight design of laminated composite plates, clamped on all sides and subjected to transverse and/or dynamic loads. To do this, we shall consider:

(i) A cross-ply laminate with the thickness of each ply as a design variable.

(ii) An anti-symmetric angle-ply laminate with the fiber orientation in each ply and thickness of each ply as the design variables.

(iii) An in-plane symmetric orthotropic laminate having an equal number of fibers at along the $(+\theta)$ and $(-\theta)$ direction in each ply with the fiber orientation of each ply as the design variable.

The equilibrium equations in terms of the displacements u, υ and w for a general laminate are given by Eqs. (11.1.1) to (11.1.3). The *objective function*

$F(t_k)$ for this case also is given by Eq. (11.1.7). The *behavior and side constraints* are given by Eqs. (11.1.8) to (11.1.11).

The *unconstrained optimization problem* is given by Eq. (11.1.12). It is solved by using the Fletcher-Powell variable metric method. The algorithm requires knowledge of behavior variables and their derivatives. For this, it is necessary to obtain the behavior constraints, namely, stresses, deflections, and frequency. Hence, we need to solve the governing equations. The governing equations and the boundary conditions have even and odd derivatives of the variables u, υ and w. No closed form solution is possible. The Galerkin technique is therefore employed to obtain the response.

For a clamped-clamped laminate subjected to a transverse load $q(x, y)$, the displacements u, υ and w are expressed as

$$u(x, y) = \sum_m \sum_n U_{mn} \phi_{mn}(x, y) \qquad (11.2.1a)$$

$$\upsilon(x, y) = \sum_m \sum_n V_{mn} \phi_{mn}(x, y) \qquad (11.2.1b)$$

$$w(x, y) = \sum_m \sum_n W_{mn} \phi_{mn}(x, y) \qquad (11.2.1c)$$

The transverse load $q(x, y)$ can be expressed as

$$q(x, y) = \sum_m \sum_n Q_{mn} \phi_{mn}(x, y) \qquad (11.2.1d)$$

The displacement function $\phi_{mn}(x, y)$ is given by

$$\phi_{mn}(x, y) = X_m(x)\, Y_n(y) \qquad (11.2.2)$$

Here, $X_m(x)$ and $Y_n(y)$ are the beam characteristic functions satisfying the boundary conditions along the X- and Y-direction of the plate, and are taken as

$$X_m(x) = \cosh(\lambda_m x) - \cos(\lambda_m x) - C_m[\sinh(\lambda_m x) - \sin(\lambda_m x)] \qquad (11.2.3)$$

$$Y_n(y) = \cosh(\lambda_n y) - \cos(\lambda_n y) - C_n[\sinh(\lambda_n y) - \sin(\lambda_n y)] \qquad (11.2.4)$$

where λ_m and λ_n are the roots of the characteristic equation

$$\cos\lambda_i\, \cosh\lambda_i = 1 \qquad (11.2.5)$$

The parameter C_i is defined as

$$C_i = \frac{\cosh(\lambda_i) - \cos(\lambda_i)}{\sinh(\lambda_i) - \sin(\lambda_i)} \qquad (11.2.6)$$

For more details, the reader may refer to Soni and Iyengar [31].

Numerical computations are carried out for the rectangular laminate of boron/epoxy composite laminate, treating the number of plies (N) and aspect ratio (p) as parameters. Details of the material properties are given in Example 11.1.

Table 11.2.1 gives the results for the optimal design of a cross-ply laminate with varying aspect ratio. Here thickness of each ply is treated as a design variable. From the table, it is observed that for a square laminate, the maximum contribution to the bending stiffness is made by the plies oriented along the X-axis. However, for a rectangular laminate, the maximum contribution to the bending stiffness is from the plies oriented along the Y-direction. In both cases, the contribution to the stiffness by the outer plies is predominant. Further, optimum weight per unit area remains fairly constant. For a square laminate, deflection is the active constraint, whereas for a rectangular laminate, frequency becomes the active constraint. Table 11.2.2 lists the results for an anti-symmetric angle-ply laminate with $\theta = 30°$. The laminate, as we find, tends to be specially orthotropic at the optimum point. Further, for the two ply laminate, i.e. square and rectangular, deflection is the active constraint.

Table 11.2.1 Optimum ply thickness (cross-ply rectangular laminate) given $q_0 = 0.0138$ N/mm^2, $W_{max} = 5$ mm, $f_{max} = 14$ Hz

p	Ply thickness (mm)								Optimum weight (N/mm^2)
	t_1	t_2	t_3	t_4	t_5	t_6	t_7	t_8	
1.0	3.788	0.088	3.818	0.010					0.000157
1.5	0.010	3.688	0.143	3.715					0.000154
2.0	0.010	3.735	0.127	3.750					0.000155
1.0	3.784	0.037	0.024	0.038	3.812	0.010			0.000157
1.5	0.010	3.652	0.069	0.076	0.066	3.732			0.000155
2.0	0.010	3.704	0.061	0.077	0.063	3.725			0.000156
1.0	3.269	0.138	0.296	0.269	0.292	0.140	3.296	0.013	0.000157

Table 11.2.2 Optimum ply thickness (cross-ply rectangular laminate) given $\theta = 30°$, $q_0 = 0.0138$ N/mm^2, $W_{max} = 5$ mm, $f_{max} = 14$ Hz

p	Ply thickness (mm)			Optimum weight (N/mm^2)	Active constraint
	t_1	t_2	t_3		
1.0	1.153	2.784		0.000161	Deflection
1.5	1.594	3.852		0.000222	Deflection
2.0	1.838	4.441		0.000256	Deflection
1.0	0.879	1.793	1.265	0.000161	Deflection
1.5	1.219	2.484	1.743	0.000222	Deflection
2.0	1.404	2.864	2.011	0.000256	Deflection

Table 11.2.3 shows the results for the optimum design of an anti-symmetric angle-ply laminate, treating the ply fiber orientation and thickness as design variables. Here, for all values of p we considered, except 1.1, the ply thickness distribution is almost uniform. Table 11.2.4 shows the results for an in-plane symmetric orthotropic laminate. Here, each ply is assumed to contain equal number of fibers in the $(+\ \theta)$ and $(-\ \theta)$ directions with respect to the X-axis. Also the objective function is taken as the maximization of the first natural frequency, treating the fiber orientation of each ply as the design variable. As can be noted from this table, the optimal design assumes a symmetric cross-ply configuration for a square laminate having the fiber orientation of its outermost plies along the X-axis.

Table 11.2.3 Minimum weight design of an anti-symmetric laminate given $N = 4$, $q_0 = 0.0098$ N/mm^2, $W_{max} = 3$ mm, $f_{max} = 15$ Hz, intial values: $\theta = 30°$, $t_i = 4$ mm

p	Fiber orientation θ (degree)	Optimum ply thickness (mm)		Optimum weight (N/mm^2)	Active constraint
		t_1	t_2		
1.0	0	2.288	2.275	0.00018618	Deflection
1.1	90	1.969	2.601	0.00018645	Deflection
1.25	90	2.274	2.271	0.00018543	Deflection
1.5	90	2.236	2.236	0.00018236	Deflection
2.0	90	2.167	2.175	0.00017716	Deflection

Table 11.2.4 Optimal values for an orthotropic symmetric laminate (total laminate thickness 8 mm)

p					
0.5	1.0	1.25	1.5	1.75	2.0
Fiber orientation (θ) and frequency (f_0)					
$[0/0]_s$	$[0/90]_s$	$[90/90]_s$	$[90/90]_s$	$[90/90]_s$	$[90/90]_s$
82.3541	21.8122	21.0944	20.8041	20.6644	20.5885
$[0/0/0]_s$	$[0/90/0]_s$	$[90/90/90]_s$	$[90/90/90]_s$	$[90/90/90]_s$	$[90/90/90]_s$
82.3541	21.8122	21.0944	20.8041	20.6644	20.5885
$[0/0/0/0]_s$	$[0/0/0/0]_s$	$[0/0/0/0]_s$	$[0/0/0/0]_s$	$[0/0/0/0]_s$	$[0/0/0/0]_s$
82.3541	21.8122	21.0944	20.8041	20.6644	20.5885

The variation of optimum weight per unit area with aspect ratio is plotted in Fig. 11.2.1 for an anti-symmetric four ply laminate. As can be seen, the weight decreases as the aspect ratio increases.

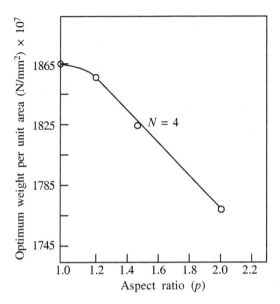

Fig. 11.2.1 Variation of optimum weight with aspect ratio (anti-symmetric four ply laminate).

Figure 11.2.2 compares the optimum weight of a cross-ply laminate with that of an anti-symmetric 30° angle-ply laminate. Here, the weight of the cross-ply laminate is minimum for each value of the aspect ratio considered. Further, for an anti-symmetric 30° laminate, the optimum weight increases with the increase in aspect ratio.

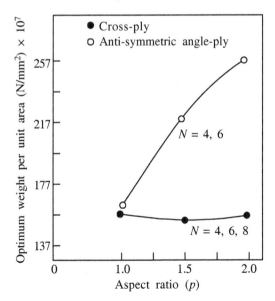

Fig. 11.2.2 Variation of optimum weight with aspect ratio (cross-ply and anti-symmetric angle-ply laminates).

Example 11.3

Find the minimum weight design of simply supported composite laminates subjected to in-plane and transverse loading.

Solution

Schmit and Farshi [29] presented a method for obtaining the minimum weight design of symmetric composite laminates subjected to multiple in-plane loading conditions. The problem is cast as a nonlinear programming problem in which the thickness of each ply is treated as design variable and ply fiber orientation is preassigned. McKeown [32] presented an algorithm based on the simplex method for the optimization of composite laminates with orthotropic layers. The analysis requires that the ply sequence be symmetric about the mid-plane of the laminate. Here we shall consider

(i) An anti-symmetric angle-ply laminate with each ply having a different thickness and a preassigned fiber orientation.
(ii) An anti-symmetric angle-ply laminate with ply thickness and fiber orientation as design variable.

The equations in terms of u, υ, w governing the behavior of an anti-symmetric angle-ply laminate are

$$A_{11}\frac{\partial^2 u}{\partial x^2} + A_{66}\frac{\partial^2 u}{\partial y^2} + (A_{12} + A_{66})\frac{\partial^2 \upsilon}{\partial x \partial y} - 3B_{16}\frac{\partial^3 w}{\partial x^2 \partial y} - B_{26}\frac{\partial^3 w}{\partial y^3} = 0 \qquad (11.3.1)$$

$$A_{11}\frac{\partial^2 u}{\partial x^2} + A_{66}\frac{\partial^2 u}{\partial y^2} + (A_{12} + A_{66})\frac{\partial^2 \upsilon}{\partial x \partial y} - 3B_{16}\frac{\partial^3 w}{\partial x^2 \partial y} - B_{26}\frac{\partial^3 w}{\partial y^3} = 0 \qquad (11.3.2)$$

$$D_{11}\frac{\partial^4 w}{\partial x^4} + 2(D_{12} + 2D_{66})\frac{\partial^4 w}{\partial x^2 \partial y^2} + D_{22}\frac{\partial^4 w}{\partial x^4} - B_{16}\left(3\frac{\partial^3 u}{\partial x^2 \partial y} + \frac{\partial^3 \upsilon}{\partial x^3}\right)$$

$$- B_{26}\left(\frac{\partial^3 u}{\partial y^3} + 3\frac{\partial^3 \upsilon}{\partial x \partial y^2}\right) = q(x, y) \qquad (11.3.3)$$

For in-plane compressive loading, $q(x, y)$ in Eq. (11.3.3)

$$\left(-N_x\frac{\partial^2 w}{\partial x^2} - N_y\frac{\partial^2 w}{\partial y^2}\right)$$

the boundary conditions are

$$w = 0, \ M_x = B_{16}\left(\frac{\partial u}{\partial y} + \frac{\partial \upsilon}{\partial x}\right) - D_{11}\frac{\partial^2 w}{\partial x^2} - D_{12}\frac{\partial^2 w}{\partial y^2} = 0$$

$$\hspace{6cm} (x = 0, \ a) \ (11.3.4a)$$

$$u = 0, \ N_{xy} = A_{66}\left(\frac{\partial u}{\partial y} + \frac{\partial \upsilon}{\partial x}\right) - B_{16}\frac{\partial^2 w}{\partial x^2} - B_{26}\frac{\partial^2 w}{\partial y^2} = 0$$

$$w = 0, \ M_y = B_{26}\left(\frac{\partial u}{\partial y} + \frac{\partial v}{\partial x}\right) - D_{12}\frac{\partial^2 w}{\partial x^2} - D_{22}\frac{\partial^2 w}{\partial y^2} = 0$$

$$(y = 0, \ b) \ (11.3.4b)$$

$$v = 0, \ N_{xy} = A_{66}\left(\frac{\partial u}{\partial y} + \frac{\partial v}{\partial x}\right) - B_{16}\frac{\partial^2 w}{\partial x^2} - B_{26}\frac{\partial^2 w}{\partial y^2} = 0$$

The objective function here is

$$\text{Minimize } \overline{F} = (F / A) = \text{Minimize} \sum_{i=1}^{N/2} t_i \qquad (11.3.5)$$

where N is the total number of plies and t_i is the thickness of ith ply. It is assumed that the material of all the plies is the same.

The behavior constraints are:

(i) The deflection W of the plate should be less than W^*, the assigned value, i.e.

$$W \leq W^* \qquad (11.3.6)$$

(ii) The minimum buckling load for the laminate must be greater than that prescribed, i.e.

$$N_x \geq N_x^*$$

(iii) The stress within the domain of the laminate should be less than that predicted by the Tsai-Hill criterion, i.e.

$$\left(\frac{\sigma_L}{\sigma_{LU}}\right)^2 + \left(\frac{\sigma_T}{\sigma_{TU}}\right)^2 + \left(\frac{\sigma_{LT}}{\sigma_{LTU}}\right)^2 - \left(\frac{\sigma_L}{\sigma_{LU}}\right)\left(\frac{\sigma_T}{\sigma_{LU}}\right) < 1 \qquad (11.3.7)$$

The side constraints are:

(i) The fiber orientation in each ply is bounded, i.e.

$$-\pi/2 \leq \theta_i \leq \pi/2 \ (i = 1, \ 2,..., \ N/2) \qquad (11.3.8)$$

(ii) The thickness of each lamina is bounded, i.e.

$$t_1 \leq t_i \leq t_2 \ (i = 1, \ 2,..., \ N/2) \qquad (11.3.9)$$

The optimization problem defined by Eqs. (11.3.5)-(11.3.9) is a constrained optimization problem. This constrained problem is transformed into a series of unconstrained problems

$$P(r,t_i) = F(t_i) - r\sum_{j=1}^{m} f_j(\theta_i, t_i) \qquad (11.3.10)$$

where $f_j(\theta_i, \ t_i) = \ln[g_j(\theta_i, \ t_i)]$ if $g_j(\theta_i, \ t_i) \geq 0$.

The constraints defined by Eqs. (11.3.6)-(11.3.9) are

$$g_i = 1 - (t_i/t_2), \ i = 1, \ 2,..., \ N/2 \qquad (11.3.11)$$

$$g_{i+N/2} = (t_i/t_1) - 1, \ i = 1, \ 2,..., \ N/2 \qquad (11.3.12)$$

$$g_{i+N} = 1 - [\theta_i/(\pi/2)], \ i = 1, \ 2,..., \ N/2 \qquad (11.3.13)$$

$$g_{i+3N/2} = 1 - [\theta_i/(-\pi/2)], \ i = 1, \ 2,..., \ N/2 \qquad (11.3.14)$$

$$g_{2N+1} = 1 - (W/W^*) \qquad (11.3.15)$$

$$g_{2N+2} = (N_x/N_x^*) - 1 \qquad (11.3.16)$$

$$g_{i+2N+2} = 1 - \left(\frac{\sigma_L}{\sigma_{LU}}\right)_i^2 - \left(\frac{\sigma_T}{\sigma_{TU}}\right)_i^2 - \left(\frac{\sigma_{LT}}{\sigma_{LTU}}\right)_i^2 + \left(\frac{\sigma_L}{\sigma_{LU}}\right)_i \left(\frac{\sigma_T}{\sigma_{LU}}\right)_i, i = 1,2...,N/2$$
$$(11.3.17)$$

For problem (i), we disregard the constraints (11.3.13) and (11.3.14) as the fiber orientation is pre-assigned. The constrained optimization problem given by Eq. (11.3.10) is then solved by the Fletcher-Powell variable metric method. This method requires the knowledge of constraints and their derivatives in terms of design variables. For this, we need to solve the governing differential equations. The displacement functions satisfying the boundary conditions exactly are given by

$$u(x, y) = \sum_m \sum_n U_{mn} \sin \frac{m\pi x}{a} \cos \frac{n\pi y}{b} \qquad (11.3.18a)$$

$$\upsilon(x, y) = \sum_m \sum_n V_{mn} \cos \frac{m\pi x}{a} \sin \frac{n\pi y}{b} \qquad (11.3.18b)$$

$$w(x, y) = \sum_m \sum_n W_{mn} \sin \frac{m\pi x}{a} \sin \frac{n\pi y}{b} \qquad (11.3.18c)$$

Substituting Eqs. (11.3.18) in governing equations and simplifying, the expression for buckling load is

$$N_x = \frac{1}{\pi^2 a^2 \left[m^2 + K^2 n^2 (a/b)^2\right]} \left[T_{33} + \frac{2T_{12}T_{23}T_{13} - T_{22}T_{13}^2 - T_{11}T_{23}^2}{T_{11}T_{22} - T_{12}^2}\right] \qquad (11.3.19)$$

in which

$$T_{11} = A_{11}m^2\pi^2a^2 + A_{66}n^2\pi^2(a/b)^2$$
$$T_{12} = (A_{12} + A_{66})\pi^2 mn(a/b)$$
$$T_{13} = -[3B_{16}\pi^2m^2 + B_{26}\pi^2n^2(a/b)^2]\pi n(a/b)$$
$$T_{22} = A_{66}\pi^2m^2 + A_{22}\pi^2n^2(a/b)^2$$
$$T_{23} = -[B_{16}\pi^2m^2 + 3B_{26}\pi^2n^2(a/b)^2]m\pi$$
$$T_{33} = D_{11}\pi^4m^4 + 2(D_{12} + 2D_{66})\pi^4m^2n^2(a/b)^2 + D_{22}\pi^4$$

$K = N_y/N_x$, and a and b are the dimensions of the laminate along the X- and Y-direction, respectively.

For a laminate subjected to a uniform transverse load $q(x, y)$, the displacement coefficients U_{mn}, V_{mn} and W_{mn} are

$$U_{mn} = \frac{q_{mn}}{a^5 \bar{D}}(T_{22}T_{13} - T_{12}T_{23}) \tag{11.3.20a}$$

$$V_{mn} = \frac{q_{mn}}{a^5 \bar{D}}(T_{11}T_{23} - T_{12}T_{13}) \tag{11.3.20b}$$

$$W_{mn} = \frac{q_{mn}}{a^5 \bar{D}}(T_{11}T_{22} - T_{12}^2) \tag{11.3.20c}$$

where

$$\bar{D} = \frac{T_{11}T_{22} - T_{12}^2}{a^8}\left[T_{33} + \frac{2T_{12}T_{23}T_{13} - T_{22}T_{13}^2 - T_{11}T_{23}^2}{T_{11}T_{22} - T_{12}^2}\right]$$

The transverse load $q(x, y)$ is expressed in terms of the Fourier series as

$$q(x, y) = \sum_m \sum_n q_{mn} \sin\frac{m\pi x}{a}\sin\frac{n\pi y}{b} \tag{11.3.21}$$

where

$$q_{mn} = 16q_0/(\pi^2 mn), \quad m, n = 1, 3, 5... \tag{11.3.22}$$

q_0 being the intensity of loading.

Numerical computations have been carried out for the rectangular laminates of boron/epoxy, treating N and p as parameters. The material properties are the same as given in earlier examples. However, the values for the transverse load and *behavior* and *side constraints* are taken as

$$q_0 = 5 \times 10^{-4} \text{ kg/mm}^2$$
$$W^* = 2.00 \text{ mm}, \quad N_x^* = 5.00 \text{ kg/mm}$$
$$t_1 = 0.125 \text{ mm}, \quad t_2 = 2.5 \text{ mm}$$

The results for an anti-symmetric four-ply laminate with $p = 0.5$ and $K = 0.5$ are shown in Table 11.3.1. It is observed that for a given fiber orientation, the weight is optimum when the ply thickness ratio is 0.414. Further, this weight is for each value of minimum when the fiber orientation is approximately $\pm 10°$.

Table 11.3.2 gives the variation of ply thickness for a square laminate with fiber orientation $\pm 30°$. The optimum weight here is the same for each value of N. Table 11.3.3 shows the variation of ply thickness for a rectangular laminate with fiber orientation $\pm 10°$. The optimum weight is less than that obtained when the fiber orientation is $\pm 30°$.

Table 11.3.1 Variation of optimum weight with fiber orientation (anti-symmetric angle-ply laminate) given $N = 4$, p (= a/b) = 0.5, $K = 0.5$

θ ($\pm$ degree)	Optimum ply thickness (mm)		Optimum weight (kg/mm^2)	Ply thickness ratio (t_1/t_2)
	t_1	t_2		
30	0.4103	0.9885	2.7976	0.415
45	0.4420	1.0676	3.0172	0.414
10	0.3928	0.9488	2.6832	0.414

Table 11.3.2 Variation of optimum ply thickness with number of plies (square laminate) given p (= a/b) = 1.0, $K = 0.0$, $\theta = \pm 30°$

N	Ply thickness ratio					Optimum weight (kg/mm^2)
	t_1/t_2	t_3/t_4	t_5/t_6	t_7/t_8	t_9/t_{10}	
4	0.6087/ 1.4692					4.1558
6	0.4667/ 0.9508	0.6603				4.1556
8	0.3824/ 0.6806	0.4554/ 0.5595				4.1558
12	0.2779/ 0.4263	0.3017/ 0.3886	0.3275/ 0.3557			4.1554
16	0.2224/ 0.3075	0.2299/ 0.2906	0.2391/ 0.2756	0.2501/ 0.2626		4.1588
20	0.1829/ 0.2395	0.1875/ 0.2307	0.1919/ 0.2228	0.1968/ 0.2151	0.2025/ 0.2084	4.1588

Table 11.3.4 shows the optimum fiber orientation and the ply thickness corresponding to it along with aspect ratio and biaxial loading ratio. Optimum weight increases with the increase in biaxial loading ratio when p is around 0.5. However, the ply thickness ratio remains fairly constant with the increase in biaxial loading ratio as the laminate tends to be a square laminate. Which of the constraints will be active when the weight is optimum depends on the aspect ratio.

Figure 11.3.1 depicts the variation of ply thickness ratio with the number of plies. The thickness ratio here of the anti-symmetric plies approaches an asymptotic value. It approaches unity if the constraint on the minimum ply thickness is active. Further, the thickness ratio of the inner plies is very close to unity and remains fairly constant.

Table 11.3.3 Variation of optimum ply thickness with number of plies given p $(= a/b) = 0.5$, $K = 0.5$, $\theta = \pm 10°$

N	Ply thickness ratio					Optimum weight (kg/mm^2)
	t_1/t_2	t_3/t_4	t_5/t_6	t_7/t_8	t_9/t_{10}	
4	0.3928/ 0.9488					2.6832
6	0.3043/ 0.6184	0.4183				2.6830
8	0.2500/ 0.4458	0.2936/ 0.3522				2.6832
12	0.1840/ 0.2806	0.1956/ 0.2491	0.2076/ 0.2247			2.6832
16	0.1468/ 0.2132	0.1522/ 0.1836	0.1552/ 0.1695	0.1597/ 0.1623		2.6850
20	0.1271/ 0.1924	0.1269/ 0.1374	0.1257/ 0.1270	0.1261/ 0.1273	0.1264/ 0.1252	2.6830

Table 11.3.4 Optimum fiber orientation and ply thickness (angle-ply laminate, $N = 4$)

p	Loading ratio $(K = N_Y/N_X)$	Ply thickness (mm)		θ (degree)		Optimum weight (kg/mm^2)	Active constraint	Ply thickness ratio	
		t_1	t_2	θ_1	θ_2			(t_1/t_2)	$-\theta_1/\theta_2$
0.5	0.0	0.6423	0.6420	0.01	– 0.04	2.57	Buckling	1.0	0.25
	0.5	0.5591	0.7821	7.56	– 15.44	2.68	Buckling	0.714	0.489
	1.0	0.4984	0.9081	15.44	– 25.44	2.81	Buckling	0.548	0.605
0.8	0.0	0.4971	1.9991	37.20	– 37.42	3.39	Deflection	0.414	0.994
	0.5	0.5043	1.2161	38.01	– 38.16	3.44	Buckling	0.414	0.996
1.0	0.0	0.5367	1.4182	45.10	– 45.14	4.01	Deflection	0.414	0.999
	0.5	0.5373	1.4176	44.90	– 44.92	4.01	Deflection	0.414	0.999
	1.0	0.5885	1.4214	44.93	– 44.87	4.02	Buckling	0.414	1.001
2.0	0.0	2.2313	0.2165	89.99	– 58.80	4.89	Deflection	10.30	1.53
	0.5	2.3225	0.1250	89.94	– 60.05	4.89	Deflection	18.58	1.497

Example 11.4
Determine the optimum parameters of a hybrid composite laminate under in-plane loading.

Solution
The performance of a hybrid laminate made of two or more types of fibers

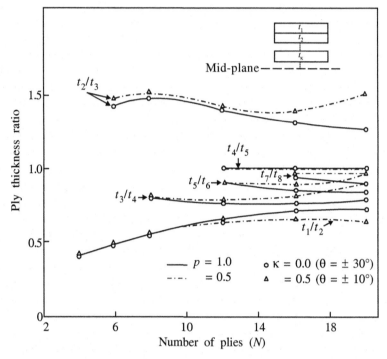

Fig. 11.3.1 Variation of ply thickness ratio with number of plies (anti-symmetric laminate).

is better than that of a laminate made of only single type of fiber. It is, however, possible to generate a hybrid laminate using a single material by varying the fiber volume fraction in each layer.

In what follows, we will optimize the first buckling load of a laminate, treating the volume fraction and fiber orientation as design variables. We shall consider the anti-symmetric composite laminate, either simply supported or clamped with a constraint on total fiber content.

The properties of the laminate depend on the properties of its constituents and the nature of distribution of the constituents. The mechanical properties of a lamina in terms of the properties of its constituents (fiber and matrix) are given in Jones [25] as

$$E_L = E_f V_f + E_m (1 - V_f) \tag{11.4.1}$$

$$E_T = \frac{1 + 2\eta V_f}{1 - \eta V_f}, \quad \eta = \frac{(E_f / E_m) - 1}{(E_f / E_m) + 2} \tag{11.4.2}$$

$$G_T = \frac{1 + 2\zeta V_f}{1 - \zeta V_f}, \quad \zeta = \frac{(E_f / E_m) - 1}{(E_f / E_m) + 2} \tag{11.4.3}$$

$$\nu_{LT} = \nu_f V_f + \nu_m (1 - V_f) \tag{11.4.4}$$

where V_f is the fiber volume fraction, and E_L, E_T, G_{LT}, ν_{LT} are the mechanical properties of the lamina along and normal to the fibers.

The equations governing the behavior of the general rectangular laminate in terms of u, υ and w are [25]

$$A_{11}\frac{\partial^2 u}{\partial x^2} + 2A_{16}\frac{\partial^2 u}{\partial x \partial y} + A_{66}\frac{\partial^2 u}{\partial y^2} + A_{16}\frac{\partial^2 \upsilon}{\partial x^2} + (A_{12}+A_{66})\frac{\partial^2 \upsilon}{\partial x \partial y} + A_{26}\frac{\partial^2 \upsilon}{\partial y^2}$$

$$- B_{11}\frac{\partial^3 w}{\partial x^3} - 3B_{16}\frac{\partial^3 w}{\partial x^2 \partial y} - (B_{12}+2B_{66})\frac{\partial^3 w}{\partial x \partial y^2} - B_{26}\frac{\partial^3 w}{\partial y^3} = 0 \tag{11.4.5}$$

$$A_{16}\frac{\partial^2 u}{\partial x^2} + 2A_{26}\frac{\partial^2 \upsilon}{\partial x \partial y} + A_{26}\frac{\partial^2 u}{\partial y^2} + A_{66}\frac{\partial^2 \upsilon}{\partial x^2} + (A_{12}+A_{66})\frac{\partial^2 u}{\partial x \partial y} + A_{22}\frac{\partial^2 \upsilon}{\partial y^2}$$

$$- B_{16}\frac{\partial^3 w}{\partial x^3} - 3B_{26}\frac{\partial^3 w}{\partial x \partial y^2} - (B_{12}+2B_{66})\frac{\partial^3 w}{\partial x^2 \partial y} - B_{22}\frac{\partial^3 w}{\partial y^3} = 0 \tag{11.4.6}$$

$$D_{11}\frac{\partial^4 w}{\partial x^4} + 4D_{16}\frac{\partial^4 w}{\partial x^3 \partial y} + 2(D_{12}+2D_{66})\frac{\partial^4 w}{\partial x^2 \partial y^2} + 4D_{26}\frac{\partial^4 w}{\partial x \partial y^3} + D_{22}\frac{\partial^4 w}{\partial y^4}$$

$$- B_{11}\frac{\partial^3 u}{\partial x^3} - 3B_{16}\frac{\partial^3 u}{\partial x^2 \partial y} - (B_{12}+2B_{66})\frac{\partial^3 u}{\partial x \partial y^2} - B_{22}\frac{\partial^3 \upsilon}{\partial y^3} - B_{16}\frac{\partial^3 \upsilon}{\partial x^3}$$

$$- (B_{12}+2B_{66})\frac{\partial^3 \upsilon}{\partial x^2 \partial y} - 3B_{26}\frac{\partial^3 \upsilon}{\partial x \partial y^2} = -\lambda\left[k_1\frac{\partial^2 w}{\partial x^2} + k_2\frac{\partial^2 w}{\partial y^2} + 2k_3\frac{\partial^2 w}{\partial x \partial y}\right] \tag{11.4.7}$$

where A_{ij}, B_{ij}, D_{ij} are the various stiffness coefficients and $\bar{N}_x = \lambda k_1$, $\bar{N}_y = \lambda k_2$, $\bar{N}_{xy} = \lambda k_3$.

The *objective is to maximize the first buckling load of the laminate.* This is written as

$$\text{Maximize } \{\phi(\theta_i, V_{fi})\}, \quad i = 1, 2, 3 \tag{11.4.8}$$

θ_i is the fiber orientation and V_{fi} is the fiber volume fraction of the ith ply and ϕ, the non-dimensional buckling load, is defined as

$$\phi = 12b^2\lambda_{cr}/(\pi^2 t^3 Q_{22}) \tag{11.4.9}$$

The *behavior constraints* are:

(i) The total fiber content of the laminate is equal to that specified, i.e.

$$2\sum_{i=1}^{3} V_{fi}t_i = \bar{V}_f \tag{11.4.10}$$

where t_i is the thickness of the ith ply and $\bar{V}_f$ is the specified total fiber content of the laminate.

(ii) The fiber volume fraction in each lamina is bounded, i.e.

$$\bar{V}_{fl} \leq \bar{V}_{fi} \leq \bar{V}_{fu} \qquad (11.4.11)$$

where the subscripts u and l stand for the upper and lower bounds, respectively.

The *side constraints* are specified as:

(i) The fiber orientation of each ply is bounded, i.e.

$$0° \leq \theta_i \leq 90° \ (i = 1, 2, 3) \qquad (11.4.12)$$

(ii) The ply thickness should be greater than the specified value, i.e.

$$t_{\min} \leq t_i \ (i = 1, 2, 3) \qquad (11.4.13)$$

(iii) The thickness of each ply is greater than zero, i.e.

$$t_1 \geq 0, \ t_2 \geq 0, \ t_3 \geq 0 \qquad (11.4.14)$$

The optimization problem as defined by Eq. (11.4.8) and Eqs. (11.4.10)-(11.4.14) is a constrained optimization problem. This is transformed to an unconstrained problem using a penalty function. The Fletcher-Powell conjugate method is then used for unconstrained minimization. The differential equation is solved to obtain the constraints and their derivatives.

For an anti-symmetric angle-ply laminate, the following stiffness coefficients are identically zero.

$$B_{11} = B_{12} = B_{22} = D_{16} = D_{26} = 0 \qquad (11.4.15)$$

To decouple the in-plane and bending behavior of such a laminate, each of the stiffness coefficients B_{16} and B_{26}, in addition, has to be made zero. If the laminate is six-layered as shown in Fig. 11.4.1, then to do this, the condition to be satisfied is [33]

$$\gamma = \left[\frac{B_3}{B_1} - \frac{A_3}{A_1} \right] \bigg/ \left[\frac{B_2}{B_1} - \frac{A_2}{A_1} \right] > 0, \ R_1 > 0, \ R_2 > 0 \qquad (11.4.16)$$

where

$$A_k = (Q_{11} - Q_{12} - 2Q_{66})_k$$
$$B_k = (Q_{12} - Q_{22} + 2Q_{66})_k, \ k = 1, 2, 3$$

$$R_1 = t_1/t_2, \ R_2 = t_2/t_3$$

The governing equation (11.4.7) for w now reduces to

$$D_{11} \frac{\partial^4 w}{\partial \xi^4} + 2(D_{12} + 2D_{66})p^2 \frac{\partial^4 w}{\partial \xi^2 \partial \eta^2} + D_{22} p^4 \frac{\partial^4 w}{\partial \eta^4}$$
$$= -\lambda a^2 \left[k_1 \frac{\partial^2 w}{\partial \xi^2} + k_2 p^2 \frac{\partial^2 w}{\partial \eta^2} + 2k_3 p \frac{\partial^2 w}{\partial \xi \partial \eta} \right] \qquad (11.4.17)$$

The non-dimensional coordinates, ξ, η are defined as

$$x = a\xi, \ y = b\eta$$

The boundary conditions for a laminate which is simply supported at all edges are

$$w = 0, \ M_\xi = D_{11}\frac{\partial^2 w}{\partial\xi^2} + D_{12}p^2\frac{\partial^2 w}{\partial\eta^2} = 0 \quad (\xi = 0,1) \qquad (11.4.18)$$

$$w = 0, \ M_\eta = D_{12}\frac{\partial^2 w}{\partial\xi^2} + D_{22}p^2\frac{\partial^2 w}{\partial\eta^2} = 0 \quad (\eta = 0,1) \qquad (11.4.19)$$

For a laminate with all edges clamped

$$w = 0, \ \partial w/\partial\xi = 0 \ (\xi = 0,1) \qquad (11.4.20)$$

$$w = 0, \ \partial w/\partial\eta = 0 \ (\eta = 0,1) \qquad (11.4.21)$$

The Galerkin method is employed to obtain the buckling load.

Numerical computations are carried out for the six-layered alternating angle-ply laminates of boron/epoxy composite with the following properties:

$$E_f = 3.97 \times 10^6 \ \text{kg/cm}^2, \ E_m = 3.38 \times 10^4 \ \text{kg/cm}^2$$

$$\nu_f = 0.2, \ \nu_m = 0.2$$

$$\bar{V}_f = 0.6, \ \bar{V}_{fl} = 0.10, \ \bar{V}_{fu} = 0.75$$

The procedure followed to obtain optimum configuration is:

(i)   With the given fiber and matrix properties, E_L, E_T, G_{LT}, ν_{LT} [Q] and [$\bar{Q}$] are calculated.

(ii) The ply thickness ratios R_1 and R_2 are taken as percentage of the total laminate thickness. Those values that render the laminate decoupled are taken.

(iii) The elements of the matrix [D] for the entire laminate are then calculated.

(iv) The minimum buckling load is obtained by solving the homogenous algebraic equation.

(v)   Steps (i)-(v) are repeated to get the values of the fiber orientation and the fiber volume fraction that yield the lowest buckling load.

Table 11.4.1 gives the optimum fiber orientation and volume fraction for a laminate simply supported on all edges and subjected to a combination of in-plane loads. For each load, as can be seen, the maximum fiber volume fraction is assigned to the outermost layer and the minimum to the innermost layer. Also, each buckling load here is found to be about 25% higher than that of an identical laminate but with uniform volume fraction.

Table 11.4.2 shows the results obtained for a clamped laminate. The distribution of ply thickness and fiber volume fraction is the same as that

Table 11.4.1 Optimum fiber volume fraction and orientation (simply supported hybrid laminate, $N = 6$)

In-plane loading ratio			p	Fiber volume fraction			Thickness as percentage of total thickness			θ (degree)	Buckling load ($\phi \times 10^{-6}$)
k_1	k_2	k_3		V_{f1}	V_{f2}	V_{f3}	t_1	t_2	t_3		
1	0	0	0.5	0.749	0.745	0.121	13.6	25.4	11.0	0.0	2.57
			1.0	0.749	0.745	0.137	12.6	25.2	11.3	45.5	5.48
			2.0	0.749	0.745	0.107	12.7	25.5	10.9	45.5	21.95
1	1	0	0.5	0.749	0.745	0.136	13.5	25.2	11.3	18.0	1.97
			1.0	0.749	0.745	0.114	13.6	25.4	11.0	45.0	2.74
			2.0	0.749	0.745	0.112	13.6	25.6	10.8	72.0	7.90
0	0	1	0.5	0.749	0.745	0.124	13.6	25.4	10.9	32.3	8.53
			1.0	0.749	0.745	0.113	13.6	25.4	10.8	45.0	12.37
			2.0	0.749	0.745	0.137	13.5	25.1	11.2	57.6	34.14
1	0	1	0.5	0.749	0.745	0.116	13.5	25.2	11.2	1.6	2.41
			1.0	0.749	0.745	0.110	13.6	25.5	10.8	45.0	4.72
			2.0	0.749	0.745	0.137	13.5	25.3	11.2	48.6	16.51

Table 11.4.2 Optimum fiber volume fraction and orientation (clamped hybrid laminate, $N = 6$)

In-plane loading ratio			p	Fiber volume fraction			Thickness as percentage of total thickness			θ (degree)	Buckling load ($\phi \times 10^{-6}$)
k_1	k_2	k_3		V_{f1}	V_{f2}	V_{f3}	t_1	t_2	t_3		
1	0	0	0.5	0.749	0.745	0.100	13.7	25.6	10.7	0.0	10.027
			1.0	0.749	0.745	0.114	13.6	25.5	10.9	0.2	11.316
			2.0	0.749	0.745	0.119	13.6	25.5	10.9	43.68	36.485
1	1	0	0.5	0.749	0.745	0.138	13.6	25.2	11.2	29.0	5.299
			1.0	0.749	0.745	0.138	13.6	25.2	11.2	56.5	5.657
			2.0	0.749	0.745	0.116	13.7	25.5	10.8	60.8	21.192
0	0	1	0.5	0.749	0.745	0.138	13.6	25.2	11.2	25.36	13.317
			1.0	0.749	0.745	0.137	13.5	25.2	11.3	45.02	17.089
			2.0	0.749	0.745	0.138	13.5	25.2	11.3	64.60	51.921
1	0	1	0.5	0.749	0.745	0.114	13.6	26.5	10.9	16.15	7.031
			1.0	0.749	0.745	0.115	13.6	25.2	11.2	36.80	8.102
			2.0	0.749	0.745	0.116	13.6	25.2	11.2	51.00	26.144

obtained for the laminate simply supported on all edges. Here too, the buckling load is 25% higher than that of an identical laminate but with uniform volume fraction.

Example 11.5
Determine the optimal design of a hybrid, laminated skew laminate under multiple constraints.

Solution
In Example 11.4, it was shown that the performances of rectangular hybrid laminates are better than rectangular regular laminates. In the design of modern high speed aircraft and missile structures, swept wing and tail surfaces are extensively employed. We shall consider here the problem of optimization of a hybrid anti-symmetric skew composite laminate with a pair of opposite edges simply supported. The other pair is clamped and subject to transverse and inertia loads. The ply thickness, fiber orientation and skew angle are treated as design variables. The constraints are imposed on maximum deflection and minimum fundamental natural frequency of the laminate.

Figure 11.5.1 shows a laminate with an orthogonal coordinate axis and the skew (oblique) coordinate axis. The relations between the two axes are

$$\eta = x - y\frac{\sin\alpha}{\cos\alpha}, \quad \zeta = y/\sin\alpha \tag{11.5.1}$$

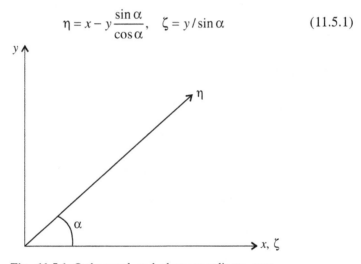

Fig. 11.5.1 Orthogonal and skew coordinate axes.

The governing equations in terms of the orthogonal coordinate system are given by Eqs. 11.4.5 to 11.4.7. The right-hand terms of Eq. 11.4.7 are replaced by the terms $q(x, y)$ and $\rho\dfrac{\partial^2 w}{\partial t^2}$. The boundary conditions along the edges of the laminate are

$$u = 0, \quad \upsilon = 0$$
$$w = 0, \quad \partial w / \partial x = 0 \quad (x = 0, a) \tag{11.5.2a}$$

$$u = 0, \quad \upsilon = 0$$

$$w = 0, \quad M_y = B_{26}\left(\frac{\partial \upsilon}{\partial x} + \frac{\partial u}{\partial y}\right) - D_{12}\frac{\partial^2 w}{\partial x^2} - D_{22}\frac{\partial^2 w}{\partial y^2} = 0 \quad (y = 0, \; b) \, (11.5.2b)$$

Using Eq. (11.5.1), we can write the relations between various derivatives in the rectangular and skew coordinate system [29] as

$$\frac{\partial^2}{\partial x^2} = \frac{\partial^2}{\partial \eta^2}$$

$$\frac{\partial^2}{\partial y^2} = \frac{\partial^2 \sin^2 \alpha}{\partial \eta^2 \cos^2 \alpha} + \frac{1}{\cos^2 \alpha}\frac{\partial^2}{\partial \zeta^2} - \frac{2\sin \alpha}{\cos^2 \alpha}\frac{\partial^2}{\partial \zeta \partial \eta}, \frac{\partial^4}{\partial x^4} = \frac{\partial^4}{\partial \eta^4}$$

$$\frac{\partial^4}{\partial y^4} = \frac{\sin^4 \alpha}{\cos^4 \alpha}\frac{\partial^4}{\partial \eta^4} - \frac{4\sin^2 \alpha}{\cos^4 \alpha}\frac{\partial^4}{\partial \eta^3 \partial \zeta} + \frac{6\sin^2 \alpha}{\cos^4 \alpha}\frac{\partial^4}{\partial \zeta^2 \partial \eta^2}$$

$$- \frac{4\sin \alpha}{\cos^4 \alpha}\frac{\partial^4}{\partial \eta \partial \zeta^3} + \frac{1}{\cos^4 \alpha}\frac{\partial^4}{\partial \zeta^4}$$

$$\frac{\partial^4}{\partial x^2 \partial y^2} = \frac{\sin^2 \alpha}{\cos^2 \alpha}\frac{\partial^4}{\partial \eta^4} + \frac{1}{\cos^2 \alpha}\frac{\partial^4}{\partial \eta^2 \partial \zeta^2} - \frac{4\sin \alpha}{\cos^2 \alpha}\frac{\partial^4}{\partial \eta^3 \partial \zeta}$$

as so on.

The *objective function*, which is the weight of the laminate, is defined as

$$\text{Minimize } F = (\rho ab \cos \alpha)\sum_{i=1}^{N} t_i \tag{11.5.3}$$

where α is the skew angle, ρ is the density of the material, t_i is the thickness of the ith lamina, and a, b are the laminate dimensions.

The *behavior constraints* are:

(i) The deflection of the laminate should be less than that prescribed, i.e.

$$g_1 = W^* - W \geq 0 \tag{11.5.4}$$

(ii) The fundamental natural frequency of the laminate should be greater than that prescribed, i.e.

$$g_2 = f - f^* \geq 0 \tag{11.5.5}$$

The *side constraints* are:

(i) The thickness of each lamina is bounded, i.e.

$$g_{2+i} = t_i - t_l \geq 0 \; (i = 1, 2..., N) \tag{11.5.6}$$

$$g_{N+2+i} = t_u - t_i \geq 0 \; (i = 1, 2..., N) \tag{11.5.7}$$

where the subscripts u and l stand for the upper and lower bounds, respectively.

(ii) The skew angle is bounded, i.e.

$$g_{2N+2} = \alpha - \alpha_l \geq 0 \qquad (11.5.8)$$

$$g_{2N+4} = \alpha_u - \alpha \geq 0 \qquad (11.5.9)$$

The optimization problem as stated above is solved by using the sequential augmented Lagrange multiplier method [33]. This technique requires the values of design variables. To obtain these, the governing equation and the boundary conditions are transformed into skew coordinates, and the transformed equations are then solved by applying the Galerkin technique.

Numerical computations have been carried out for the hybrid anti-symmetric skew laminate with boron/epoxy and Kevlar/epoxy composite constituting its plies.

The properties of boron/epoxy are: $E_L = 2.1 \times 10^6$ kg/cm², $E_T = 0.21 \times 10^6$ kg/cm², $v_{LT} = 0.3$ and those of the Kevlar/epoxy composite are: $E_L = 8.3 \times 10^5$ kg/cm², $E_T = 0.56 \times 10^5$ kg/cm, $v_{LT} = 0.34$.

The density ρ of the hybrid composite is taken as 0.00169 kg/cm³ and the other dimensional and side constraints are

$$W^* = 0.3 \text{ cm}, f^* = 5.0 \text{ Hz}$$
$$t_l = 0.016 \text{ cm}, t_u = 0.8 \text{ cm}$$
$$\alpha_u = 60°, \alpha_u = 15°$$

Table 11.5.1 lists the optimum values of the skew angle, ply fiber orientation and thickness along with the active behavior constraints. It is observed that, at the optimum point, the skew angle attains the maximum prescribed value for each aspect ratio. We further observe that the thickness of Kevlar/epoxy laminae is lower bounded. This is expected since Kevlar/epoxy is more flexible than boron/epoxy. Further, for the aspect ratios considered, the optimum fiber orientation with a constraint on maximum deflection lies between 75° and 90° whereas that for a laminate with a constraint on the lowest natural frequency lies between 70° and 80°.

Table 11.5.1 Optimum ply thickness, fiber orientation and skew angle (orthotropic hybrid laminate)

p	θ	α	Ply thickness (cm)		Optimum weight (kg/cm²)	Behavior constraint
			t_1	t_2		
0.5	79.63	60	0.04945	0.016	0.000221	Max. def.
1.0	82.95	60	0.06044	0.016	0.000258	Max. def.
1.5	86.47	60	0.06661	0.016	0.000279	Max. def.
0.5	71.97	60	0.09666	0.016	0.000381	Fund. freq.
1.0	74.30	60	0.18706	0.016	0.000686	Fund. freq.
1.5	78.69	60	0.23569	0.016	0.000851	Fund. freq.

For the hybrid skew laminate subjected to a uniformly distributed load, Fig. 11.5.2 shows the variation of central deflection with fiber orientation for various values of aspect ratio and skew angle and Fig. 11.5.3, the variation of first natural frequency with fiber orientation for diverse values of skew angle and aspect ratio. From the two figures, it is observed that the effects of skew angle, fiber orientation, and aspect ratio on deflection and frequency are very much interrelated.

Genetic Algorithm Technique (Applications)

As stated earlier, the simple elements of natural genetics, namely, reproduction, crossover and mutation are adopted in genetic algorithms to simulate a biological evolution process which in turn can be effectively utilized as a search procedure which is applicable across a spectrum of problems.

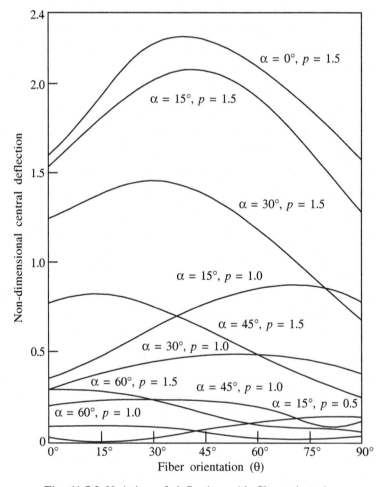

Fig. 11.5.2 Variation of deflection with fiber orientation.

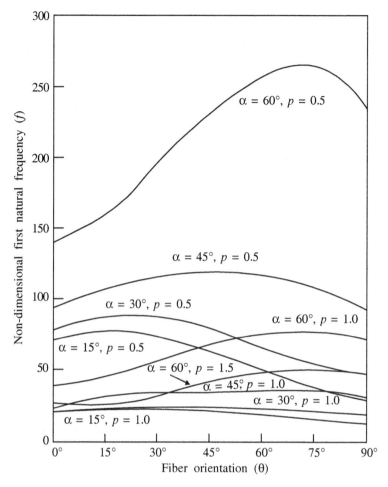

Fig. 11.5.3 Variation of frequency with fiber orientation.

A genetic algorithm differs from other search methods used in engineering optimization in the following ways:

(i)   A genetic algorithm uses a population of points at a time in contrast to the single-point approach used by the traditional optimization methods. The initial population generated is random and hence likely to cover the entire search space. Therefore, a genetic algorithm solution is unlikely to get trapped in a local minimum.

(ii) It only uses the objective function information. No derivatives are required.

(iii) A genetic algorithm works with coding of the parameters rather than the parameters themselves.

The implementation of a simple genetic algorithm consists of the following three operators:

(i) Reproduction
(ii) Crossover
(iii) Mutation

Reproduction is the first operator applied and it constitutes the first step in a genetic algorithm. It involves selection of individuals based on their fitness values relative to those of the population. Goldberg and Deb [34] have discussed in detail the various selection schemes. The two most commonly used selection procedures are proportionate selection and tournament selection.

The crossover operation proceeds in two steps. First, two individual strings are selected at random from the mating pool generated by the reproduction operator. A crossover site is selected at random on the probability of crossover, along the string length and the alleles are swapped between the two strings following the crossover site. The resulting strings are placed in the new population which is filled with new progenies.

The mutation operator is applied to the new strings with a specified mutation probability. A mutation is the occasional random alteration of an allele's value. The application of the mutation operator introduces an element of diversity. It may lead to entirely new areas of search space, thus preventing the genetic algorithm from getting trapped in some local minima. However, the probability is kept very low to prevent loss of genetic information.

We shall consider some examples illustrating the application of genetic algorithms for obtaining the optimal design of composite laminates.

Example 11.6
Determine the optimal design of composite laminates with and without cutout undergoing free vibration.

Solution
It is not possible to get a closed form solution for a laminate with cutout. Finite element formulation is, therefore, employed to obtain the fundamental frequency. Four noded rectangular elements are used to represent the total domain of the plate. Due to the presence of cutout, a uniform mesh will not work and hence a graded mesh is employed in the neighborhood of the cutout. A C^1 continuous shear flexible element based on the higher order shear deformation theory is developed using the Hermite interpolation formulae as indicated by Bogner et al. [35]. For details of the derivation of the finite element characteristic equation for frequency, the reader may refer to Reddy and Chao [36].

The *objective function* is written as

$$\text{Maximize } \lambda \qquad (11.6.1)$$

$$\text{Subject to } \theta \in [0°, \pm 30°, \pm 45°, \pm 60°, 90°] \qquad (11.6.2)$$

$$\sum_{i=1}^{nl} h_i = h \qquad (11.6.3)$$

$$h_l \leq h_i \leq h_u \qquad (11.6.4)$$

Evaluation of the Fitness

If the constraint is not violated, then the fitness is the obtained frequency value from the finite element solution. If the constraint is violated, a very high penalty value of the order of 100,000 is imposed as

$$\text{Fitness value} = \frac{\text{Frequency}}{10^5 \times (h - \text{Total thickness})} \qquad (11.6.5)$$

Coding in the Genetic Algorithm

The variables, namely, ply angles (θ) and group ply thickness (h) are coded in binary string. A ten layer laminate is considered for the optimization study. Only practical fiber orientations $0°, \pm 30°, \pm 45°, \pm 60°, 90°$ are considered for design. The ply angles are coded as 1, 2, 3, 4, 5, 6, 7 and 8 which stand for the possible stacks, $60°, 45°, -30°, 0°, 30°, 45°, 60°$ and $90°$, respectively. Laminate thickness is considered directly in multiples of ten and generated within the specified bounds on individual layer thickness.

Convergence Criteria

The convergence of the optimum solution is decided by

(i)   The values of the constraint at the final solution.
(ii) Repetition of the same optimum for different runs.
(iii) Distribution of the population in final generation.

Table 11.6.1 shows the maximum non-dimensional fundamental frequency and optimum fiber orientations for a ten layer thick laminate without cutout for a composite with the properties

$E_L = 130$ GPa, $E_L/E_T = 13$, $G_{LT}/E_T = 0.5$, $G_{LT'}/E_T = 0.5$, $G_{TT'} = 0.33$, $v_{LT} = 0.35$.

Table 11.6.1 Optimum fiber orientations with fundamental frequency (thick laminate)

a/h	a/b	Fiber angle	Ply thickness	Frequency	λ (uniform)
5	1	$[-45/-45/$ $45/45/-45]_s$	$[0.2/0.3/0.3/$ $0.1/0.2]_s$	11.9403	11.9338
5	2	$[0/0/0/0/0]_s$	$[0.2/0.3/0.1/$ $0.2/0.2]_s$	20.1163	20.1028

From the table, it is observed that for rectangular plates ($a/b = 2.0$), for thick laminates, the optimum fiber orientation is $0°$ for all the plies. However, for square laminates, the fiber orientation is $\pm 45°$. For both the cases, the thickness of each ply is different. The last column indicates the frequency obtained when the thickness of all plies is equal and is taken as 0.2.

Table 11.6.2 shows the results for a thin laminate with the same material property.

Table 11.6.2 Optimum fiber orientations with fundamental frequency (thin laminate)

a/h	a/b	Fiber angle	Ply thickness	Frequency	λ (uniform)
100	1	[− 45/45/− 45/ − 45/− 45]$_s$	[0.1/0.3/0.3/ 0.1/0.2]$_s$	25.8217	24.9995
100	2	[0/0/0/0/0]$_s$	[0.3/0.2/0.1/ 0.2/0.2]$_s$	72.8507	72.7780

Table 11.6.3 shows the results for a simply supported square laminate with square cutout. The two runs give the results for two different initial populations. The convergence history is shown in Fig. 11.6.1.

Table 11.6.3 Optimum fiber orientation with frequency for a simply supported square laminate with square cutout

a/h	Run	Fiber angle	Ply thickness	Frequency
100	1	[− 45/45/45/ − 45/45]$_s$	[0.2/0.3/0.2/ 0.1/0.2]$_s$	31.5317
100	2	[− 45/45/− 60/ − 45/45]$_s$	[0.1/0.3/0.1/ 0.3/0.2]$_s$	31.4588

Table 11.6.4 shows the effect of initial population studies for a symmetric, ten layer simply supported laminate without cutout for $a/h = 5$.

Table 11.6.4 Effect of initial population on optimum frequency for a supported square laminate ($a/h = 5$)

a/h	Run	Fiber angle	Ply thickness	Frequency
5	1	[− 45/− 45/45/ − 45/− 45]$_s$	[0.2/0.3/0.2/ 0.1/0.2]$_s$	11.9403
5	2	[− 45/− 45/45/ − 45/45]$_s$	[0.2/0.2/0.2/ 0.2/0.2]$_s$	11.9409

The convergence runs for the laminate with and without cutout are shown in Fig. 11.6.1 and Fig. 11.6.2, respectively. It is observed that as the

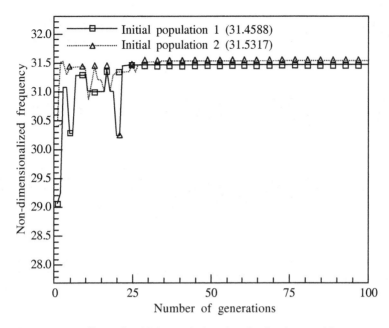

Fig. 11.6.1 Effect of initial population for the laminate with cutout.

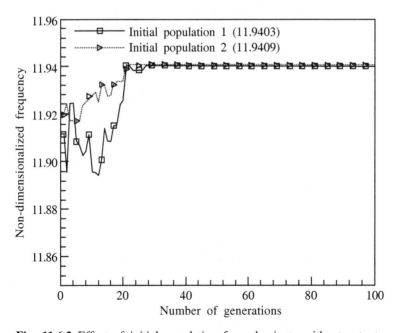

Fig. 11.6.2 Effect of initial population for a laminate without cutout.

numbers of generations are increased, the solution converges to a global minimum. Further, initial population has very little effect on the global solution.

Example 11.7

Determine the optimal design of laminated composite plates undergoing large amplitude oscillations.

Solution

Dynamic analysis is an important field where the designer has to concentrate while designing structures. The structures include specifically spacecraft or aircraft structures where thin skins are used. In practical applications, the assumption that the laminates vibrate at small amplitude may not always be valid. It is, therefore, necessary to understand the effect of large amplitude vibrations in the optimum design of laminates for many applications. Reddy and Chao [36] presented a finite element analysis with large deflection theory including transverse shear, governing moderately thick laminated anisotropic composite plates. Large deflection of shear deformable composite plates using a simple higher-order theory with a higher-order C^1 continuous refined element is reported by Gajbir et al. [37]. Tenneti and Chandrashekhara [38] studied the nonlinear vibration of laminated plates using a refined shear flexible finite element. In this example, we make an attempt to determine the optimum design of composite structures under dynamic constraints.

Formulation

The problem is formulated using a plate of thickness h composed of orthotropic layers of thickness h_i with fibers oriented at angles $\pm\ \theta$ as shown in Fig. 11.7.1. The higher-order displacement model which gives the parabolic variation of shear stresses over the thickness of the laminate, used for the present analysis is given as [39]

$$u(x, y, z, t) = u_0(x, y, t) + f_1(z)\psi_x(x, y, t) + f_2(z)\theta_x(x, y, t)$$
$$\upsilon(x, y, z, t) = \upsilon_0(x, y, t) + f_1(z)\psi_y(x, y, t) + f_2(z)\theta_y(x, y, t) \qquad (11.7.1)$$
$$w(x, y, z, t) = w_0(x, y, t)$$

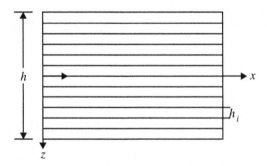

Fig. 11.7.1 Composite laminate with orthotropic layers.

where

$$f_1(z) = C_1 z - C_2 z^3$$
$$f_2(z) = -C_4 z^3$$

u_0, υ_0 and w_0 are displacements of the middle plane of the laminate and θ_x, θ_y, ψ_x and ψ_y are the rotations.

The nonlinearity due to large amplitude is taken care of by using the Von Karman strain-displacement relation. The procedure adopted by Ganapathi and Varadan [40] is used to obtain the strain energy expression for the laminate.

In the present study, a C^0 nine-noded isoparametric quadrilateral finite element with 7 DOF per node (u, υ, w, ψ_x, ψ_y, θ_x, θ_y) is employed. The full plate is idealized by 2 × 2 meshes for finding the frequencies in the large amplitude range. Reduced integration is used to evaluate the transverse shear stress while full integration is carried out for bending and stretching. For details, the reader may refer to Sivakumar et al. [41]. Going through the procedure, the governing equation for a nonlinear eigen value problem is obtained as

$$[M]\{\ddot{\delta}\} + \left[[K_{MB}] + \frac{1}{2}[K_{NL1}] + \frac{1}{3}[K_{NL2}] + [K_s] \right]\{\delta\} = 0 \quad (11.7.2)$$

where $[K_{NL1}]$, $[K_{NL2}]$ are the nonlinear stiffness matrices corresponding to cubic, quartic functions of the displacement vector. Equation (11.7.2) is solved by using the solution procedure for the direct iteration method suggested by Tenneti and Chandrashekhara [38] and Kanaka Raju and Hinton [42].

Optimization

Two problems are considered for optimization:

(i) The objective here is to maximize the fundamental frequency of the laminate for a given amplitude of oscillation. No behavioral constraints are specified. The material property is taken for computations as

$E_L/E_T = 40$, $G_{LT}/E_T = 0.6$, $G_{LT'}/E_T = 0.6$, $G_{TT'}/E_T = 0.5$, $\nu_{LT} = 0.25$, $\rho = 1500$ kg/m^3

The optimization problem is written as

Maximize λ_{1NL}
subject to $\theta \in 0°$, $45°$, $-45°$, $90°$

A genetic algorithm is used to obtain the optimum value. Figure 11.7.2 shows the variation of maximum fundamental frequency parameter with amplitude ratio. It also shows the optimum ply orientation (θ_{opt}) corresponding to the maximum frequency parameter for eight layer symmetric and anti-symmetric laminates. In the linear range, the optimum ply orientations corresponding to the maximum fundamental frequency

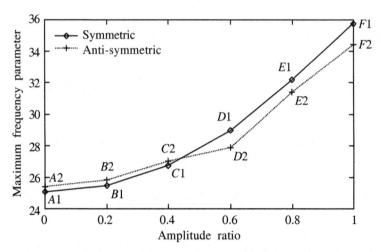

Fig. 11.7.2 Variation of maximum fundamental frequency with amplitude ratio. A1: $[45°/- 45°/45°/- 45°]_s$, C1: $[45°/- 45°/45°/90°]_s$, D1: $[45°/- 45°/90°/0°]_s$, E1: $[- 45°/45°/90°/0°]_s$, A2: $[45°/- 45°/45°/- 45°]_2$, F2: $[45°/- 45°/- 45°/45°]_2$ (Material-1, $a/h = 100$, $a/b = 1$, simply supported with movable ends).

parameter for both symmetric and anti-symmetric laminates are $[45°/- 45°/45°/- 45°]_{s,2}$. When the amplitude ratio increases, the ply orientations corresponding to the maximum fundamental frequency change. The optimum ply orientations obtained at an amplitude ratio of 1.0 for a symmetric laminate are $[- 45°/45°/90°/0°]_s$ and for an anti-symmetric laminate, they are $[45°/- 45°/- 45°/45°]_2$. Therefore, it is seen that the ply orientation which is optimum for a small amplitude range does not remain the same at the higher amplitude ranges. It is, therefore, desirable to take this factor into account when designing laminates undergoing large amplitude vibrations.

(ii) The second problem deals with the optimum weight design of a laminated plate in the large amplitude range. Ply thickness, ply angles and material of the plies are considered as the variables. A ten layer square symmetric laminate of 0.3 m with its edges simply supported is used in this optimization problem.

The material properties are chosen according to Table 11.7.1. Only 45°, – 45°, 0°, 90° are considered for ply angles. A total of 15 variables with a string length of 40 is considered for the analysis. The problem is formulated as

Maximize W

$$2\pi\omega_{1NL} \geq 15000$$
$$0.0005 \leq h_i \leq 0.008$$

Table 11.7.1 Composite material properties

Material code, *mc*	Material	E_L/E_T	G_{LT}/E_T	$G_{LT'}/E_T$	$G_{TT'}/E_T$	ν_{LT}	ρ (kg/m³)
1	Gr/epoxy	40	0.6	0.6	0.5	0.25	1500
2	G/epoxy	3.0	0.5	0.5	0.4	0.26	1800
3	K/epoxy	14.82	0.375	0.375	0.375	0.34	1460
4	B/epoxy	10.0	0.3	0.3	0.275	0.23	2000

Subject to

$$\theta_i \in 0°, \ 90°, \ 45°, \ -45°$$
$$mc \in 1, \ 2, \ 3, \ 4, \ i = 1, \ 2,..., \ nl/2$$

The design is obtained for different values of amplitude to thickness ratio. Due to the hardening effect in the large amplitude range, the stiffness of the material increases in some orientations. Selection of these orientations allows reduced weight designs for the given frequency constraints. Further, the design of the laminate also changes according to the amplitude of vibration. Table 11.7.2 gives the optimum weight design of laminated composite plates in the large amplitude range.

Table 11.7.2 Optimum design of a ten layer symmetric laminate with simply supported edges in the large amplitude range ($a = 0.3$)

Amplitude ratio A/h	Optimum ply thickness (*m*)	Ply angle (degree)	Material code	Optimum weight (kg)	Constraint values $2\pi\omega_{1NL}$
0.0	[0.0025/0.0055/ 0.0015/0.0005/ 0.0065]$_s$	[45.0/− 45.0/ 90.0/45.0/ − 45.0]$_s$	[1/1/3/3/1]$_s$	4.441	15891.470
0.25	[0.0055/0.0050/ 0.0050/0.0025/ 0.0060]$_s$	[− 45.0/− 45.0/ − 45.0/45.0/ 45.0]$_s$	[1/3/3/1/1]$_s$	4.0428	15205.70
0.5	[0.0025/0.0020/ 0.0005/0.0045/ 0.0020]	[− 45.0/45.0/ 90.0/90.0/ 45.0]$_s$	[1/1/3/1/1]$_s$	3.236	15380.79

Summary

• The various techniques for optimization for composite laminates were discussed.
• A large number of worked out examples are presented.
• The advantages and disadvantages of the techniques are also discussed.

References

1. Fletcher, R. and Reeves, C.M., 1964, "Function Minimization by Conjugate Gradients", *Computer J.*, 7, 2, pp. 149-154.

2. Davidon, W.C., 1959, "Variable Metric Method for Minimization", *Technical Report No. ANL-599*.
3. Fletcher, R. and Powell, M.J.D., 1963, "A Rapidly Converging Decent Method for Minimization", *Computer J.*, 6, pp. 163-168.
4. Fiacco, A.V. and McCormick, G.P., 1968, *Nonlinear Programming: Sequential Unconstrained Minimization Techniques*, Wiley, New York.
5. Hooke, R. and Jeeves, T.A., 1961, "Direct Search Solution of Numerical and Statistical Problems", *J. ACM*, 8, pp. 212-221.
6. Powell, M.J.D., 1964, "An Efficient Method for Finding the Minimum of a Function of Several Variables without Calculating Derivatives", *Computer J.*, 7, pp. 155-162.
7. Holland, J.H., *Adaptation in Natural and Artificial Systems*, University of Michigan Press, Ann Arbor, Michigan, USA.
8. Goldberg, D.E., 1989, *Genetic Algorithm in Search, Optimization and Machine Learning*, Addison-Wesley.
9. Fox, R.L., 1971, *Optimization Method for Engineering Design*, Addison-Wesley, Massachusetts.
10. Haug, E.J. and Arora, J.S., 1979, *Applied Optimal Design*, Wiley, New York.
11. Reklaitis, G.V., Ravindran, A. and Ragsdell, K.M., 1983, *Engineering Optimization: Methods and Applications*, Wiley.
12. Haftka, R.T., 1992, *Elements of Structural Optimization*, 3rd Edition, Kluwer Academic Publishers.
13. Rao, S.S., 1966, *Engineering Optimization*, 3rd Edition, Wiley.
14. Iyengar, N.G.R. and Gupta, S.K., 1997, *Structural Design Optimization*, East West Press, New Delhi.
15. Deb, K., 2001, *Multi Objective Optimization Using Evolutionary Algorithms*, John Wiley and Sons.
16. Goldberg, D.E. and Samtani, M.P., 1986, "Engineering Optimization via Genetic Algorithm", *Proceedings of the Ninth Conference on Electronic Computation*, pp. 471-482.
17. Rao, S.S., Pan, T.S. and Venkayya, V.B., 1990, "Optimal Placement of Actuators in Actively Controlled Structures Using Genetic Algorithm", *J. AIAA*, 29, pp. 942-943.
18. Chaturvedi, D., Deb, K. and Chakrabarty, S.K., 1995, "Structural Optimization Using Real Coded Genetic Algorithm", *Proceedings of Symposium on Genetic Algorithm*, Computer Society of India, pp. 73-81.
19. Nagendra, S., Haftka, R.T. and Gurdal, Z., 1991, "Buckling Optimization of Laminates Stacking Sequence with Strain Constraints", *Proceedings of the Tenth Conference on Electronic Computation*, Indianapolis, pp. 205-215.
20. Callahan, J.K. and Weeks, E.G., 1992, "Optimum Design of Composite Laminates Using Genetic Algorithms", *Composite Engineering*, 2, pp. 149-160.
21. Kogiso, M., Watson, L.T., Gurdal, L., Haftka, R.T. and Nagendra, S., 1994, "Design of Composite Laminates by a Genetic Algorithm with Memory", *Mechanics of Composite Materials and Structures*, 1, pp. 95-117.
22. Mahesh, K., Kishore, N.N. and Deb, K., 1996, "Optimum Design of Composite Turbine Blade Using Genetic Algorithm, *Adv. Composite Materials*, 5, pp. 87-98.
23. Muc, A. and Gurba, W., 2001, "Genetic Algorithms and Finite Element Analysis in Optimization of Composite Structures", *Composite Structures*, 54, pp. 275-281.

24. Lin, C.Y. and Hajela, P., 1992, "Genetic Algorithms in Optimization Problems with Discrete and Integer Design Variables", *Engineering Optimization*, 19, pp. 309-327.
25. Jones, R.M., 1975, *Mechanics of Composite Materials*, Scripta Book Co., Washington.
26. Hirano, Y., 1979, "Optimum Design of Laminated Plates under Axial Compression", *J. AIAA*, 17, pp. 1017-1019.
27. Khot, N.S., Venkayya, V.B. and Berke, L., 1976, "Optimum Design of Composite Structures with Stress and Displacement Constraints", *J. AIAA*, 14, pp. 131-132.
28. Khot, N.S., Venkayya, V.B., Johnson, C.D. and Tischler, V.A., 1973, "Optimization of Fiber Reinforced Composite Structures", *Int. J. Solids and Structures*, 9, pp. 1225-1236.
29. Schmit, L.A., Jr. and Farshi, B., 1977, "Optimum Design of Laminated Fiber Composite Plates", *Int. J. Num. Meth. Engg.*, 11, pp. 623-640.
30. Singh, K., 1976, "Optimum Design of Fiber Reinforced Composite Structural Elements", *M.Tech. Thesis*, Department of Aerospace Engineering, Indian Institute of Technology Kanpur.
31. Soni, P.J. and Iyengar, N.G.R., 1983, "Optimal Design of Clamped Laminated Composite Plates", *Fiber Science and Technology*, 19, pp. 281-296.
32. McKeown, J.J., 1975, "A Quasi Linear Programming Algorithm for Optimizing Fiber Reinforced Structure", *Comp. Meth. Appl. Mech. Eng.*, 6, pp. 123-154.
33. Gill, P.E. and Murry, W. (Eds.), 1974, *Numerical Methods for Constrained Minimization*, Academic Press, New York.
34. Goldberg, B.E. and Deb, K., 1991, "A Comparison of Selection Schemes Used in Genetic Algorithms", *Foundations of Genetic Algorithms*, I, pp. 69-93.
35. Bogner, F.K., Fox, R.L. and Schmit, L.A., 1996, "The Generation of Inter Element Compatible Stiffness and Matrices by the Use of Interpolation Formulas", *Proc. Conf. Matrix Methods in Structural Mechanics*, AFFDL-TR-66-80, Wright Patterson A.F.B., pp. 397-444.
36. Reddy, J.N. and Chao, W.C., 1981, "Large Deflection and Large Amplitude Free Vibrations of Laminated Composite Material Plates", *Composite Struct.*, 13, pp. 341-347.
37. Gajbir, S., Venkateswara Rao, G. and Iyengar, N.G.R., 1993, "Large Deflection of Shear Deformable Composite Plates Using a Simple Higher Order Theory", *Composite Engineering*, 3, pp. 507-525.
38. Tenneti, R. and Chandrashekhara, K., 1994, "Nonlinear Vibration of Laminated Plates Using Refined Shear Flexible Finite Element", *Adv. Composite Mat.*, 4, pp. 145-158.
39. Shankara, C.A. and Iyengar, N.G.R., 1992, "A C^0 Element for the Free Vibration Analysis of Laminated Composite Plates", *J. Sound and Vib.*, 191, pp. 721-738.
40. Ganapathi, M. and Varadan, T.K., 1991, "Nonlinear Flexural Vibrations of Laminated Orthotropic Plates, *Comput. Struct.*, 39, pp. 685-688.
41. Sivakumar, K., Iyengar, N.G.R. and Kalyanmoy, D., 1999, "Optimum Design of Laminated Composite Plates Undergoing Large Amplitude Oscillations", *App. Composite Mat.*, 6, pp. 87-98.
42. Kanaka Raju, K. and Hinton, E., 1980, "Nonlinear Vibrations of Thick Plates Using Mindlin Plate Elements", *Int. J. Num. Meth. Engg.*, 16, pp. 247-257.

12. Manufacturing Techniques for Composites

In this chapter, we will discuss:

• Different manufacturing techniques employed to make composite parts or products.
• The advantages and disadvantages of various methods generally employed in the laboratories to make test specimens.

12.1 Introduction

The procedure most common to all polymeric composite processes is the combination of a resin, a curing agent, reinforcing fiber, and in some cases a solvent. Heat and pressure are generally used to "cure" the mixture into a finished product or part. The method of manufacturing depends on the type of fiber. They could be continuous fibers, chopped fibers, prepreg tapes or fiber cloth.

Prepregging is the process where the resin and curing agent mixture are impregnated into the reinforcing fiber. Such impregnated reinforcements (fibers), also known as prepregs, take three main forms: woven fabrics, roving and unidirectional tapes. Fabrics and tapes are provided as continuous rolls in widths up to 600 mm and lengths up to several hundred meters. The fabric or tape thickness constitutes one ply in the construction of a multi-ply lay-up. Impregnated roving is wound onto cores or bobbins and is used for filament winding. Once the resin mixture has been impregnated onto the fibers, the prepreg must be stored in a refrigerator or freezer (temperature of − 18°C) until ready for use in the manufacturing process. This cold storage prevents the chemical reaction from occurring prematurely and retains its tackiness. Prepreg tapes are used widely in the advanced composite industry, particularly in aircraft and aerospace.

A large number of well established processing methods are available for fiber reinforced composites. The selection of the fabrication process depends on the type of fiber (continuous, short, chopped strand, prepreg etc.), matrix material and the temperature required to form, melt or cure the matrix. As stated in earlier chapters, in the case of composites, the material is created and the product is formed at the same time unlike in metals. In metals, the material is created first and then processed to make the product. The processes in the case of composites with polymer matrix, metal matrix and ceramic matrix are different. Further, with polymer matrix, the process is different for thermosetting and thermoplastic polymers. The choice of the process also depends on the volume and number of items of the same product to be fabricated. The various processes employed are described below.

12.2 Fabrication Process for Thermosetting Polymer Composites

As stated earlier, a large number of techniques are available for fabrication. These are: (i) hand lay-up, (ii) spray-up, (iii) vacuum bag molding, (iv) pressure bag molding, (v) autoclave molding, (vi) sheet molding compound (SMC), (vii) bulk molding compound (BMC), (viii) filament winding, (ix) pultrusion and (x) liquid composite molding.

We now discuss some of these techniques in detail.

Hand Lay-up Technique

This is one of the simplest and most commonly used methods for fabrication of composite products of small volumes or large size. The open mold is made of wood, plastic, metal, composite or plaster of Paris. The accuracy and finish of the product depends on how well the mold has been prepared. The mold surface is then coated with the mold release to prevent the bonding of the resin material to the mold. If a smooth surface is required, a gel coat is applied to the mold. One layer of chopped strand mat, woven fibers or fiber cloth is placed on the mold. Resin and catalyst mixed separately is then applied on the surface. A roller is then used for proper wetting of the fibers. This process also removes excess resin and trapped air bubbles. Alternate layers of fibers and resin are used till the required thickness is attained (Fig. 12.1). At this stage, the final product is still wet. Curing is performed at high temperature for completion of the reaction. Once the curing is complete, the resin hardens. The product is then removed from the mold.

Spray-up Technique

This is once again an open mold process. The mold is prepared in the same manner as described above. However, in this case, chopped glass fibers are used. The chopped fibers are mixed with the resin and catalyst. The mixture is sprayed on the mold [1]. The surface is rolled by hand to remove excess resin and air bubbles. The quality of the product depends on the skill of the operator (Fig. 12.2).

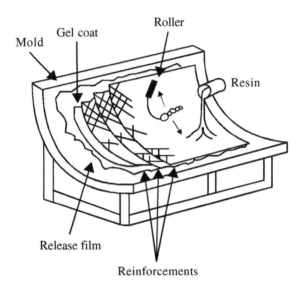

Fig. 12.1 Hand lay-up technique.

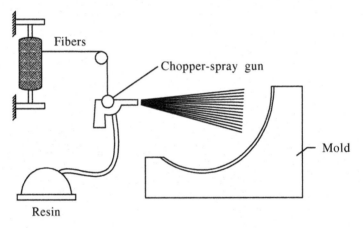

Fig. 12.2 Spray-up technique for chopped fiber composites [1].

Bagging Technique

There are three types of bagging techniques: (a) vacuum bagging technique, (b) pressure bagging technique and (c) autoclave technique. The products realized by the autoclave technique are of very good quality, however, it is an expensive technique.

Vacuum Bag Molding [2]

Vacuum bagging is a technique employed to create mechanical pressure on a laminate during its cure cycle. Pressurizing a composite lamination serves several functions:

(i)   It removes trapped air between layers.
(ii) It compacts the fiber layers for efficient force transmission among fiber bundles and prevents shifting of fiber orientation during curing.
(iii) It reduces humidity.
(iv) Finally, and most important, the vacuum bagging technique optimizes the fiber-to-resin ratio in the composite part.

A typical wet laminate and corresponding female mold may be placed inside a vacuum bagging assembly. First, a release film or peel-ply material must be draped over any portion in contact with the resin. If the part is too complex to have the release draped evenly, sections may be cut and applied individually to the contours. However, the layers must overlap at all joints to ensure adequate protection. The type of release layer varies depending upon the desired surface texture and resin content that the component is to have. Peel-ply will also allow excess resin to pass through it and become absorbed in the bleeder material. If a smooth, finished surface is desired, a release film should be used. Perforated release film will allow excess resin to pass through, while non-perforated release film holds the resin in the molded part. The next layer to be added is the bleeder/breather material. These are porous, high temperatures fabrics. This material provides two important functions. First, it absorbs excess resin from the laminate during processing. Second, it is this layer which ensures that the vacuum is distributed evenly within the bag. The breather layer can also be used to apply pressure in hard to reach areas of a mold by placing rolled or folded sections deep into sharp angles or troughs. It also helps in the removal of air and volatiles during curing. Figure 12.3 shows a representative sketch.

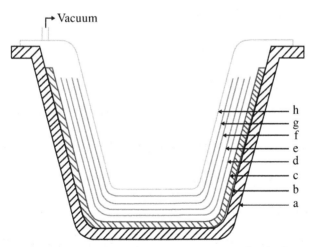

a: Mold, b: Release agent applied to mold, c: Lay-up, d: Bleeder, e: Breather, f: Porous release film, g: Pressure pad and h: Bag

Fig. 12.3 Vacuum bag technique.

The quality of the product developed depends on the process of laying up and bagging. The size of the component that can be molded using the bagging process is limited by the curing equipment.

Pressure Bag Molding

The molding technique is similar to the vacuum bag technique. However, in this case, pressures in excess of atmospheric pressure are applied on the laminate inside the closed mold for shaping. A flexible bag is placed over the lay-up in the mold. Inflation of the bag by compressed air enables the lay-up to take the shape of the mold. Generally, one need not use higher temperature for curing. However, the curing process may be accelerated by placing the component in an oven or by heating the mold in an oven (Fig. 12.4).

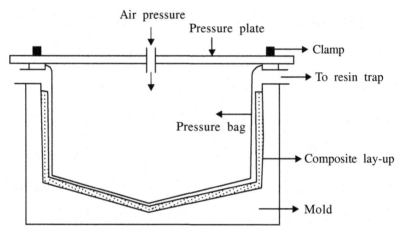

Fig. 12.4 Pressure bag molding technique.

Autoclave Molding

This technique is an extension of the pressure bag molding technique. Though expensive, it is largely used in the aircraft industry to manufacture parts which require high accuracy and reliability. Autoclave technique involves a heated pressure vessel in which the mold is placed with the lay-up of the desired product. The whole mold is sealed in a vacuum bag. The air is drawn out of the bag. The vacuum pressed mold and bag are placed in an airtight heating chamber (or an autoclave). Usually an inert atmosphere is provided inside the autoclave through the introduction of nitrogen or carbon dioxide. Exotherms may occur if the curing step is not done properly. Figure 12.5 shows a simple sketch of the technique [1].

The door of the autoclave is closed and heat is applied (usually 120-180°C). As a result of heating, the resin starts to liquefy. This is when pressure is introduced (usually 3-8 atm). The pressure expels the air and excess resin out of the component. After a few minutes in these conditions, the component is cured. The pressure is released from the machine and the

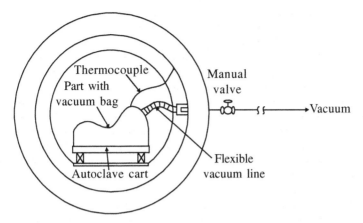

Fig. 12.5 Autoclave technique.

cooling cycle starts. The pressure and temperature are controlled as required for complete curing. One advantage of the autoclave technique is that one can control temperature and pressure in different segments of the pressure vessel; thereby components of unlimited variety can be manufactured at the same time.

Sheet Molding Compound
This technique is largely used in the automobile industry. Chopped fibers are mixed thoroughly with the matrix to make panels and doors of cabins etc. The compound so prepared is in an uncured state. It is then pressed between two heated matched dies. The pressure required depends on the flow characteristics. SMC benefits from very high volume production ability and excellent part reproducibility. It is cost effective because of low labor requirements and industry scrap is reduced substantially. Weight reduction, due to lower dimensional requirements and because of the ability to consolidate many parts into one, is also advantageous. The level of flexibility also exceeds many counterpart processes.

Bulk Molding Compound
Bulk molding compound (BMC) is a blend of various inert fillers, fiber reinforcement, catalysts, stabilizers, matrix and pigments that form a viscous, 'puttylike' compound for compression or injection molding. BMC is highly filled and reinforced with short fibers. The volume fraction of reinforcement is of the order 10 to 30% and with fiber length typically up to 12.5 mm.

Filament Winding
It is a capital-intensive process used mainly to manufacture small and large diameter pipes, tubes, cylinders and pressure tanks in medium to high production volume. As the name implies, it involves the winding of continuous,

pre-saturated reinforcing filaments around a rotating mandrel, until the whole surface is covered at the desired depth. Continuous reinforcements in the form of rovings are fed from a number of *creels*. The filaments from the *creel* are saturated as they pass through a resin bath just before they meet the mandrel. They are then gathered into a band of given width and wound over the mandrel. The winding, depending on the complexity of the machine, can be performed in two or more angles. A pre-impregnated, B stage filament can be used (prepreg winding). The winding angles and the placements of the filaments are controlled by machines traversing at speeds synchronized with mandrel rotation. After curing, at the final stage of production, the mandrel has to be removed, usually with the help of a hydraulic extractor. The quality of the finished products is usually very good, as the filament reinforcement is continuous. However, most of the wound products are somewhat resin rich and lack longitudinal reinforcement (mandrel rotation prohibits reinforcement to be placed along the longitudinal axis of the component). The winding angle employed for making pipes etc., depends on the strength requirements and may vary from helical to circumferential. Generally, a combination of different winding patterns is employed to give optimal performance. Figure 12.6 shows a typical filament winding technique.

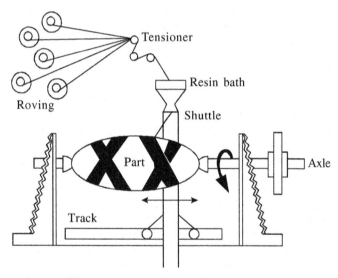

Fig. 12.6 Filament winding technique.

Pultrusion

It is a continuous, high capital and material intensive process for the manufacture of composite products of constant cross-section. Unlike filament winding which mounts reinforcement in the transverse (or circumferential) direction of the mandrel, pultrusion places the primary reinforcement in the longitudinal direction. It is performed by pulling continuous filaments together with chopped strand mat tapes through a resin bath to a heated metal die cavity of the desired cross-section and shape. This die serves as the mold and the curing oven at the same time. The higher the temperature of the die, the larger is the speed of pulling. It is also possible, instead of using a resin bath for saturation, to inject the resin directly into the die cavity. As the profile is pulled out already cured, a saw at the end of the production line cuts it to the desired length. A large number of profiles such as rods, tubes and many complex structural shapes can be manufactured by using proper dies. Length of the product is unlimited. Quality of pultruted parts is very good. Their main disadvantage is the lack of transverse reinforcement as the pulling mechanism makes its placement very difficult. Generally, one can obtain a fiber volume fraction as high as 65%. Figure 12.7 shows a typical layout for pultrusion technique.

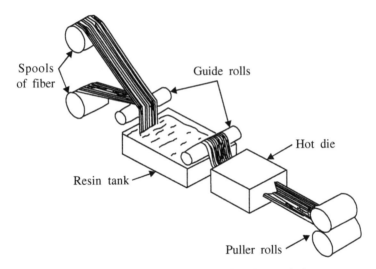

Fig. 12.7 A typical layout for pultrusion technique.

The pultrusion process is suitable for thermosetting resins. Thermoplastic resins may be used, but special impregnation equipment is required to soften the resin so that the reinforcement is completely wetted and cured as its passes through the die.

Liquid Composite Molding [3]
There are a number of variants of liquid composite molding (LCM) techniques. Notable among them are: resin transfer molding (RTM), reaction injection molding (RIM), vacuum assisted resin transfer molding (VARTM), resin film infusion (RFI), structural reaction injection molding (SRIM) and vacuum assisted reaction injection molding (VARIM). The processes mentioned differ in (i) how the resin is delivered, (ii) how the air bubbles are removed,

(iii) nature of tooling and (iv) whether vacuum is used or not. This is an expensive process which employs a coupling (male and a female) metal mold that is heated. The reinforcement is cut with precision and placed in the mold cavity. Generally, a pre-form is used (many different layers of reinforcement are pre-cut and held together in particular pattern, according to the shape of the mold, with the help of a "binder". After placing the reinforcement, the two matching molds are closed tightly and liquid resin is pushed inside through the carefully positioned openings or injection "gates". The air is expelled through other carefully positioned openings, the "vents", and the reinforcement is saturated. The whole process can be assisted by vacuum (VARTM). When full cure is reached, the component is ejected from the mold cavity. RTM is used for high production volumes. Figure 12.8 shows the RTM technique employed for typical composites.

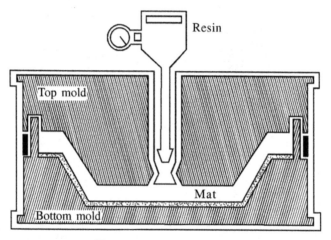

Fig. 12.8 RTM technique.

Liquid transfer molding is generally employed when autoclave molding becomes expensive. Some of the other advantages are: (i) fully repeatable production, (ii) parts with smooth finish on both sides are obtained, (iii) complex parts can be fabricated with ease, (iv) short turn around time, (v) less waste and (vi) reduction in emission of volatiles.

12.3 Manufacturing Process for Thermoplastic Polymer Composites

Matched/Unmatched Die Molding

In the thermoplastic matched/unmatched die molding process, an uncured laminate consisting of thermoplastic prepreg tape layers is passed through an infrared oven. It is heated to a temperature close to the melting point of the thermoplastic resin. The heated laminate is then quickly placed in a

(a) Unmatched mold

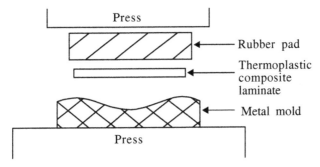

(b) Matched mold

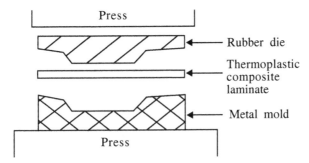

Fig. 12.9 Unmatched and matched mold techniques.

matched/unmatched metal mold for final forming. Figure 12.9 shows matched/unmatched molding techniques.

Injection Molding

In injection molding of reinforced thermoplastics, a compound of resin and fibers are in pelletized form. The pellets can be either short fibers or long fibers. The pellets are formed separately and are fed into the hopper of the injection machine. Thermoplastic granules are fed via a hopper into a screw-like plasticating barrel where melting occurs. The melted plastic is injected into a heated mold where the part is formed. This process is often fully automated. Figure 12.10 shows a typical injection molding technique.

Thermoforming

Thermoplastic sheets are also produced by the lamination of chopped strand mats in a thermoplastic matrix. The thermoforming of thermoplastic organic sheets is due to the unlimited shelf life and the simple handling is a very promising method for the manufacturing of components of high-strength and high-stiffness. Thermoforming uses heat and pressure to transform

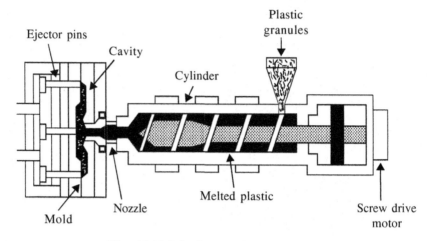

Fig. 12.10 Injection molding equipment.

sheet thermoplastics into the desired shape. In simplest terms, the sheet is preheated, then transferred to a temperature-controlled mold and conformed to its surface until cooled.

Summary

- In this chapter, we discussed some of the techniques for composite manufacturing.
- The method chosen depends on the number of units of the same product to be manufactured.

References

1. Gibson, R.F., 1994, *Principles of Composite Material Mechanics*, McGraw-Hill Book Co., International Edition.
2. Agarwal, B.D., Broutman, L.J. and Chandrashekhara, K., 2006, *Analysis and Performance of Fiber Composites*, 3rd Edition, John Wiley and Sons, New York.
3. Mukhopadhyay, M., 2004, *Mechanics of Composite Materials and Structures*, Universities Press (India) Pvt. Ltd., Hyderabad.

13. Experimental Evaluation of Composite Properties

In this chapter, we will discuss:

- The various testing techniques for polymer composites.
- The limitations, approximations and accuracy of composites.

13.1 Introduction

Mechanical property data are essential in the design process if structures have to perform as intended – reliably and in a cost effective manner during their design life. What tests should be carried out to give the required and accurate data? How precisely do the tests have to be conducted? Many questions arise: Are data obtained from small specimens meaningful when large structures are being designed? What effect will the operating environment have? Some of these have been well answered in the case of isotropic materials. What is the effect of testing machines on the data obtained? Do the support conditions of the testing machine have any effect on the data obtained? For composites, which are in general anisotropic or in a special case, orthotropic, the situation is very much different in comparison to the tests to be carried out for isotropic specimens. How does the data obtained from one testing machine differ from other machines? How does the data differ from one operator to the other? Are the data so obtained sensitive? We shall discuss the tests generally conducted to evaluate the composite material developed.

The methods generally used for evaluating isotropic materials such as aluminum cannot be directly used for composites as they are anisotropic and also have various types for coupling depending on the lamina sequencing. Further, the presence of fiber also renders it more difficult as the failure modes are different. Unlike metallic materials, there are a number of standards for the characterization of composites. These standards vary from country to country. The test specimens also vary depending on the standards.

Generally, straight edge specimens are employed for the determination of mechanical properties of composites. It is difficult to make a dog bone specimen as the fibers in the composite may be broken, may be misaligned or separate from the matrix. The accuracy of the test results depends on various parameters such as the size of the specimen, fiber orientation and fiber distribution. It has been observed that there is a lot of variation in properties in the tests carried out. Figure 13.1 shows the test results in actual specimens carried out in a particular organization. They are a result of the various factors mentioned above.

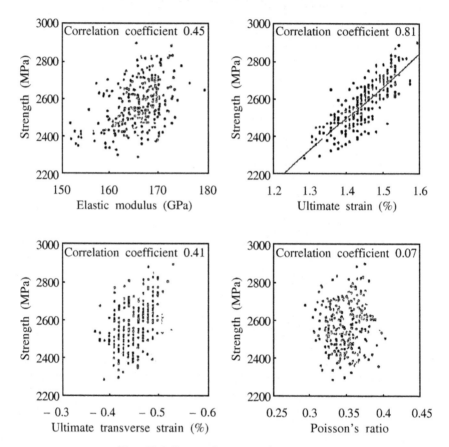

Fig. 13.1 Scatter in composite test data.

Properties such as tensile strength, tensile modulus and Poisson's ratio of the straight edge unidirectional composites are determined by the tensile test in accordance with the standards. Table 13.1 shows the variations in material properties from tests carried out by various organizations.

These figures bring home the point that the test results for composites are sensitive to the machines in which testing is carried out. It depends on grips, end tabs etc.

Table 13.1 Variation in material properties

Property	1	2*	3*	4	5	Max. diff. %
Elastic constant ($\times 10^6$ psi)						
Longitudinal tensile modulus	20.8	18.1	21	20.6	18.5	16
Longitudinal compressive modulus	18.6	14.5	21	19.8	18.5	45
Transverse tensile modulus	1.9	1.8	1.7	1.3	1.6	46
In-plane shear modulus	0.85	-	0.65	0.8	0.65	31
Poisson's ratio	0.30	-	-	0.32	0.25	28
Strength properties ($\times 10^3$ psi)						
Longitudinal tension	274	190	180	164	169	67
Longitudinal compression	280	126	180	126	162	122
Transverse tension	9.5	5.2	8	5.4	6.0	83
Transverse compression	39	-	30	21	25	86
In-plane shear	17.3	-	12	8.4	-	106
Interlaminar shear	-	13.5	13	-	7.1	90

*Divisions of the same company

ASTM 3039 is generally used for tensile testing. The basic mechanical properties essential for characterizing the composites are:

(i)   Tensile modulus (uniaxial)
(ii)   Compressive modulus (uniaxial)
(iii) Flexural modulus
(iv)   In-plane shear modulus
(v)   Poisson's ratio
(vi)   Tensile strength
(vii) Compressive strength
(viii) Flexural strength
(ix)   Interlaminar shear strength
(x)   Fracture toughness (various modes)

As stated earlier, the mechanical properties of composites are affected by a large number of parameters. They can be grouped as (i) properties of the fiber, (ii) surface characteristics of fibers, (iii) properties of the matrix material, (iv) properties of any other phase present in the composite, (v) volume fraction, (vi) spatial distribution and alignment of fibers, (vii) nature of contact, (viii) processing parameters such as degree of adhesion between fibers and matrix, and (ix) physical integrity of fibers.

The fiber alignment in the test coupons and the actual product generally deviate from the ideal state assumed for a mathematical model. Coupling between various phases is an influential factor. Good coupling is desirable where composites with high modulus and high strength are required. Poor coupling is sometimes desirable since local separation between fiber and matrix can arrest the crack propagation. The decoupled fiber may also give rise to stress concentration and initiate failure.

Thin sheets with 1 to 2 mm thickness for coupon specimens are manufactured from layers of fibers which are pre-impregnated by partially cured resin. A single prepreg is generally 0.125 or 0.25 thick. A number of layers are put together to get the required thickness. Unidirectional specimens are generally employed for obtaining the fundamental properties of the composite. Specimens that are too thin may give unrealistic results. If the specimen is too thick, its strength may exceed that of the end tabs giving rise to the failure of end tabs.

The purpose of the end tab is to provide a compatible gripping surface to transfer the load to the specimen and to protect the outer fibers of the specimen.

Before testing, the quality of the specimen needs to be checked. Fabrication techniques are employed for manufacturing laminates. If not properly controlled, they may introduce defects in the specimen. These defects will bring down the mechanical properties. During the manufacturing processes, voids are formed in the specimen if the technique is not properly controlled.

De-lamination may be caused by the entrapped air which will bring down the strength of the composite. A number of standards are available for testing composite specimens. Some of them are: (i) ASTM 3039 for 0° and 90° specimens, (ii) ISO 927 for 0° specimen and (iii) CRAG methods for 0° and 90° specimens. Figure 13.2 shows the various specimens depending on the standards selected for testing.

Tensile Properties

The tensile specimen is straight edged and is of constant cross-section. The end tabs are adhesively attached to the specimen. The material of the end tab is chosen to transfer the load to the specimen without resulting in the stress concentration at the ends. Further, the grips of the testing machine should not result in the fracture of the specimen. The strength of the end tabs must be higher than that of the specimen. This will ensure that the end tab does not fail before the specimen fails. The specimen is held by the grips generally of the wedge type. The speed of loading is kept generally low so that it simulates static loading. Generally, strain gauges are employed to measure the strain either in longitudinal or transverse loading. The strain gauges are bonded to the middle portion of the specimen, since the stress distribution is uniform across the cross-section. Depending on the requirements, one can use rectangular rosettes to measure the strains. One can also use an extensometer for measuring strains. The applied load is generally measured by a load cell. From the test data obtained, one draws the stress/strain curve for the material from which longitudinal tensile modulus E_{11} and the major Poisson's ratio v_{12} are obtained from the tensile data of 0° unidirectional laminates. Transverse tensile modulus E_{22} and the minor

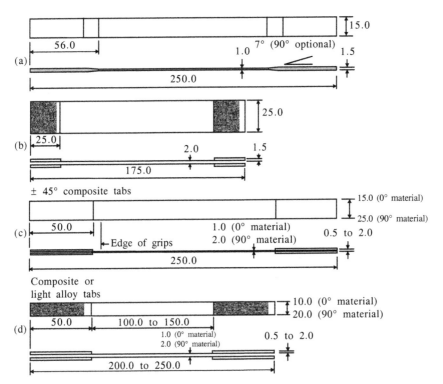

Fig. 13.2 Tensile specimens for use with aligned (0° to 90°) fiber-reinforced materials: (a) ASTM D 3039 (0°), (b) ASTM D 3039 (90°), (c) ISO 527 (0°) and (d) CRAG methods 300 (0°) and 301 (90°).

Poisson's ratio v_{21} are determined from the tensile test of 90° unidirectional laminates. If the loads are applied till the failure of the specimen, we can find the ultimate longitudinal and transverse tensile strengths.

Generally, clamped ends are used for the tensile test. This results in the development of both normal and shear stresses [1]. The stresses near the support can be written as

$$\varepsilon_x \neq 0, \ \varepsilon_y = \gamma_{xy} = 0 \tag{13.1}$$

The boundary conditions at the clamped edges are

$$\sigma_x \neq 0 \ \text{and} \ \tau_{xy} \neq 0 \tag{13.2}$$

In the presence of τ_{xy}, the stress-strain curve does not give E_x but gives rise to $\overline{Q}_{11}$. This effect can be eliminated if the specimen is large and the ends are free.

If one employs an off-axis specimen, a uniaxial tensile load results in both extension and shear deformations. Since the ends of the specimen cannot move, as they are constrained by grips, shear and bending moments are induced which result in a non-uniform deformation. As a consequence, the true modulus E_{true} is obtained by incorporating the corrections to the experimental modulus [2].

$$E_{\text{true}} = (1 - \eta)_{\text{expt}}$$

where

$$\eta = \frac{3\bar{S}_{16}^2}{\bar{S}_{11}^2 \left[3(\bar{S}_{66} / \bar{S}_{11}) + 2\left(\dfrac{L}{W}\right)^2 \right]} \qquad (13.3)$$

in which L is the specimen length between grips, W is the width of the specimen and $\bar{S}_{16}$, $\bar{S}_{11}$ and $\bar{S}_{66}$ are the elements of the compliance matrix.

For large values of length-to-width ratio, the value of η tends to zero. Rizzo [3], on the basis of experiments, has suggested that for tensile testing of an off-axis specimen, the ratio of length to width should be greater than 10.

Compressive Properties

The specimen used for testing compressive properties is very different from that used for testing tensile properties. Under compressive load, a tensile specimen tends to buckle at loads which are much lower than the compressive specimen. Hence, the specimen has to be such that it can withstand compressive load without geometric buckling. Generally, for compressive tests, one uses block specimens or rod specimens. Fabricating a block specimen is generally difficult. When such a specimen is subjected to a compressive load, there is likelihood of local failing called brooming. However, this can be overcome by encasing the ends in metal [4, 5]. A good number of loading fixtures and specimen types are available in literature [6]. The test set-up developed by the Illinois Institute of Technology Research Institute (IITRI) [7] for composite materials is shown in Fig. 13.3. In this test set-up, a flat specimen is employed. As shown in the figure, the exposed test specimen is very small as compared to overall length of the specimen. As in the case of the tension test, in the compression test also, the data is recorded by the strain gauge embedded in the test area along the loading direction. The strain in the perpendicular direction is also measured by strain gauges. One obtains the elastic modulus from the stress-strain and the Poisson's ratio from the data obtained from the strain gauges. If the specimen is loaded up to failure, ensuring there is no bucking of the specimen, it is the ultimate strength in compression.

One can also determine the tensile and compressive properties by employing sandwich beam specimens [1]. A specimen is generally manufactured by having a thin composite layer at the top and bottom of the sandwich beam with a thick core. The specimen is subjected to a pure bending moment. This results in the top layer having tensile stresses while the bottom layer has compressive stresses. Thus, a single specimen can be used for testing both tensile and compressive properties. In the actual case, the axial stresses vary across the thickness in the top and bottom surfaces.

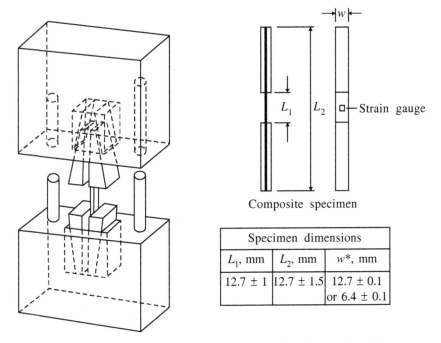

Specimen dimensions		
L_1, mm	L_2, mm	w^*, mm
12.7 ± 1	12.7 ± 1.5	12.7 ± 0.1 or 6.4 ± 0.1

Fig. 13.3 IITRI compression test fixture and specimen dimensions.

If, however, the top and bottom surfaces are quite thin, then one can fairly assume that the tensile and compressive stresses are uniform across the thickness.

In-plane Shear Properties
The mechanical properties that are determined by this test are the in-plane shear modulus and the in-plane shear strength. The types of specimen and test employed result in pure shear strength and modulus. Some of the tests are: (i) torsion tube test, (ii) rails shear test and (iii) Iosipescu in-plane shear test. The choice of the technique depends on the ease with which the specimen can be fabricated.

Torsion Tube Test
One of the simplest ways of getting pure shear state in the specimen is to apply a pure torque to a thin walled circular tube. This state of stress occurs only far away from the supported ends.

$$\tau_{xy} = \frac{T}{2\pi r^2 t} \tag{13.4}$$

where r is the mean radius, t is the thickness and T is the applied pure torque. It is assumed that the thickness of the tube is much smaller than the radius. In view of this, it can be fairly assumed that the shear stress is constant across the thickness. Pagano and Whitney [8] have shown that the torsion is the most desirable test specimen for in-plane shear test. However, some drawbacks exist. The specimen should be of constant thickness throughout the axial length and the support should be free to move axially so that bending moment and axial forces are not generated in the specimen. Strain is measured with the help of strain gauges. Strain gauges measure only axial strains. Hence, the strains are measured along the axis and 45° to the longitudinal axis. The shear strain is then obtained by drawing Mohr's circle for strains.

Rail Shear Test

The rail shear is employed to obtain the in-plane shear properties. ASTM standard D 4255/D4255M-01 describes the test technique. The rail shear type of test is shown schematically in Fig. 13.4. Two rail tests can be used for obtaining shear modulus and ultimate shear strength [9]. In the two rails shear test, the rails are fastened along the long edges usually by bolts on either side. A tensile load is applied as shown in the figure. The movement

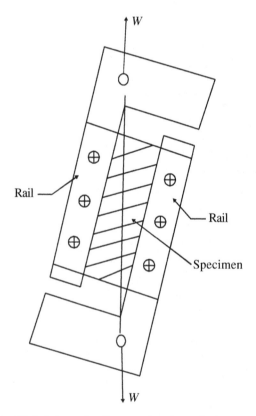

Fig. 13.4 Schematic diagram for the rail shear test.

of rails introduces shear stresses in the specimen. The in-plane shear stress is obtained as

$$\tau_{xy} = \frac{W}{Lh} \qquad (13.5)$$

As stated earlier, the shear strains are obtained with the help of strain gauges mounted at 45° to the longitudinal axis. The shear strain is given by

$$\gamma_{xy} = 2\varepsilon_{45°} \qquad (13.6)$$

The specimen length-to-width ratio should be greater than 10 for uniform shear stress.

± 45° Shear Test

This test is carried out with a symmetric laminate according to ASTM D3518 by applying uniaxial tensile loading. Stress-strain diagram is plotted from the measured data as

$$\tau_{12} = \frac{\sigma_{xx}}{2} \qquad (13.7)$$

$$\gamma_{12} = \varepsilon_{xx} - \varepsilon_{yy}$$

where σ_{xx}, ε_{xx} and ε_{yy} represent uniaxial tensile stress, longitudinal strain and transverse strain, respectively. Figure 13.5 shows the schematic diagram for the shear test.

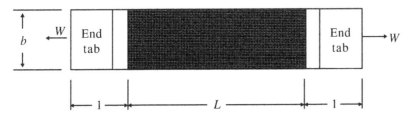

Fig. 13.5 Schematic diagram for [± 45]ₛ shear test.

Iosipescu Shear Test

This test, originally developed for isotropic materials, was later applied by Walrath and Adams [10] to fiber reinforced composites for obtaining the shear strength and shear modulus of composite specimens. It is essentially a four point bending test which develops pure shear at the middle of the notch plane. At the tip of the notch, there is stress concentration which can be reduced by increasing the notch angle. Generally, a 90° notch angle and notch depth equal to 20% of the specimen width and notch root radius of 1.3 mm is used. The schematic diagram is shown in Fig. 13.6. The in-plane shear stress is given by

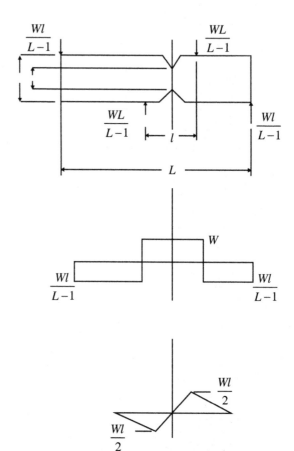

Fig. 13.6 Schematic diagram of the Iosipescu shear test and associated shear force and bending moment diagram.

$$\tau = \frac{W}{dt} \qquad (13.8)$$

where W is the applied transverse load, d is the distance between notches and t is the thickness of the specimen.

Shear strain in the specimen is measured by obtaining longitudinal strains from a ± 45° rosette embedded in the test specimen.

$$\gamma = \varepsilon_{+45°} - \varepsilon_{-45°} \qquad (13.9)$$

Flexural Properties

Flexural modulus and flexural strength for composite specimens are determined by a 4 point bending or a 3 point bending test on beam type specimens. Figures 13.7 and 13.8 show the schematic diagrams of the 3 point and 4 point bending test along with the shear force and bending

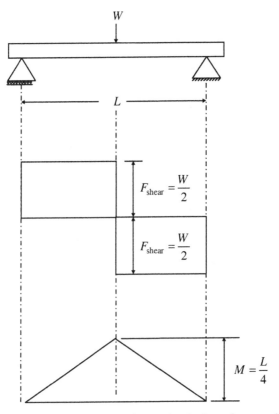

Fig. 13.7 3 point bending test and associated shear force and bending moment diagrams.

moment diagrams. In the 3 point bending test, the bending moment varies along the length. However, it has a peak value at the middle of the beam. The normal stress varies across the depth of the beam with a maximum at the top and bottom of the cross-section. The maximum normal stress is given by

$$\sigma_{max} = \frac{6M}{bt^2} \tag{13.10}$$

where M is the bending moment across the cross-section, b is the width and t is the thickness of the specimen. Maximum shear stress is given by

$$\tau = \frac{3F_{shear}}{2bt} \tag{13.11}$$

In the case of a composite specimen made-up of lamina oriented differently, the maximum fiber stress will not occur on the top surface. It may occur at any point across the section. In such cases, lamination, discussed in earlier chapters, has to be used to obtain the maximum stress.

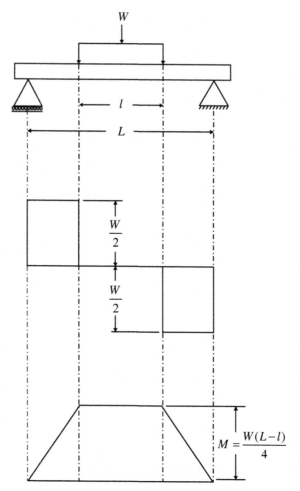

Fig. 13.8 4 point bending test and associated shear force and bending moment diagrams.

However, if the specimen is made up of lamina with same fiber orientation, then the stress is highest at the top surface.

Interlaminar Shear Strength

Interlaminar shear stress is developed when the laminates have plies which are oriented differently. This results mainly because of a mismatch of Poisson's ratio between the adjacent plies and also because of the cross coupling coefficient. Stress exists at the interface of the adjacent lamina. It is also large at the free edges. Large interlamina shear stresses may result in delamination failure in the laminated composite structures. This in turn will reduce the strength of the composite. It is, therefore, necessary to evaluate the interlamina shear stresses. Carlsson et al. [11] have shown that the interlamina stresses depend very much on the stacking sequence. Interlaminar shear strength (ILSS) refers to the shear strength parallel to the lamina surface. In the three point bending test on homogenous materials, as shown by Eqs. (13.10) and (13.11), the shear stress does not depend on the length of the specimen. However, the bending stress depends on the L/t ratio. Hence, it is possible in homogenous materials that the shear stress may exceed the normal stress and the shear strength may reach the ultimate value for the material. If the length of the three point bending beam is short enough, then the failure is initiated by interlaminar shear stress. Delamination in the composite will result in distribution of the stresses within the laminate which may result in fiber breakage. A number of test methods [11] have been developed to measure the static interlaminar fracture toughness of composites.

The interlaminar fracture toughness of the unidirectional composite is measured in terms of the critical strain energy release rate. Failure may occur in three different modes: Mode I is the opening mode, Mode II is associated with the in-plane shear mode and Mode III is referred to as tearing mode which is a combination of Mode I and Mode II.

The strain energy release rate in Mode I is determined by the double cantilever specimen made up of a unidirectional composite with even layers. Figure 13.9 shows the schematic diagram of the test. The ASTM standard D 5528-01 provides the details of the method. The initial crack is introduced in the specimen during fabrication itself at the end of the specimen in the middle of the thickness of the laminate. Displacement of the end point of the beam is measured during the test as shown in Fig. 13.9. The displacements are measured by a linear voltage differential transformer (LVDT) or an extensometer. As the loading is increased, the change in the crack length is

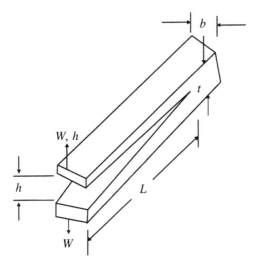

Fig. 13.9 Schematic diagram of double cantilever specimen.

measured. Load is applied slowly till the initial crack grows to a predetermined length. The specimen is unloaded and loaded again till the crack grows to a defined length. This process is repeated to obtain a series of load displacement curves. The compliance k is given by

$$k = \frac{2L^3}{3EI} \tag{13.12}$$

The strain energy release rate per unit width of the specimen is given by

$$G_1 = \frac{W^2 L^2}{bEI} \tag{13.13}$$

The critical strain energy release rate G_{1c} is determined when $W = W_{cr}$.

The strain energy release rate for Mode II is determined by the three point beam bending test. A crack is introduced at one end of the specimen during the fabrication stage itself by introducing a thin teflon film. The crack is initiated in the middle of the thickness of the beam. It is difficult to obtain a stable crack growth in the case of three point bending test. One has to fabricate a number of specimens with a different initial crack length to draw compliance and crack growth length [9]. The strain energy release rate is given by

$$G_{II} = \frac{W^2}{2b} \frac{dk}{da} \tag{13.14}$$

where W is the applied load, b is the width of the beam, a is the crack length at very loading, k is the specimen compliance and $\frac{dk}{da}$ is the slope of the compliance versus the crack length curve.

Summary

In this chapter, we discussed some of the techniques for evaluating the properties of the composite specimens. However, there are many more tests generally conducted to critically evaluate the properties. It is to be pointed out that in view of the variation of the properties, one has to conduct tests on a number of specimens.

References

1. Agarwal, B.D., Broutman, L.J. and Chandrashekhara, K., 2006, *Analysis and Performance of Fiber Composites*, John Wiley and Sons.
2. Pagano, N.J. and Halpin, J.C., 1968, "Influence of End Constraint in the Testing of Anisotropic Bodies", *J. Comps. Mater.*, 2, pp. 18-31.
3. Rizzo, R.R., 1969, "More on the Influence of End Constraint on Off-axis Tensile Test", *J. Comp. Mater.*, 3, pp. 202-219.

4. Fried, N. and Winans, R.R., 1963, in *Symposium on Standard Filament Wound Reinforced Plastics*, ASTM STP 327, American Society of Testing and Materials, Philadelphia, pp. 83-95.

5. Hadcock, R.N. and Whiteside, J.B., 1969, *Composite Materials: Testing and Design*, STP460, American Society of Testing and Materials, Philadelphia.

6. Whitney, J.M., Daniel, I.M. and Pipes, R.B., 1981, "Experimental Mechanics for Fiber Reinforced Composite Materials", *SESA Monograph No. 4*, Society of Experimental Stress Analysis, Brookfield Centre.

7. Hofer, Jr., K.E. and Rao, P.N., 1977, "A New Static Compression Fixture for Advanced Composite Materials", *J. Testing and Eval.*, 5, pp. 278-283.

8. Pagano, N.J. and Whitney, J.M., 1970, "Geometric Design of Composite Cylindrical Specimen Characterization Specimens", *J. Comp. Mater.*, 4, pp. 360-378.

9. Mallick, P.K., 2008, *Fiber-reinforced Composites: Materials, Manufacturing and Design*, Third Edition, CRC Press.

10. Walrath, D.E. and Adams, D.F., 1983, "The Iosipescu Shear as Applied to Composite Materials", *Exp. Mech.*, 23, pp. 105-110.

11. Carlsson, L.A., Adams, D.F. and Pipes, R.B., 2003, *Experimental Characterization of Advanced Composite Materials*, 3rd Edition, CRC Press, Florida.

Index